BASIC ELECTRONICS

FOR +2 SC STUDENTS OF ELECTRONICS

Volume I

SADASIVA BISWAL

ATLANTIC

PUBLISHERS & DISTRIBUTORS (P) LTD

Published by
ATLANTIC
PUBLISHERS & DISTRIBUTORS (P) LTD
7/22, Ansari Road, Darya Ganj, New Delhi 110002
Phones : +91-11-40775252, 40775214, 23273880, 23275880
Fax : +91-11-23285873
Web : www.atlanticbooks.com
E-mail : orders@atlanticbooks.com

Branch Office: Chennai
Phones : +91-44-48531784, 28291383
E-mail : chennai@atlanticbooks.com

Printed & bound in India by Atlantic Print Services

DEDICATED
TO
MY PUPILS

PREFACE

Electronics and Electronic Appliances have become the way of life in this age. Every body of the present society is more or less crazy after the subject and its applications. With the advancement of science and technology, *Electronics* has been introduced into industry in a large scale to meet the challenge of the modern man. Starting from the fabrication of vacuum tubes to transistors, the science of *Electronics* has moved rapidly to fabricate integrated circuits which in turn have paved the way for generating micro-processors and super computers. The facilities of internet have been felt enormously in the field of our communication network. Ostensibly, this has surpassed all the far and near distances of the world by bringing every corner of land, water and space into its fold.

Feeling the importance of *Electronics* in the present-day society, more and more people are taking interest in learning, teaching, researching and practising *Electronics*. Consequently the subject has been introduced in the courses of study starting from secondary to higher secondary and degree-science classes. In order to facilitate the students to have specialisation in this subject, some states have introduced it as a separate subject of study starting from +2 stage to under-graduate and post-graduate levels.

Understanding the need of our pupils of *Electronics*, the present book *Basic Electronics* has been planned in two volumes. Volume I covers the topics : *Resistors, Capacitors, Inductors, Electromagnetic Induction, A.C. Fundamentals, Thermionic Emission and Vacuum Tubes, Semiconductors and Transistors, Rectifiers, Transistors, Amplifiers, Feedback Amplifiers* and *Oscillators*. Volume II deals with *Modulation and Transmission, Demodulation and Reception, Measuring Instruments, Microphones and Loudspeakers, Integrated Circuits, Antenna Theory and Propagation of Radiowaves, Television and Radio, Power Electronics and Digital Electronics*. Each topic has been cast in the form of questions and answers. At the end of each chapter, some numerical problems have been included with their solutions. This will no doubt help the students in preparing for their examinations. In spite of all our efforts, some gap might be there in the book to be fulfilled. We shall be highly obliged to those teachers and students who bring those wants to our notice and we shall try our best to incorporate those in the next edition of the work.

I am highly indebted to both my teachers and students who have really inspired me to go for a work like this. Their inestimable help and prolonged initiation enable me to learn more and more by reading, writing and investigating.

I am ever indebted to my parents for their eternal blessings on every walk of my life. Their sweet compassion, love and affection always provide me enough strength to work through all odds. All of my achievements go to them.

The inestimable help and untiring efforts rendered by my two *putra ratnas*, Saumya and Shaunak and *Sukanya* Shashwatee Swagatica in preparing the manuscript are truly unforgettable. I solemnly acknowledge their sincere co-operation and benevolent deeds and for which I am greatly thankful to them.

I feel incompetent in acknowledging the role of my wife Dr. Bidyut Lata Ray, who has constantly given me encouragement and support in every step of the work. Even her knowledge of the extent of my indebtedness to her is only approximate.

I render my sincere thanks to M/s Atlantic Publishers and Distributors, New Delhi for bringing out the book. I am highly indebted to Dr. K.R. Gupta, Managing Director of the Company for his ungrudging help and co-operation in this matter.

Lastly, I am grateful to many authors and publishers of several excellent works from which I have drawn the material.

In spite of our several efforts, some errors might have been crept in the book. If such errors will be brought to our notice by our learned readers, due attempts will be made to correct those in the next edition of the book.

Knowledge dispels all ignorance.

SADASIVA BISWAL

CONTENTS

VOLUME II

1
Resistors

Q. 1. What is a resistor?

Ans. *Resistor* : Resistor is an electrical device that offers resistance to the flow of current through it.

Q. 2. What are different types of resistors?

Ans. *Types of resistors* : Resistors can be classified according to different methods, such as :

(i) According to material used
(ii) According to the nature or type
(iii) According to power rating

(i) Classification according to materials used

The resistors are classified into four categories, according to the materials they have :

(a) Wire wound resistors
(b) Carbon resistors
(c) Carbon film resistors
(d) Metal film resistors

(a) Wire wound resistors. These resistors are made of materials like copper and aluminium. They have low resistance. But other wire wound resistors are available which have high resistivity and zero temperature coefficient, *i.e.*, there resistance doesn't vary with temperature. Such resistors are made of silver, nichrome, german, manganin etc.

When wire made of these substances is wound on a tubular ceramic form, the result is a wire wound resistor (fig. 1.1).

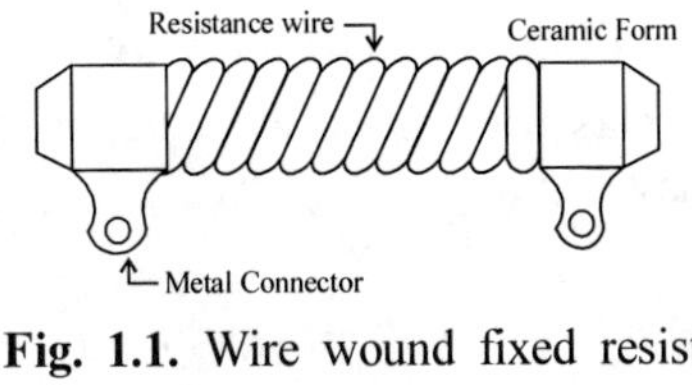

Fig. 1.1. Wire wound fixed resistor.

A wire wound resistor shown above is usually a multilayer coil, each turn and each layer is insulated from each other. The alternate layers are wound in opposite direction to avoid induction effect. These resistors are usually covered with a hard, vitreous (insulating), protective coating. High wattage resistors are usually of this type. But the wire size limits the maximum value of resistance to about 50 kΩ.

(b) Carbon resistor (Composition resistor). It is made from powdered Carbon mixed with a day binding material and baked into a small, hard cylinder with a wire attached to each end (fig. 1.2).

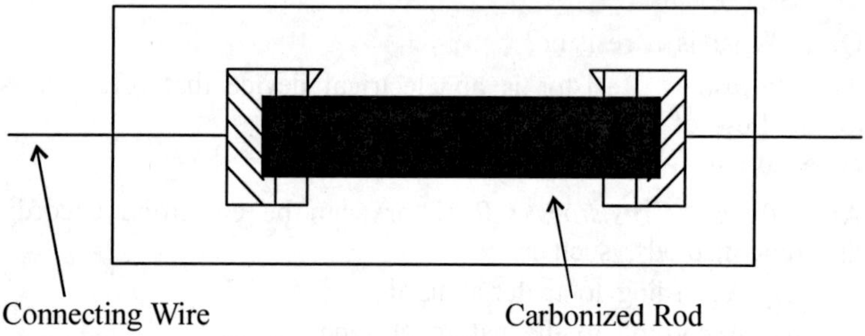

Fig. 1.2. Fixed type carbon resistor.

The value of resistance depends on the proportion of carbon in the mixture used. Carbon resistors range in value from a fraction of an ohm to several mega ohms. Size has no bearing on resistance value affecting only the power rating of the resistor. Carbon resistors have a positive temperature coefficient *i.e.*, the resistance increases with increase of temperature. Common power ratings of carbon resistors are $\frac{1}{8}$, $\frac{1}{4}$, $\frac{1}{2}$, 1, 2, 3 watts.

(c) Carbon film resistor. It consists of a glass or ceramic rod coated with a carbonized layer, which in turn is covered with a ceramic nonconductive coating.

(d) Metal-film resistors. It has a metal such as nichrome deposited on a glass or ceramic form. A spiral groove is cut with a lathe or laser beam through the deposit to produce a spiral wound resistor, which is then covered with a protective layer.

(ii) Classification according to nature or type

The resistors are classified into two categories according to the nature or type they have

(a) Fixed resistor

(b) Variable resistor

(a) Fixed resistor. Its value is constant and cannot be varied. Its circuit symbol is given in fig. 1.3.

Fig. 1.3

(b) Variable resistor. The resistance between the terminals is varied by changing a sliding contact. Various circuit symbols are shown in fig. 1.4 (a), (b), (c).

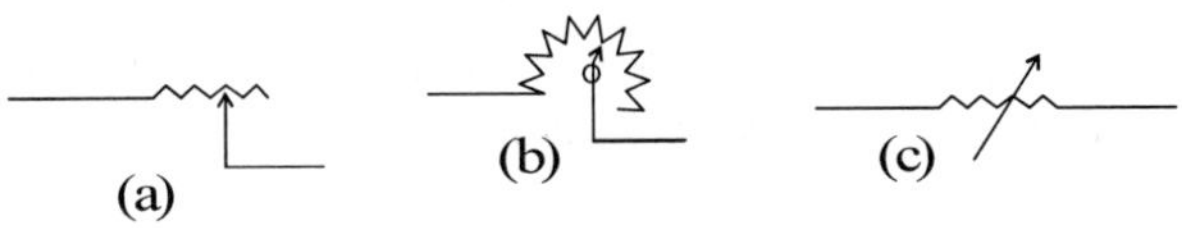

Fig. 1.4

A common form of variable resistor is a rheostat. It is an adjustable wire wound resistor with a sliding contact that can be fixed at the position when the desired value of resistance is obtained. A resistor of this type is partly covered with a vitreous coating but part of the wire will be bare to allow the slider to make contact.

An adjustable resistor is shown in fig. 1.5 in which the sliding contact can be tightened by a machine screw at the desired position. Other types of variable resistors usually known as potentiometer are shown in the fig. 1.5. Here the resistor is made into the form of an

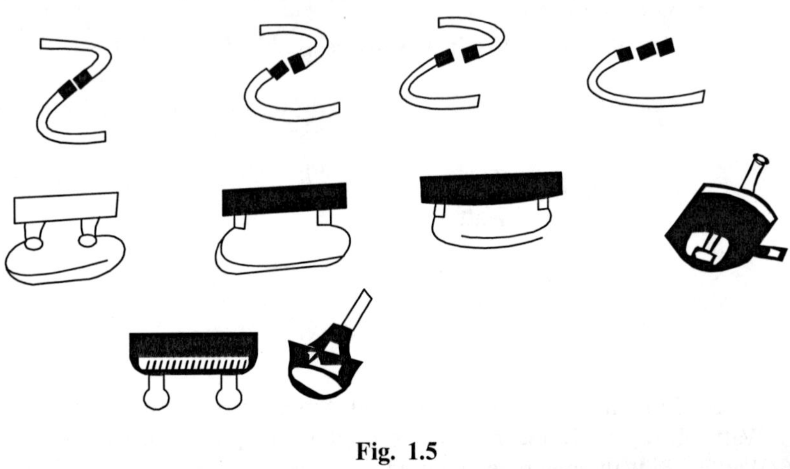

Fig. 1.5

arc with two fixed terminals at the two ends. The variable contact moves along the arc controlled by an axle which is insulated from the resistor. The resistor in this case may be wire wound type (for low value) or carbon-type (Megohm or even higher value).

(iii) Classification according to power rating

Resistors, whether they are carbon or wirewound, have a resistance and power rating. The power rating indicates how much heat the resistor is capable of dissipating without getting unduly heated under normal conditions. If cooled by blowing air over it, the resistor may be capable of considerably greater power dissipation. If enclosed in an unventilated area, it may become excessively hot and burn out when dissipating its rated power or even less.

Usually the power rating required is computed by one of the power formulae and a resistor of twice the computed power rating is used.

Example

A 10000 Ω resistor is connected across 200 V. What should be the power rating of the resistor?

The power dissipated by resistor is given by any of the three formulae :

$$P = VI \qquad ...(1)$$

$$P = I^2R \qquad ...(2)$$

$$P = \frac{V^2}{R} \qquad ...(3)$$

Using the equn. (3) in the above problem

$$P = \frac{V^2}{R}$$

$$= \frac{(200)^2}{10000} \text{ Watt}$$

$$= \frac{40000}{10000} \text{ Watt}$$

$$= 4 \text{ Watt}$$

Therefore, the resistor to be used should have power rating of 4 Watt. However, if the resistor is used in a tightly enclosed space, 6 Watt to 8 Watt might be required.

The maximum permissible current through a resistor is calculated using the equn. (2) in the form,

$$P = I^2 R$$

$$\Rightarrow I = \sqrt{\frac{P}{R}} \qquad ...(4)$$

Similarly, the max^{m} permissible voltage is calculated using equn. (3) in the form,

$$P = \frac{V^2}{R}$$

$$\Rightarrow V = \sqrt{PR} \qquad ...(5)$$

The resistors are available normally with the ratings 1/8W, ¼W, ½W, 1W, 2W, 3W, 5W, 10W, 20W, 30W etc. The low power rating resistors are usually of carbon type and higher ones are normally wirewound type. The low-rating resistors are of small size and resistors' size increases with increase in rating. A ½W carbon resistor is about 3 mm in diameter and 1.25 cm long. Fig. 1.5 shows some resistors of different power rating.

Q. 3. Write down the Colour code for resistor.

Ans.

Colour code for resistor

Sl. No.	Colour	Ist colour	2nd colour	Multiplier (3rd colour)	Tolerance (4th colour)
1	Black	—	0	10^0	—
2	Brown	1	1	10^1	1%
3	Red	2	2	10^2	2%
4	Orange	3	3	10^3	3%
5	Yellow	4	4	10^4	4%
6	Green	5	5	10^5	—
7	Blue	6	6	10^6	—
8	Purple (violet)	7	7	10^7	—
9	Grey	8	8	10^8	—
10	White	9	9	10^9	—
11	Silver	—	—	—	10%
12	Gold	—	—	—	5%
13	Body Colour	—	—	—	20%

Q. 4. Define resistance. Mention its unit in MKSa.

Ans. *Resistance :* The resistance of a resistor is defined from ohm's law which states that for a conductor at constant temperature the current is directly proportional to the potential difference between the two ends of the conductor. Mathematically, $V \propto I$

$$\text{or,} \quad \frac{V}{I} = \text{constant } R$$

This constant R is known as the resistance of the conductor, V is the potential difference between the two ends of the conductor and I is the current flowing through the conductor.

Unit of resistance. The MKS unit of resistance is called ohm.

Defn. of ohm. One ohm is the resistance of a conductor through which a steady current of one ampere flows when a steady potential difference of one volt is maintained across the two ends of the conductor.

Q. 5. Define specific resistance. What are its units?

Ans. *Resistivity or Specific resistance :* The resistance of a conductor varies directly as its length and inversely as its area of cross section. So, mathematically we can write :

$$R \propto l, \text{ when } A \text{ and } T \text{ constant}$$

$$R \propto \frac{1}{A}, \text{ where } l \text{ and } T \text{ constant}$$

Thus, $R \propto \dfrac{l}{A}$, when both l and A vary, but T remains constant.

$$\text{or,} \quad R = \rho\frac{l}{A}$$

$$\text{or,} \quad \rho = \frac{RA}{l} \qquad \text{...(6)}$$

where 'ρ' is called the resistivity or the specific resistance of the conductor.

If A and l are given values of unity, it is seen that $\rho = R$ ...(7)

Thus, the resistivity or the specific resistance of the material of the conductor is numerically equal to the resistance of the conductor having unit cross section and unit length.

Units of ρ :

MKS → ohm-meter

CGS → ohm-centimeter

Q. 6. What are the factors upon which the resistance of a conductor depends? Explain.

Ans. *Factors upon which the resistance of a resistor depends :* The resistance of a resistor depends upon the following factors :

(a) The length of the conductor
(b) The area of cross section of the conductor
(c) The material of the conductor
(d) The temperature of the conductor

(a) The length of the conductor

The resistance of a conductor is directly proportional to the length of the conductor where the area of cross section, the temperature and the material remain constant.

Thus, $R \alpha l$, where A and T and the material are kept constant.

(b) The area of cross section of the conductor

The resistance of a conductor is inversely proportional to its area of cross section, when l, T and the material remain constant. Thus,

$R \propto \frac{1}{A}$, when 1, T and the material remain constant.

(c) The material of the conductor

The resistance of a conductor varies with the material of the conductor. Two wires of different material having same length and same area of cross section at same temperature will have different resistances. For instance, an iron wire and a copper wire of equal length and cross section will have different resistances at the same temperature. This means that they will conduct electric current differently. Therefore, the resistance of a resistor depends upon the nature of the material.

(d) The temperature of the conductor

The resistance of a conductor depends upon the temperature of the conductor. It is found that the resistance of all substances changes more or less with the changes of temperature. There are three types of changes observed.

(i) The resistance may increase with increasing temperature.

This is true of all pure metals and most alloys.

(ii) The resistance may decrease with increase of temperature.

This is true of a semiconductor like carbon and of glass and many electrolytes.

(iii) The resistance may be independent of temperature.

This is approximately true of many special alloys, such as manganin (copper → 0.84. Ni → 0.12, Mn → 0.04)

Experiments have shown that, for a moderate temperature range, the change of resistance with temperature of metallic conductors can be given by the following equation.

$$R_\theta = R_0 \ (1+\alpha\theta) \qquad ...(8)$$

Where R_θ is the resistance at θ°C; R_θ is the resistance at temperature 0°C and α is known as the temperature co-efficient of resistance.

From equn. (8) α is given by :

$$\alpha = \frac{R_\theta - R_0}{R_0\theta} \qquad ...(9)$$

Defn. of α → The temperature co-efficient of resistance is thus defined as the change in resistance per unit resistance per degree rise in temp., based upon the resistance at 0°C.

Units of α

The units of α are /°C or /°F

Determination of α from the graph between R_θ and θ →.

A graph may be plotted between R_θ and θ. For a pure metal the graph becomes a straight line not passing through the origin. This shows that the resistance isn't zero at 0°C. The slope of this curve

$\frac{\Delta R}{\Delta \theta}$ is a constant. Since

$$\alpha = \frac{\Delta R / \Delta\theta}{R_0} = \frac{Slope}{R_0}$$

It is clear that the value of α depends upon the base temperature chosen for R_o. In computations involving temperature variation of resistance, the value of R_o must first be obtained by using equn. (8).

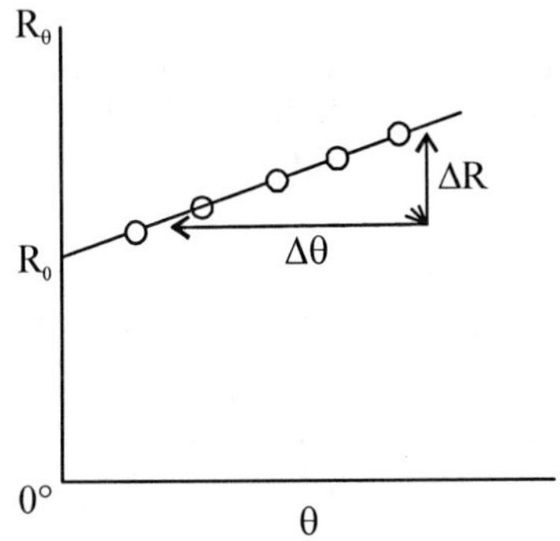

Fig. 1.6. Variation of resistance with temperature.

Q. 7. Describe how does temperature affect resistivity?

Ans. *Changes of resistivity with temperature :* The resistivity of a conductor changes with temperature according to the following relation :

$$\rho_\theta = \rho_0 (1 + \alpha \, \theta) \qquad \text{...(10)}$$

It has been found that the resistivity of a conductor decreases with decrease of temperature. This is also valid according to the above relation. α is known as the temperature co-efficient of resistivity which is given by,

$$\alpha = \frac{\rho_\theta - \rho_0}{\rho_0} \qquad \text{...(11)}$$

Where ρ_θ is the resistivity at temperature θ and ρ_0 is the resistivity at temp. 0°C. The resistivity of some metals at very low temperatures drops suddenly and becomes practically zero. This phenomenon is called super conductivity.

It is interesting to note the fact that the value of α for all pure metals is roughly the same, namely, $\frac{1}{273}$, or about 0.004, per celsius degree. Observe that this value is the same as the coefficient of expansion of an ideal gas. It suggests that, at the absolute zero of temperature (– 273°C), a conductor would have zero resistance, *i.e.*, a current once started would continue indefinitely without the expenditure of any energy to keep it going.

Q. 8. What is critical temperature of a conductor?

Ans. *Critical temperature* : Critical temperature of a conductor is defined as that temperature below which the resistivity of the conductor falls to zero, so that the conductor becomes a super conductor.

Q. 9. Graphically show the variation of resistivity of conductors, semi-conductors and super conductors with temperature.

Ans. The variation of resistivity of conductors, semi-conductors and super conductors with temperature :

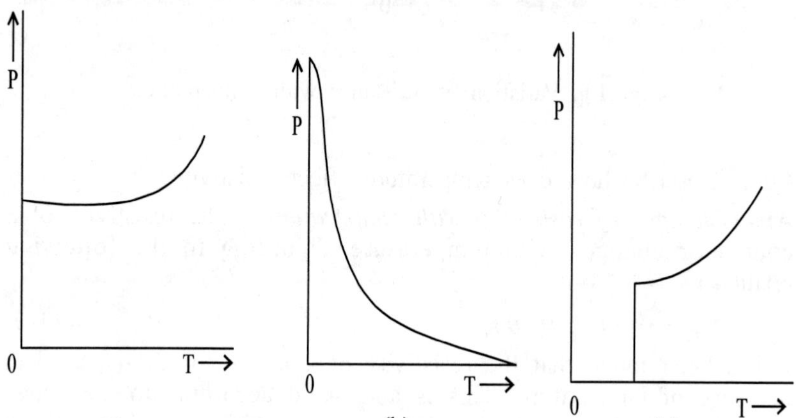

(a) Conductor (b) Semi-conductor (c) Super conductor

Fig. 1.7. Variation of resistivity with temperature.

Q. 10. Deduce the law of combination of resistors in series and in parallel.

Ans. *Combination of resistors* : Resistors may be combined in series or in parallel.

(a) Resistors in series

Suppose that a box contains three coils of wire whose resistances are R_1, R_2 and R_3 which are connected in series as shown in fig. 1.8. If one were asked to determine the resistance of whatever is inside the box without opening it, he might place it in the circuit shown and measure the current I in the box and the voltage V across it. He would then write,

$$R = \frac{V}{I} \qquad \text{...(12)}$$

where R is the resistance of the part of the circuit inside the box.

Let us now determine the relation of R, the combined resistance, to the individual resistances R_1, R_2 and R_3. The current in each of the resistors is I, since the current isn't divided in the box. The voltage across the individual resistors are, $V_1 = IR_1$, $V_2 = IR_2$, $V_3 = IR_3$

The sum of these three voltages must be equal to V, the voltage across the box; thus :

$$\begin{aligned} V &= V_1 + V_2 + V_3 \\ &= IR_1 + IR_2 + IR_3 \\ &= I\,(R_1 + R_2 + R_3) \end{aligned} \qquad \text{...(13)}$$

$$R_1 + R_2 + R_3 = \frac{V}{I}$$

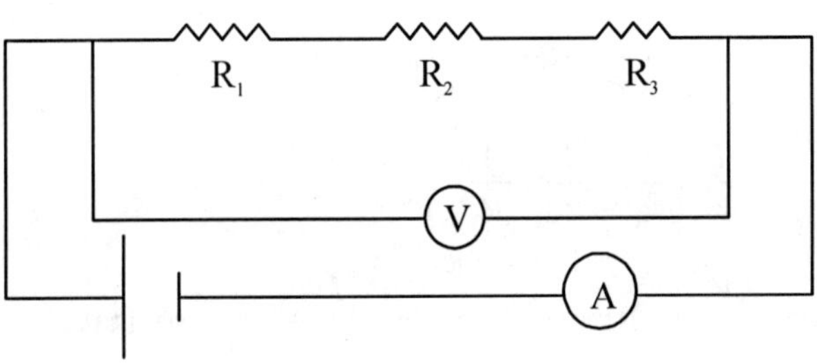

Fig. 1.8. Resistors in series.

but this is identical with $R = \dfrac{V}{I}$, so that

$$R = R_1 + R_2 + R_3 \qquad \text{...(14)}$$

The following facts may therefore be noted for series connection of resistors :

(1) The current in all parts of a series circuit is the same.

(2) The voltage across a group of resistors connected in series is equal to the sum of the voltages across the individual resistors.

(3) The total resistance of a group of conductors connected in series is equal to the sum of the individual resistances.

(b) Resistors in parallel

Suppose that a box contains a group of three resistors in parallel, of resistances R_1, R_2 and R_3, as shown in fig. 1.9. The resistance of the combination will be $R = \frac{V}{I}$, where V is the voltage across the terminals of the box and I is the total current in it. Since the voltage across each of the resistors is V, the voltage across the terminals of the box, the currents in the individual resistors are :

$$I_1 = \frac{V}{R_1}, \quad I_2 = \frac{V}{R_2}, \quad I_3 = \frac{V}{R_3}$$

The sum of these three currents must be the total current I, so that :

$$I = I_1 + I_2 + I_3$$

or, $$I_1 = \frac{V}{R_1} + \frac{V}{R_2} + \frac{V}{R_3}$$

This can be written,

$$I = V\left(\frac{I}{R_1} + \frac{I}{R_2} + \frac{I}{R_3}\right)$$

or, $$\frac{I}{V} = \frac{1}{R_1} + \frac{1}{R_2} + \frac{1}{R_3}$$

Since $\frac{V}{I}$ = R, we know that $\frac{I}{V} = \frac{1}{R}$, so that

$$\frac{1}{R} = \frac{1}{R_1} + \frac{1}{R_2} + \frac{1}{R_3} \qquad ...(15)$$

For parallel connections of resistors, the following conditions are obtained :

(1) The currents in the various resistors are different and are inversely proportional to the resistances. The total current is the sum of the separate currents.

(2) The voltage across each resistor of a parallel combination is the same as the voltage across any other resistor. Moreover, the voltage across each separate resistor is identical with the voltage across the whole group considered as a unit.

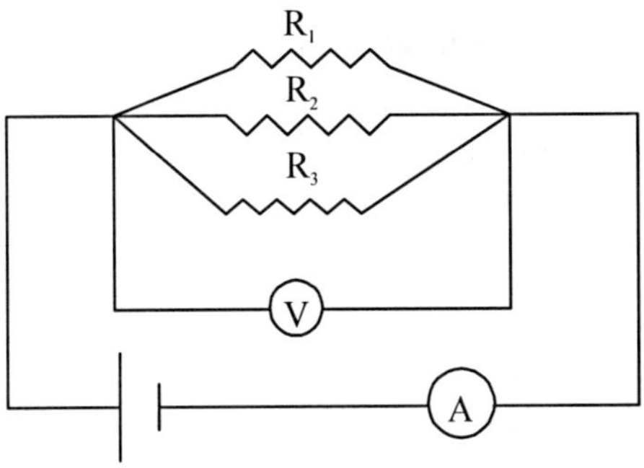

Fig. 1.9. Resistors in parallel.

(3) The reciprocal of the total resistance of a number of resistors connected in parallel is equal to the sum of the reciprocals of the separate resistances.

Note. It is important to note that connecting additional resistors in series increases the total resistance, while connecting additional resistors in parallel decreases the total resistance. For example, in an ordinary house installation, when we "turn on" more lamps, we are inserting additional resistors in parallel. We thus reduce the total resistance of the house circuit, and hence (since the voltage is constant) we increase the current in the mains.

EXERCISES

Numerical Problems with Solutions

1. Find the value of the resistor whose colour band is :
   (a) Orange, red, black
   (b) Red, red, yellow, silver
   (c) Yellow, violet, yellow, gold
   (d) Brown, black, brown, red
   (e) Brown, black, black
   (f) Grey, blue, green
   (g) Brown, black, black, silver
   (h) Red, red, gold

(i) Red, red, silver

(j) Yellow, violet, green, gold

Solution

(a) 1st stripe — Orange = 3
2nd stripe — Red = 2
3rd stripe — Black = 10^0
Since, there is no 4th stripe, the tolerance is 20%

So, $R = 32 \times 10^0\Omega \pm 20\%$
$= 32\ \Omega \pm 20\%$

(b) 1st stripe — Red = 2
2nd stripe — Red = 2
3rd stripe — Yellow = 10^4
4th stripe — Silver = 10% tolerance

So, $R = 22 \times 10^4\Omega \pm 10\%$

(c) 1st stripe — Yellow = 4
2nd stripe — Violet = 7
3rd stripe — Yellow = 10^4
4th stripe — Gold = 5% tolerance

So, $R = 47 \times 10^4\Omega \pm 5\%$

(d) 1st stripe — Brown = 1
2nd stripe — Black = 0
3rd stripe — Brown = 10^1
4th stripe — Red = 2% tolerance

So, $R = 10 \times 10^1\Omega \pm 2\%$

(e) 1st stripe — Brown = 1
2nd stripe — Black = 0
3rd stripe — Black = 10^0
No 4th stripe, so tolerance is 20%

Thus, $R = 10 \times 10^0\Omega \pm 20\%$
$= 10\ \Omega \pm 20\%$

(f) 1st stripe — Grey = 8
2nd stripe — Blue = 6
3rd stripe — Green = 10^5
No 4th stripe, so tolerance is 20%

Thus, $R = 86 \times 10^5\Omega \pm 20\%$

(g) 1st stripe — Brown = 1
2nd stripe — Black = 0
3rd stripe — Black = 10^0
4th stripe — Silver = 10% tolerance

Thus, $R = 10 \times 10^0\Omega \pm 10\%$
$= 10\Omega \pm 10\%$

(h) 1st stripe — Red = 2
2nd stripe — Red = 2
3rd stripe — Gold = 10^{-1}

Thus, $R = 22 \times 10^{-1}\Omega = 2.2\Omega$

(i) 1st stripe — Red = 2
2nd stripe — Red = 2
3rd stripe — Silver = 10^{-2}

Thus, $R = 22 \times 10^{-2}\Omega = 0.22\Omega$

(j) 1st stripe — Yellow = 4
2nd stripe — Violet = 7
3rd stripe — Green = 10^5
4th stripe — Gold = 5% tolerance

Thus, $R = 47 \times 10^5\Omega \pm 5\%$

2. Find the colour code for the following resistors :
   (a) 22 KΩ ± 10%
   (b) 560 KΩ ± 5%
   (c) 470 KΩ ± 20%
   (d) 820 KΩ ± 10%
   (e) 8Ω ± 2%
   (f) 33Ω ± 5%
   (g) 0.63Ω ± 2%
   (h) 0.47Ω ± 2%
   (i) 0.56Ω ± 2%
   (j) 8.2 MΩ ± 10%

Solution

(a) $22\ K\Omega = 22 \times 10^3\Omega$
The colour code for 2 = Red
The colour code for 3 = Orange
10% tolerance represents Silver

So, the colour code of the resistor is Red, Red, Orange, Silver.

(b) $560\ K\Omega = 56 \times 10^4\Omega$
The colour code for 5 = Green
The colour code for 6 = Blue
The colour code for 4 = Yellow
The colour code for 2% tolerance = Gold

So, the colour code of the resistor is Green, Blue, Yellow, Gold.

(c) $470\ K\Omega = 470 \times 10^3\Omega$
$= 47 \times 10^4\Omega$
The colour code for 4 = Yellow
The colour code for 7 = Violet
The colour code for 4 = Yellow
The colour code for 20% tolerance = No colour
So, the colour code of the resistor is Yellow, Violet, Yellow.

(d) $820\ K\Omega = 820 \times 10^3\Omega = 82 \times 10^4\Omega$
The colour code for 8 = Grey
The colour code for 2 = Red
The colour code for 4 = Yellow
The colour code for 2% tolerance = Silver
So, the colour code of the resistor is Grey, Red, Yellow, Silver.

(e) $8\Omega = 8.0\Omega \times = 80 \times 10^{-1}\Omega$
The colour code for 8 = Grey
The colour code for 0 = Black
The colour code for 10^{-1} = Gold
The colour code for 2% tolerance = Red
So, the colour code of the resistor is Grey, Black, Gold, Red.

(f) $33\Omega = 33 \times 10^0\Omega$
The colour code for 3 = Orange
The colour code for 0 = Black
The colour code for 5% tolerance = Gold
So, the colour code of the resistor is Orange, Orange, Black, Gold.

(g) $0.63\Omega = 63 \times 10^{-2}\Omega$
The colour code for 6 = Blue
The colour code for 3 = Orange
The colour code for 10^{-2} = Silver
The colour code for 2% tolerance = Red
So, the colour code of the resistor is Blue, Orange, Silver, Red.

(h) $0.47\Omega = 47 \times 10^{-2}\Omega$
The colour code for 4 = Yellow
The colour code for 7 = Violet
The colour code for 10^{-2} = Silver
The colour code for 2% tolerance = Red
So, the colour code of the resistor is Yellow, Violet, Silver, Red.

(i) $0.56\Omega = 56 \times 10^{-2}\Omega$

The colour code for 5 = Green

The colour code for 6 = Blue

The colour code for 10^{-2} = Silver

The colour code for 2% tolerance = Red

So, the colour code of the resistor is Green, Blue, Silver, Red.

(j) $8.2M\Omega = 8.2 \times 10^{6}\Omega \times = 82 \times 10^{5}\Omega$

The colour code for 8 = Grey

The colour code for 2 = Red

The colour code for 5 = Green

The colour code for 10% tolerance = Silver

So, the colour code of the resistor is Grey, Red, Green, Silver.

3. Three resistors of resistances 20Ω, 30Ω and 40Ω are joined in series. What is the combined resistance of the group?

Solution

In series grouping of resistors, the combined resistance R is given by

$$R = R_1 + R_2 + R_3$$

Here, $R_1 = 20\Omega$, $R_2 = 30\Omega$, $R_3 = 40\Omega$

So, $R = (20 + 30 + 40)\ \Omega$

$= 90\Omega$

Thus, the combined resistance of the group is 90Ω.

4. Three conductors whose resistances are 20Ω, 30Ω and 40Ω respectively, were connected in parallel. What is the joint resistance of this group.

Solution

$R_1 = 20\Omega$, $R_2 = 30\Omega$, $R_3 = 40\Omega$

$R = ?$

In parallel grouping of resistors, the combined resistance 'R' of the group is given by

$$\frac{1}{R} = \frac{1}{R_1} + \frac{1}{R_2} + \frac{1}{R_3}$$

$$= \frac{1}{20} + \frac{1}{30} + \frac{1}{40}$$

$$= \frac{6+4+3}{120} = \frac{13}{120}$$

So, $$R = \frac{120}{13}\Omega = 9.23\Omega$$

5. A battery of 1.5Ω internal resistance is connected to two resistors in series of 2Ω and 3Ω. What is the effective resistance of the network?

Solution

$$R_1 = 1.5\Omega,\ R_2 = 2\Omega,\ R_3 = 3\Omega$$

In series combination of resistors, the effective resistance 'R' of the group is given by

$$R = R_1 + R_2 + R_3$$
$$= 1.5\Omega + 2\Omega + 3\Omega$$
$$= 6.5\Omega$$

6. Three resistors of resistances 4Ω, 6Ω and 8Ω respectively form a triangle. Find the effective resistance across each resistor separately.

Solution

$$R_1 = 4\Omega,\ R_2 = 6\Omega,\ R_3 = 8\Omega$$

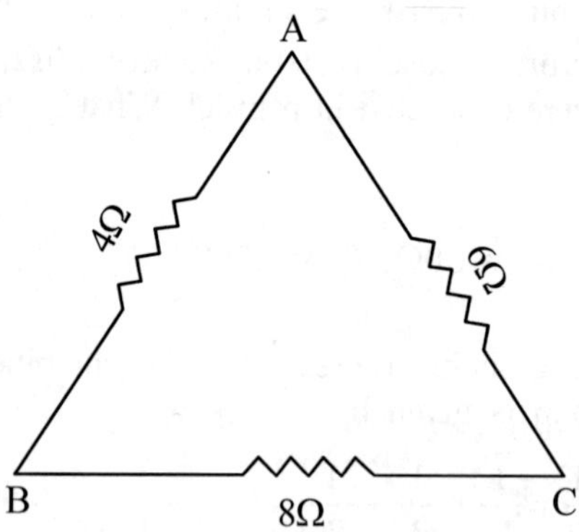

Fig. 1.10

The effective resistance across $AB = R_{AB}$. The resistances 6Ω and 8Ω are in series. So, the combined resistance R^1

$$= (6 + 8)\Omega$$
$$= 14\Omega$$

$$\therefore \; R_{AB} = \frac{R_1 R^1}{R_1 + R^1} = \frac{4 \times 14}{4 + 14} = \frac{56}{18}$$

$$= \frac{28}{9} = 3.11\Omega$$

The effective resistance across $BC = R_{BC}$

The combined resistance of 4Ω and 6Ω in series is $R^{//}$.

$$R^{//} = (4 + 6) = 10\Omega$$

$$\therefore \; R_{BC} = \frac{R_3 R^{//}}{R_3 + R^{//}} = \frac{8 \times 10}{8 + 10}$$

$$= \frac{80}{18} = \frac{40}{9} = 4.44\Omega$$

The effective resistance across $CA = R_{CA}$

The combined resistance of 8Ω and 4Ω in series is $R^{///}$.

$$R^{///} = (4 + 8) = 12\Omega$$

$$\therefore \; R_{CA} = \frac{R_2 \times R^{///}}{R_2 + R^{///}}$$

$$= \frac{6 \times 12}{6 + 12} = \frac{72}{18} = 4.0\Omega$$

7. Sketch the outline of a cube. Imagine that each of the 12 edges represents a conductor having a resistance of 1.0Ω. What is the resistance between corners diagonally opposite each other?

Solution

Let us take a cube whose sides have resistance of 1.0Ω each (Fig. 1.11).

Let a current of $6I$ pass through the junction P. According to Kirchhoff's first law (Current law), the current $6I$ is distributed among three branches as $2I$, $2I$ and $2I$. Likewise, the current meeting at any junction is divided equally among the sides of the cube as the resistance of each side is the 1.0 Ω.

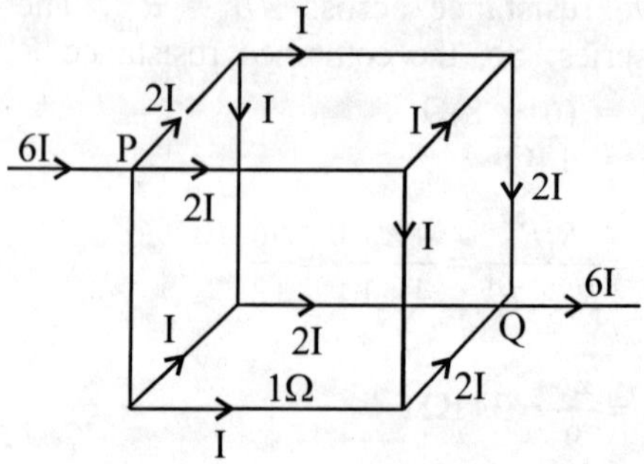

Fig. 1.11

If 'R' is the resistance between P and Q, which are diagonally opposite corners, then

$$6I \times R = 2I \times 1\Omega + I \times 1\Omega + 2I \times 1\Omega$$

$$= 5I\Omega$$

So, $R = \frac{5}{6}\ \Omega$

8. Twelve wires, each of resistance r ohms are joined in the form of a skeleton cube. Find the equivalent resistance of the cube when a cell is joined across any one of the wires forming the cube.

Solution

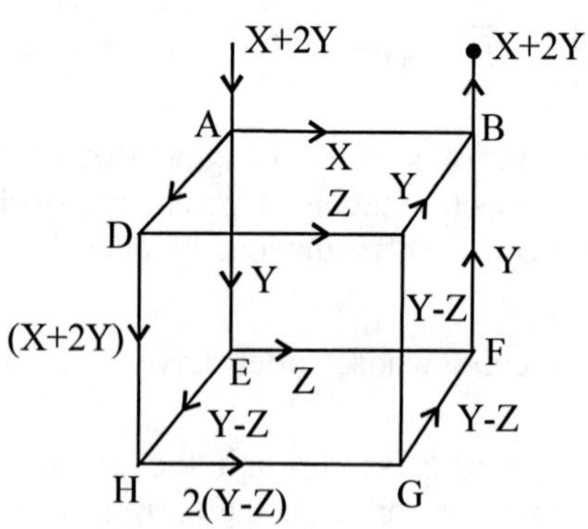

Fig. 1.12

Applying Kirchhoff's 2nd law (Voltage law) to the mesh DCGHD and ABCDA, we get

$$Zr - (y - z)\, r - 2\, (y - z)\, r - (y - z)\, r = 0 \qquad ...(1)$$

and $\quad Xr - yr - zr - yr = 0 \qquad ...(2)$

From equn. (1), we have

$$5z - 4y = 0$$

or, $\quad z = \frac{4}{5}y \qquad ...(3)$

From equn. (2), we have

$$x - 2y - z = 0$$

or, $\quad x - 2y - \frac{4}{5}y = 0$

or, $\quad 2y + \frac{4}{5}y = x$

or, $\quad \frac{14}{5}y = x$

or, $\quad y = \frac{5}{14}x \qquad ...(4)$

Let the equivalent resistance be R.

Then, P.D. across $AB = (x + 2y)\, R = xr$

or, $\quad R = \left(\frac{x}{x+2y}\right)r$

$$= \left(\frac{x}{x + 2.\frac{5}{14}x}\right)r$$

$$= \frac{1}{1+\frac{5}{7}}r$$

$$R = \frac{7}{12}r$$

9. Calculate the value of R in the following network which makes the total resistance of the circuit also equal to R.

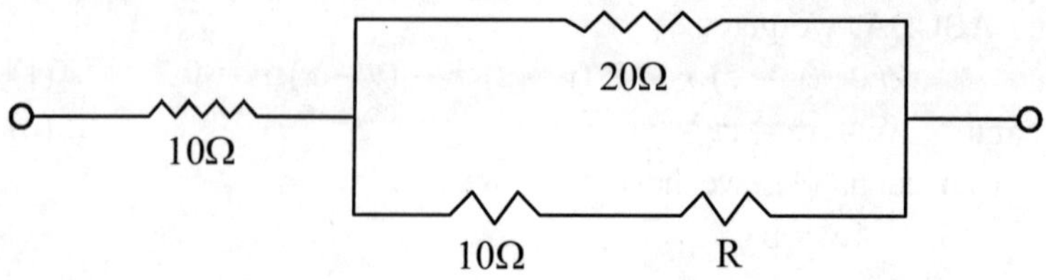

Fig. 1.13

Solution

Since 10Ω and R are in series, the combined resistance of 10Ω and R is $(10 + R)$ ohms.

Now, 20Ωs and $(10 + R)$ ohms are in parallel. So, the effective resistance is

$$R_{eff} = \frac{20 \times (10 + R)}{10 + 10 + R}$$

$$= \frac{200 + 20R}{30 + R}$$

Again, 10Ωs and R_{eff} are in series. So, the combined resistance of these two is $(10 + R_{eff})$ ohms.

According to question, this is equal to R. Hence

$$R = 10 + \frac{200 + 20R}{30 + R}$$

$$= \frac{300 + 10R + 200 + 20R}{30 + R}$$

or, $30R + R^2 = 500 + 30R$

or, $R^2 = 500$

or, $R = \sqrt{500} = 22$ ohms

10. Find the effective resistance across AB of the group of resistors as shown in the following network.

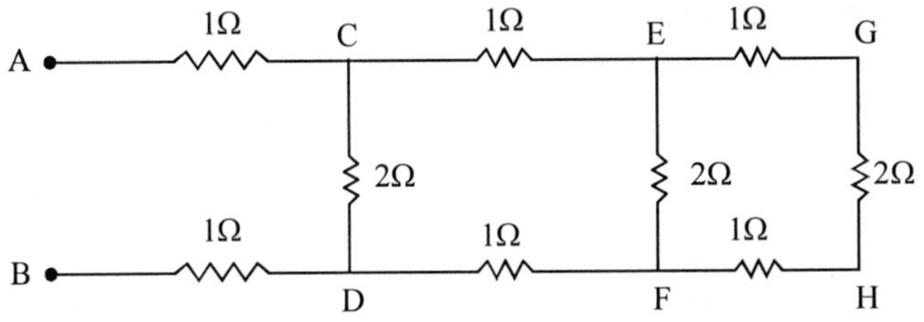

Fig. 1.14

Solution

Resistance in the branches EG, GH and the HF are in series. The effective resistance is

$$R_1 = 1 + 2 + 1 = 4 \text{ ohms}$$

This 4 ohms is parallel with 2 ohms resistance in branch EF. So, resistance across EF is

$$R_2 = \frac{4 \times 2}{4+2} = \frac{8}{6} = \frac{4}{3} \text{ ohms}$$

Now, resistances in the branches CE, EF and FD are in series. The effective resistance is

$$R_3 = 1 + \frac{4}{3} + 1 = \frac{10}{3} \text{ ohms}$$

This R_3 resistance is in parallel with 20 ohms in the branch CD. So, resistance across CD is

$$R_4 = \frac{2 \times R_3}{2 + R_3}$$

$$= \frac{2 \times \frac{10}{3}}{2 + \frac{10}{3}} = \frac{20}{16} = \frac{5}{4} \text{ ohms}$$

Now, the resistance across AB is R which is the series combination of 1Ω, $R_4\Omega$ and 1Ω. So,

$$R = 1 + R_4 + 1$$

$$= 1 + \frac{5}{4} + 1 = 2 + \frac{5}{4} = \frac{13}{4}$$

$$= 3.25 \text{ ohms}$$

Thus, the effective resistance across AB is 3.25 ohms.

❑ ❑ ❑

2

Capacitors

Q. 1. What is a capacitor?

Ans. *Capacitor :* Capacitor is a device which stores electrical charge.

The function of the capacitor is to store a large amount of charge for a given potential difference or to store a given charge at low potential difference. The relation of charge and potential difference in a capacitor is used to define capacitance.

Q. 2. Define capacitance. Mention its MKSa Unit.

Ans. *Capacitance :* The capacitance of a capacitor is defined as the ratio of the charge transferred from one plate to the other of the capacitor, to the potential difference produced between the plates. Mathematically we can write :

$$C = \frac{Q}{V} \quad \text{...(16)}$$

where C is the capacitance of the capacitor, Q is the amount of charge transferred from one plate to the other and V is the potential difference between the two plates.

Units. The MKS unit of capacitance is farad.

1 Farad = 1 coul/1 volt

Other units used for capacitance are micro farad (μF) and pico farad (PF)

$1\ \mu F = 10^{-6}$ F
1 PF $= 10^{-12}$ F

Defn. of Farad. A farad is the capacitance of a capacitor which acquires a potential difference of 1 volt when it receives a charge of 1 coulomb.

Q. 3. How can a capacitor be formed? Give its symbols.

Ans. *Formation of a capacitor :* A simple capacitor can be formed by taking two separate metallic plates with a non-conducting substance (Dielectric) placed in the space between them (fig. 2.1).

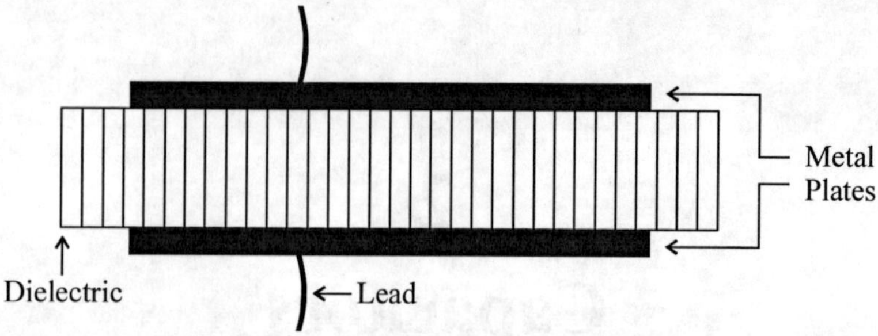

Fig. 2.1

Symbols of capacitor

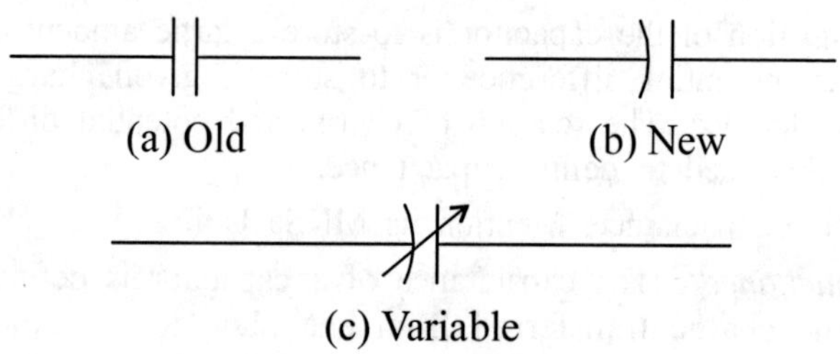

Fig. 2.2

Q. 4. Describe the function of material dielectrics in a capacitor ?

Ans. *Function of material dielectrics in a capacitor* : Air is used as a medium between the two plates of many types of capacitors. Like those in other capacitors, solid and liquid materials such as mica, paraffined paper or oil are used as the material medium between the plates. These materials have good insulating properties as well as they also respond to the presence of the electric field between the plates of the capacitor. As a result of which, the capacitance increases largely. These substances introduced in between the two plates of a capacitor are called dielectrics. The presence of a dielectric between the two plates of a capacitor increases the capacitance of the capacitor.

Q. 5. Deduce the capacitance of a parallel plate capacitor ?

Ans. *Capacitance of a parallel plate capacitor* : Let us consider a parallel plate capacitor whose two plates are charged uniformly to $+Q$ and $-Q$ and separated by a dielectric. The potential difference developed between the two plates is 'V'. Each plate has a surface area 'A'.

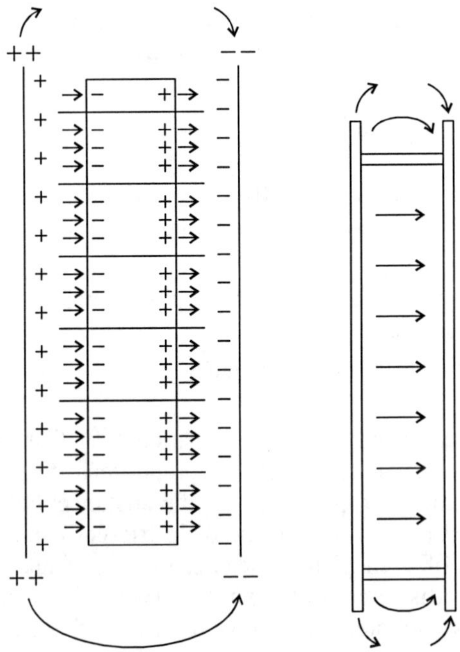

(a) (b)

Fig. 2.3. Parallel plate capacitor.

If the plates are large and their separation small, the electric field intensity between them is every where uniform. The electric field intensity is given by

$$E = \frac{\Psi}{\varepsilon q} = \frac{Q}{\varepsilon A} \qquad ...(17)$$

[$\because$ $\psi = Q$, *i.e.*, the total no. of lines of force emerging from charge is equal to the charge itself.]

From the definition of electric field intensity we have $E = \dfrac{F}{q}$

The potential difference between the two plates is given by,

$$V = \frac{W}{q} = \frac{FS}{q} = \frac{Eq}{q} S = ES$$

$$= \frac{Q}{\varepsilon A} S \qquad ...(18)$$

where S is the separation between the two plates. Now, the capacitance 'C' of the capacitor is given by

$$C = \frac{Q}{V} = \frac{Q}{\frac{Q}{\in A}S} = \frac{\in A}{S}$$

$$C = \varepsilon_r \varepsilon_o \; A/S \qquad \text{...(19)}$$

where ε_r is the relative permittivity or dielectric constant of the dielectric employed between the two plates, ε_o is the permittivity of empty space.

Q. 6. Describe the factors on which the capacitance of a parallel plate capacitor depends.

Ans. Factors on which the capacitance of a parallel plate capacitor depends.

It is now clear from the above equation (19) that the capacitance of a parallel plate capacitor is directly proportional to the area of each plate, 'A'; inversely proportional to the distance between the two plates, S; directly proportional to the dielectric constant, ε_r. This shows that as the area of each plate increases, the capacitance will increase. Increasing the distance between the two plates decreases the capacitance. As the dielectric constant rises, the capacitance also increases.

Q. 7. What are commercial capacitors ?

Ans. *Commercial capacitors :* A multi plate capacitor is formed by joining a number of parallel plate capacitors. That means a large number of plates, each of large area are used to make a commercial capacitor. This capacitor has large capacitance. In such high grade capacitors mica is used as dielectric. Low cost capacitors of capacitance upto 10 μF are usually made of alternate layers of tin or aluminium foil and waxed paper. These are frequently wound into rolls under pressure and sealed into moisture-resisting metal containers. A multi plate capacitor is shown in fig. 2.4.

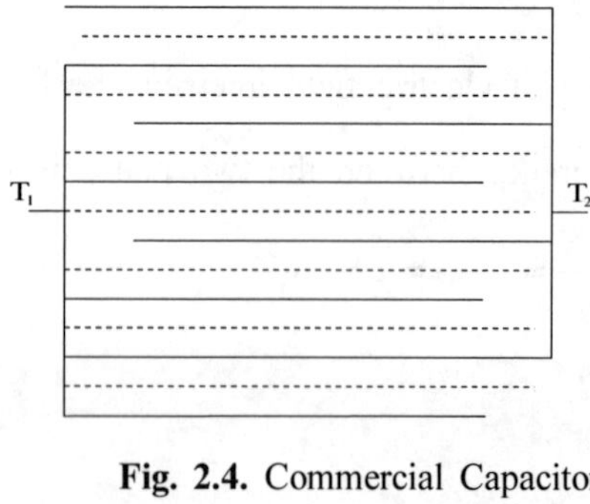

Fig. 2.4. Commercial Capacitor

Q. 8. What do you mean by dielectric strength ?

Ans. *Dielectric strength :* The dielectric strength or breakdown voltage is really a measure of the insulating quality of the material. Dielectric strength is defined as that property of the insulator which determines the maxm. potential gradient which can be applied to the material before its insulating properties are destroyed by a disruptive discharge of electricity through the insulator. The capacitor will withstand certain voltage or potential gradient. After a critical potential gradient, the dielectric strength will breakdown and the capacitor will be destroyed with sparks through the dielectric material. The dielectric strength is always measured in terms of Kilo-volt/centimeter.

TABLE

Dielectric Strength (Average values)

Dielectric material	Breakdown voltage in KV/cm
Air	30
Transformer oil	75
Turpentine	110
Paraffin oil	160
Kerosene	160
Paraffin, solid	250-450
Paraffined paper	300-500
Polystyrene	250-1500
Mica	300-700
Ebonite	300-1000
Glass	300-1600
Vacuum	> 4000

Q. 9. Derive expressions for energy stored in a capacitor ?

Ans. *Energy stored in a capacitor :* When a capacitor is charged energy is stored in the capacitor. When the capacitor is discharged, the energy is released from the capacitor. Therefore during charging of a capacitor, the current in the circuit falls and during discharging of the capacitor, the current in the circuit rises.

Let's consider one uncharged capacitor. This capacitor is gradually charged by transferring small charges 'dq' from one plate to the other. This is done continuously by steps. So, at each step the work done is vdq. Since, initially the potential difference between the two plates is zero (uncharged capacitor) and the potential increases uniformity to a final value 'v' (fully charged capacitor), the average potential difference between the two plates of the capacitor is thus $\frac{O+V}{2} = \frac{V}{2}$. Hence the work done in transferring the total charge Q from one plate to the other is given by

$$W = FS = EQS = \frac{\overline{V}}{S} QS = \frac{1}{2} VQ \qquad ...(20)$$

From the definition of capacitance we have

$$Q = CV$$

Substituting this value of Q in equn. (20) we have,

$$W = \frac{1}{2} V(CV) = \frac{1}{2} CV^2 \qquad ...(21)$$

Substituting the value of V in equn. (20) we have,

$$W = \frac{1}{2}\left(\frac{Q}{C}\right) \times Q = \frac{1}{2}\frac{Q^2}{C} \qquad ...(22)$$

In the above relations 'C' is measured in Farads, 'V' in volts and Q in coulombs and hence the energy is in joules.

Q. 10. What is power factor in capacitors ?

Ans. *Power factor in capacitors :* Power factor of a capacitor represents its percentage rating of losses.

$$\textit{Power Factor} = \frac{\text{loss of energy}}{\text{energy supplied}} \times 100\%$$

With little or no leakage or losses, a capacitor has zero power factor.

Q. 11. What is inductance in capacitors ?

Ans. *Inductance in capacitors :* In all capacitors there is a small value of inductance from the counter emf developed in the leads of the capacitor and the current travelling along the plates. In low frequency this inductance is negligible. In high frequency above 10MHz, the inductance can be taken into account.

Q. 12. What do you mean by variable capacitors ? Where such capacitors are used ?

Ans. *Variable capacitors :* Two types of capacitors are there which can be adjusted or varied. These adjustable capacitors are generally constructed of two or more flat metal plates separated by sheets of mica or plastic. The adjustable capacitors are also called padders or trimmers. These capacitors have capacitance from 30 PF to 1000 PF. The working voltage ranges from 100 to 600 volts.

Variable capacitors are formed of two sets of plates, one stator or stationary plates and the other rotor or rotatable plates. In these capacitors, air is usually used as dielectric. Sometimes mica sheet, plastic or ceramic metals are also used as dielectric. When the shafts are rotated the rotor plates mesh into the spaces between stator plates, without touching. Consequently, the area of the plates vary and hence the capacitance changes. Some special variable capacitors are also available which are called vacuum-dielectric variable capacitors. These capacitors are used in tuning circuits of radio receivers.

Q. 13. Discuss various types of capacitors.

Ans. *Types of capacitors :* Various types of capacitors are in use. These are :

(a) Vacuum capacitor
(b) Air capacitor
(c) Mica capacitor
(d) Ceramic capacitor
(e) Paper capacitor
(f) Plastic capacitor
(g) Electrolytic capacitor
(h) Oil-filled capacitor

(a) Vacuum capacitor. These capacitors are available in both fixed and variable type. They have no leakage. They are used in 5000-50,000 volt service. They have capacitances ranging from 5 to 250 PF. They are efficient for frequency 1000 MH_Z. These capacitors are used in transmitters.

(b) Air capacitor. These capacitors are made in fixed, adjustable and variable type. They have little leakage except through insulators. They are used in both low and high voltage applications. Specifically these capacitors are suitable for receivers and transmitters. Fixed and variable type of air capacitors have capacitance below 500 PF.

(c) Mica capacitor. These capacitors are available as fixed and adjustable type. They have little leakage except through some material between the plates. The working voltage for them ranges from 350 to several thousand volts. Their capacitance varies from 1.5 PF to 0.1 μF. Fixed types are used as RF by pass capacitors. They are used in the frequency 300 MH_Z and more. The efficiency decreases as the frequency drops to 10MH_Z. Adjustable types are used as padders and trimmers.

(d) Ceramic capacitor. Both fixed type and adjustable types are available. They have more low leakage. Their capacitance ranges from

1.5 PF to 0.01 μF for fixed types and upto 100 PF for adjustable types. Their working voltage is about 500 volt and frequency is 300 MH_z or more.

(e) Paper capacitor. Paper covered with oil, wax, poly chlorinated. Biphenyly (PCB) or Ester is used. They have low leakage. This leakage becomes very high with old once due to contamination with moisture. Fixed types of these paper capacitors have frequencies from 10 PF to 10 μF. Working voltage ranges from 150 to several thousand volts. They are useful in the frequency 1 or 2 MH_z. Beyond this frequency, they are less effective because there will be dielectric fatigue and hysteresis loss.

(f) Plastic capacitor. Instead of paper sheets, plastics are also used as dielectrics. These capacitors are available in tubular forms. They work effectively in high frequencies. Their capacitance is about 2 μF. Their working voltage ranges from 200 volt to 600 volt.

(g) Electrolytic capacitors. They are available in fixed types only. They have leakage if used beyond rated voltage. They work in the frequency range from few μF to 50,000 μF or even more. The working voltage is 6 to about 700 volts. These are useful only in d.c. circuits. Tantalum electrolytic capacitors operate at a higher temperatures and hence are used the most. An electrolytic capacitor consists of an Aluminium foil emersed in a solution of electrolyte. The Aluminium-foil serves as the +ve plate and the electrolyte as the –ve plate. To make an electrical connection with the liquid, another Aluminium foil is placed in the solution. To prevent the two foils from touching each other, a piece of gauze is placed between them. Then the combination is rolled into tubular shape and sealed to make air tight. A wire lead from one end connected to one of the foil plates and a wire lead from the other end connected to the other foil plate serve as the connecting leads of the capacitor.

The electrolytic capacitors are normally polarised. A special type of non-polarised electrolytic capacitor may be used during starting periods in a.c. electric motors. These capacitors are relatively small and less expensive in regard to the high value of capacitance.

(h) Oil filled capacitor. In these capacitors, insulating oil is used as the dielectric.

Q. 14. Discuss on the colour code for capacitance.

Ans. *Capacitance colour code :* The colour codes for capacitance are same as for resistors, *i.e.*, 0, 1, 2....9 are represented by the colours black, brown, red, orange... etc. The capacitance obtained from colour code is always in pico farad. In some capacitors, colours are printed and in some other capacitors numbers are printed. In fixed capacitors, however, their capacitance and working voltage are printed.

There are different types of codes, namely :

(a) Three-dot code

(b) Six-dot code

(c) Marked with number of coloured dots or bands.

(a) Three-dot code. This is the simplest code used with 500 volt mica capacitors with a tolerance 20%. A 580 PF colour-coded mica capacitor is shown below in fig. 2.5.

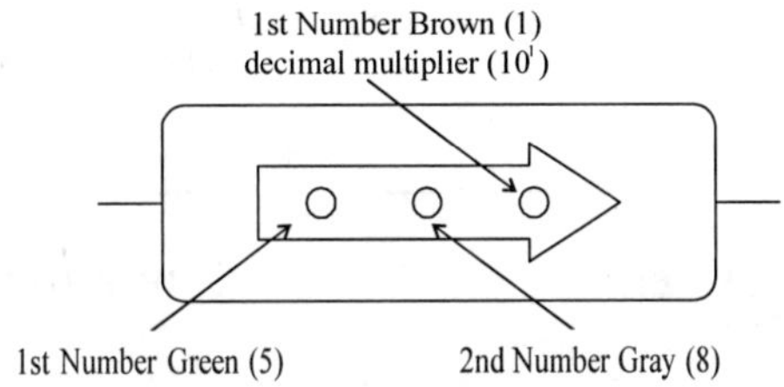

Fig. 2.5. Three-dot code.

(b) Six-dot code. A six-dot coded capacitor is shown in fig. 2.6. There are six-dots which are read clockwise from top left. White indicates mica, the next two dots indicate capacitance numbers, the fourth dot is the capacitance multiplier, the fifth dot gives the tolerance and the sixth dot gives the temperature co-efficient in parts per million (ppm). This increases as the numbers decrease from four. Such type of capacitor has voltage ratings from 300 volt to 500 volt.

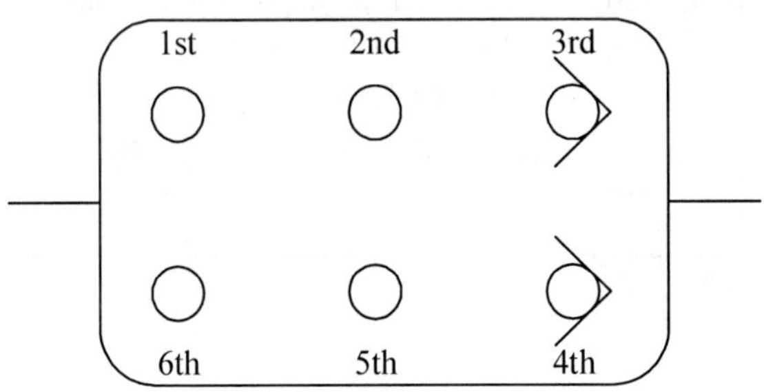

Fig. 2.6. Six-dot code.

Note. The three-dot and six-dot coded capacitors are known as EIA (Electronic Industries Association) code.

The military standard is similar to EIA except that a black first dot indicates a mica dielectric and silver a paper dielectric.

(c) A capacitor having a series of colour dots. There are some ceramic capacitors which are cylindrical in shape and are marked with a number of coloured dots or bands. One example of such a colour coded capacitor are shown below in fig. 2.7.

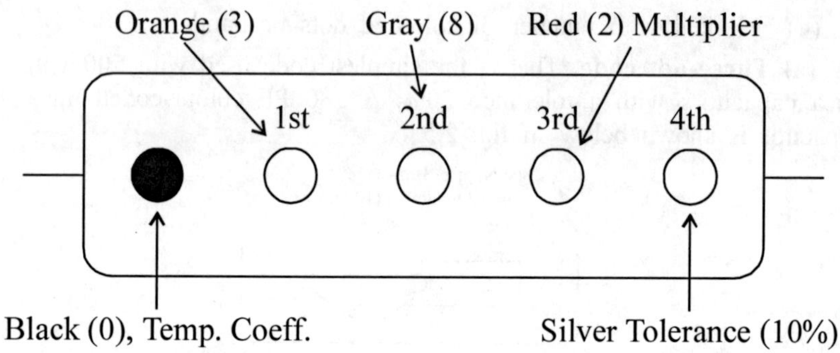

Fig. 2.7

The capacitance of the capacitor is 3800 ± 10%.

Q. 15. What is temperature co-efficient of capacitance ? Explain.

Ans. *Temperature co-efficient of capacitance* : The temperature co-efficient of capacitance is the degree by which the capacitor will change its capacitance with a change in temperature. A capacitance has zero temperature coefficient if it doesn't change with temperature. If the capacitance decreases with temperature it has negative temperature co-efficient. If the temp. co-efficient is –220, then the capacitance will be decreased by 220 ppm per degree centigrade increase in temperature. For ceramic capacitors, temp. co-efficients according to colour codes are given below :

TABLE

Temperature coefficient for capacitor

Colour	TC/°C
Black	0
Brown	– 33
Red	– 75
Orange	– 150
Yellow	– 220

Colour	TC/°C
Green	– 330
Blue	– 470
Violet	– 750
Gray	+ 30
White	+ 500

Q. 16. Derive the laws of series and parallel combination of capacitors.

Ans. *Combinations of capacitors :* Two or more capacitors can be combined in two ways :

(a) In series

(b) In parallel

(a) Series combination of capacitors

Let's consider three capacitors joined in series. A potential difference 'V' is applied across the group of the capacitors. Let C_1, C_2, C_3 be the individual capacitances of the capacitors and V_1, V_2 and V_3 represent the potential differences over the capacitors. Each of these capacitors, holds the same quantity of electricity. This follows from the fact that if the capacitors are initially uncharged, a charge upon one plate always attracts upon the other plate, a charge equal in magnitude and opposite in size.

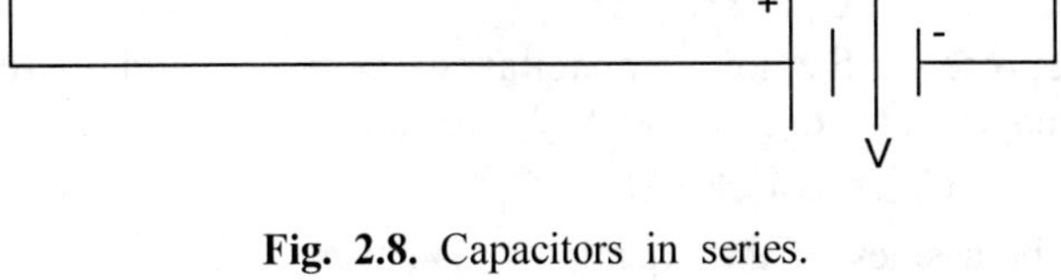

Fig. 2.8. Capacitors in series.

For the series connection it follows that :

$$V = V_1 + V_2 + V_3$$

If we substitute in this equation the various values,

$$V = \frac{Q}{C}, \quad V_1 = \frac{Q}{C_1}, \quad V_2 = \frac{Q}{C_2}, \quad V_3 = \frac{Q}{C_3}$$

We have

$$\frac{Q}{C} = \frac{Q}{C_1} + \frac{Q}{C_2} + \frac{Q}{C_3}$$

Dividing both sides of the equation by Q, we get

$$\frac{1}{C} = \frac{1}{C_1} + \frac{1}{C_2} + \frac{1}{C_3} \quad ...(23)$$

That is, for capacitors connected in series, the reciprocal of the joint capacitance is equal to the sum of the reciprocals of the several capacitances.

(b) Parallel combination of capacitors

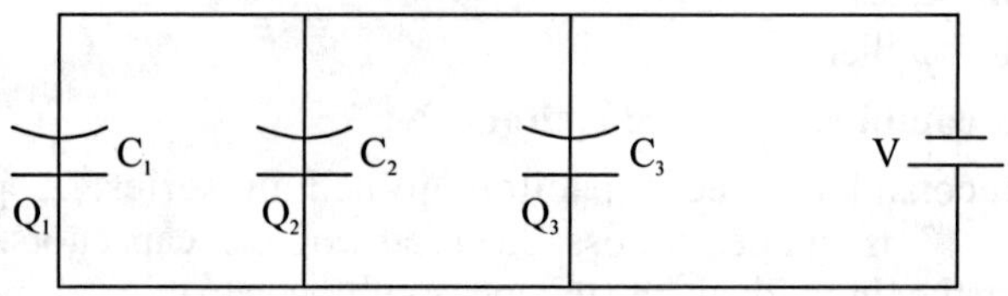

Fig. 2.9. Capacitors in parallel.

In fig. 2.9 are shown three capacitors of separate capacitances, C_1, C_2 and C_3, connected in parallel and joined to a device for establishing a potential difference, such as a cell. Obviously all the capacitors are charged to the same potential, since they are all connected directly to the same source of potential difference. By defn. of capacitance, the charge on each capacitor is,

$$Q_1 = C_1V, \; Q_2 = C_2V, \; Q_3 = C_3V$$

the potential difference over each being the same. The total charge Q is equal to the sum of the separate charges,

$$Q = Q_1 + Q_2 + Q_3$$

If 'C' represents the joint capacitance, $Q = CV$ and therefore by substitution for Q, Q_1, Q_2 and Q_3 we get

$$CV = C_1V + C_2V + C_3V$$

Dividing both sides of the equation by V, we obtain

$$C = C_1 + C_2 + C_3 \quad ...(24)$$

That is, for capacitors connected in parallel, the joint capacitance is the sum of several capacitances.

Q. 17. How can the capacitors be selected ?

Ans. *Selection of capacitors* : Capacitors can be selected by considering the following factors :

(i) **Capacitance.** A capacitor can be replaced by another having the same capacitance.

(ii) **Working voltage.** A capacitor can be chosen considering its working voltage. The rated voltage should be 10 to 20% higher than the max^{m} voltage of the circuit.

(iii) **Type of dielectric.** For radio frequency, the capacitors having mica, air, vacuum or ceramic dielectric should be used. For audio frequency, capacitors having mica, ceramic, paper or plastic dielectric should be used. For d.c. filter circuits, electrolytic capacitors are commonly used.

(iv) **Variable, adjustable or fixed.** A similar capacitor is chosen as the original one.

(v) **Physical size.** Ceramic capacitors are generally smaller than equivalent mica and paper capacitors. Electrolytic capacitors are much smaller than paper capacitors.

(vi) **Cost.** Probable cost per μF increases in the order of electrolytic, ceramic, paper, plastic, mica air, and vacuum capacitors.

Q. 18. What do you mean by a.c. and d.c. resistance of a capacitor ?

Ans. *A.C. and d.c. resistance of a capacitor :* When a capacitor is connected in a d.c. circuit, the plate gets charged up and the electron current drops to zero. The capacitor, thereafter, acts as though it had an infinite resistance.

When a capacitor is connected to an alternating source of emf, its action is different. The reversing of emf reverses the flow of electrons to and from the plates of the capacitor and the alternating flow of charge builds an alternating current. The resistance offered by the capacitor is called capacitive reactance. In an a.c. circuit, it is given by,

$$X_c = \frac{1}{j\omega C}$$

Where '*C*' is the capacity of the capacitor 'ω' is the angular frequency given by $\omega = 2\pi f$, f is the frequency, X_c is the capacitive reactance of the capacitor and j is the imaginary quantity. The magnitude of X_c is $\frac{1}{\omega C}$. The capacitive reactance in an a.c. circuit plays the role of resistance in d.c. circuit. The current in an a.c. circuit having capacitance *C* only is given by

$$i = \frac{E_0}{X_c} e^{j\omega t} \quad ...(25)$$

where $X_c = \frac{1}{j\omega C}$ is called the capacitive reactance of the circuit.

Q. 19. How does capacitance vary with a.c. emf ?

Ans. *Variation of capacitance with a.c. emf :*

In an a.c. circuit

$$E = E_o e^{jwt} \quad ...(26)$$

Where $E = q/C$ ...(27)

Then, we have $C = q/E$ ...(28)

It shows that as a.c. emf increases the capacitance 'C' falls keeping q, the instantaneous charge on the capacitor constant.

EXERCISES

Numerical problems with Solutions

1. Represent the colour code for the following capacitors :
   (a) 470 pf ± 10%
   (b) 620 pf ± 20%
   (c) 33 pf ± 5%
   (d) 66 pf ± 20%
   (e) 82 pf ± 5%

Solution

(a) 470 pf = 47 × 10^1 pf
The colour code of 4 is Yellow
The colour code of 7 is Violet
The colour code of 1 is Brown
The colour code of 10% tolerance is Silver

So, the colour code of the capacitor is Yellow, Violet, Brown, Silver.

(b) 620 pf = 62 × 10^1 pf
The colour code of 6 is Blue
The colour code of 2 is Red
The colour code of 1 is Brown
The colour code of 20% tolerance is No colour

So, the colour code of the capacitor is Blue, Red, Brown.

(c) 33 pf = 33 × 10^0 pf
The colour code of 3 is Orange

The colour code of 3 is Orange
The colour code of 0 is Black
The colour code of 5% tolerance is Gold

So, the colour code of the capacitor is Orange, Orange, Black, Gold.

(d) $66 \text{ pf} = 66 \times 10^0 \text{ pf}$
The colour code of 6 is Blue
The colour code of 6 is Blue
The colour code of 0 is Black
The colour code of 20% tolerance is No colour

So, the colour code of the capacitor is Blue, Blue, Black.

(e) $82 \text{ pf} = 82 \times 10^0 \text{ pf}$
The colour code of 8 is Grey
The colour code of 2 is Red
The colour code of 0 is Black
The colour code of 5% tolerance is Gold

So, the colour code of the capacitor is Grey, Red, Black, Gold.

2. Find the value of the capacitance from the following colour code of the capacitors :
   (a) Green, Grey, Brown
   (b) Red, Green, Red, Silver
   (c) Orange, Grey, Red, Silver
   (d) Red, Red, Brown, Gold
   (e) Violet, Orange, Black

Solution

(a) Green, Grey, Brown

First colour Green represents the number 5, second colour Grey represents the number 8, third colour Brown represents 10^1. As there is no fourth colour, tolerance is 20%.

Hence, the capacitance is $58 \times 10^1 \text{ pf} \pm 20\%$
$= 580 \text{ pf} \pm 20\%$

(b) Red, Green, Red, Silver
First colour Red represents the number 2
Second colour Green represents the number 5
Third colour Red represents 10^2
Fourth colour Silver represents 10% tolerance

Hence, the capacitance is
$25 \times 10^2 \text{ pf} \pm 10\%$
$= 2500 \text{ pf} \pm 10\%$

(c) Orange, Grey, Red, Silver
First colour Orange represents the number 3

Second colour Grey represents the number 8
Third colour Red represents 10^2
Fourth colour silver represents 10% tolerance

Hence, the capacitance is

$$38 \times 10^2 \text{ pf} \pm 10\%$$
$$= 3800 \text{ pf} \pm 10\%$$

(d) Red, Red, Brown, Gold
First colour Red represents the number 2
Second colour Red represents the number 2
Third colour Brown represents 10^1
Fourth colour Gold represents 5% tolerance

Hence, the capacitance is

$$22 \times 10^1 \text{ pf} \pm 5\%$$
$$= 220 \text{ pf} \pm 5\%$$

(e) Violet, Orange, Black
First colour violet represents the number 7
Second colour Orange represents the number 3
Third colour Black represents 10°
As there is no fourth colour, the tolerance is 20%.

Hence, the capacitance is

$$73 \times 10^o \text{ of} \pm 20\%$$
$$= 73 \text{ pf} \pm 20\%$$

3. Three capacitors have capacitances 0.5, 0.3 and 0.2 µf. What is their joint capacitance when arranged to give (a) a minimum capacitance and (b) a maximum capacitance?

Solution

In series,

$$\frac{1}{C} = \frac{1}{0.5\mu f} + \frac{1}{0.3\mu f} + \frac{1}{0.2\mu f}$$

$$= \frac{10}{5} + \frac{10}{3} + \frac{10}{2}$$

$$= 2 + 3.33 + 5$$
$$= 10.33$$

$$C = \frac{1}{10.33} = 0.0968 \ \mu f$$

In parallel,

$$C = C_1 + C_2 + C_3$$
$$= (0.5 + 0.3 + 0.2) \ \mu f$$
$$= 1.0 \ \mu f$$

4. Six 1/2 μf capacitors are connected first in series and then in parallel. What are the respective joint capacitances of the combination.

Solution

In series,

$$\frac{1}{C}=\frac{1}{C_1}+\frac{1}{C_1}+\frac{1}{C_1}+\frac{1}{C_1}+\frac{1}{C_1}+\frac{1}{C_1}$$

$$=\frac{6}{C_1}=\frac{6}{\frac{1}{2}\mu f}=\frac{12}{\mu f} \text{ or } C=\frac{1}{12}\mu f=0.083\ \mu f$$

In parallel,

$$C=C_1+C_1+C_1+C_1+C_1+C_1=6C_1$$

$$C=6\times\frac{1}{2}=3\mu f$$

5. Three capacitors of capacitance 0.1, 0.2, and 0.5 μf, respectively, are connected in parallel and the group then connected in series with another group of 0.1, 0.2 and 0.5 μf, respectively, connected is series. Calculate the total capacitance.

Solution

In parallel,

$$C = C_1 + C_2 + C_3$$
$$= (0.1 + 0.2 + 0.5)\ \mu f$$
$$= 0.8\ \mu f$$

In series,

$$\frac{1}{C^1}=\frac{1}{C}+\frac{1}{C_1}+\frac{1}{C_2}+\frac{1}{C_3}$$

$$=\frac{1}{0.8}+\frac{1}{0.1}+\frac{1}{0.2}+\frac{1}{0.5}$$

$$=\frac{10}{8}+\frac{10}{1}+\frac{10}{2}+\frac{10}{5}$$

$$=\frac{5}{4}+10+5++2$$

$$=17+\frac{5}{4}=\frac{73}{4}$$

$$\therefore C^1=\frac{4}{73}\mu f=0.054\ \mu f$$

6. A conductor is charged to a potential of 100 volts and has the electrical capacity of 2 microfarad (μf). What is the charge on the conductor?

Solution

$V = 100$ volts, $C = 2\ \mu f$

$C = Q/V \quad \Rightarrow \quad Q = CV$

$= 2 \times 100\ \mu C$

$= 200\ \mu C$

7. Find the capacity of a spherical charged conductor in vacuum whose radius is equal to the radius of the earth.

Solution

$\varepsilon_0 = 8.85 \times 10^{-12}$ coul²/N.m², $r = 6.5 \times 10^6$m

$C = 4\pi\varepsilon_0 r$

$= 4 \times 3.142 \times 8.85 \times 10^{-12} \times 6.5 \times 10^6$

$= 723.17 \times 10^{-6}$ farad

$= 723.17\ \mu f$

8. A capacitor is charged to 100 V. What is charge on its positive plate if its capacitance is 2 μf ?

Solution

$C = 2\ \mu f, \quad V = 100\ V$

$Q = CV = 2\ \mu f \times 100\ V$

$= 200\ \mu C$

9. 'n' identical oil drops, each of capacity 'C' are charged to a potential of V volts each. The radius of each drop is 'R', and the charge of each drop is 'Q' coulomb. These 'n' oil drops are merged to form a bigger drop of radius 'R_1'. Find the capacitance of this bigger drop.

Solution

$$\text{Volume of bigger drop} = \frac{4}{3}\pi R_1^3$$

$$\text{Volume of small drops} = n \times \frac{4}{3}\pi R^3$$

So, $$\frac{4}{3}\pi R_1^3 = n \times \frac{4}{3}\pi R^3$$

or, $$R_1 = n^{1/3} R$$

We know that the capacity of a sphere is proportional to its radius.

Hence, $\frac{C_1}{C} = \frac{n^{1/3}R}{R} = n^{1/3}$

or, $C_1 = n^{1/3}C$

10. The area of each plate of a capacitor is $1m^2$ and are separated by 0.1 cm. Find the capacitance considering air as the dielectric.

Solution

$A = 1\ m^2,\ S = 0.1 \times 10^{-2}\ m$

$C = \varepsilon_0\ A/S$

$= \frac{8.85 \times 10^{-12} \times 1}{0.1 \times 10^{-2}}$

$= 8.85 \times 10^{-9}\ f$

11. A multiplate commercial capacitor is made of 350 plates separated by fine sheets of mica ($\varepsilon_r = 2.5$) of thickness 10^{-5} m. The effective area of the plates is 45×10^{-3} m^2. Find the capacitance.

Solution

$2N = 350$ plates

$\varepsilon_r = 2.5$

$t = 10^{-5}\ m, \quad S = 10^{-5}/350\ m$

$A_{eff} = 45 \times 10^{-3}\ m^2, \quad A = \frac{45 \times 10^{-3}}{350} m^2$

$C_{eff} = ?$

$C = \varepsilon_r\varepsilon_o \times A/S$

$= \frac{2.5 \times 8.85 \times 10^{-12} \times 45 \times 10^{-3}}{350 \times 10^{-5}/350}$

$= 2.5 \times 8.85 \times 45 \times 10^{-10} = 995.625 \times 10^{-10}\ f$

$C_{eff} = (2N - 1)\ C$

$= (350 - 1) \times 995.625 \times 10^{-10} f$

$= 349 \times 995.625 \times 10^{-10}\ f$

$= 347473 \times 10^{-10} f$

$= 34.7473 \times 10^{-6}\ f$

$= 34.7473\ \mu f$

$\simeq 35\ \mu f$

12. A 10 μf capacitor is charged to a potential of 400 volts and is then discharged through a conductor. Find the energy given to the conductor due to discharge of the capacitor.

Solution

$C = 10\ \mu f$

$V = 400$ volts

Energy of the charged capacitor

$$= \frac{1}{2}CV^2$$

Energy given to the conductor

$$= \frac{1}{2} \times 10 \times 10^{-6} \times \left(4 \times 10^2\right)^2 \text{ Joule}$$

$$= \frac{1}{2} \times 10^{-5} \times 16 \times 10^4 \text{ Joule}$$

$= 8 \times 10^{-1}$ Joule

$= 0.8$ Joule.

13. A 2000 μf capacitor is charged at a steady rate of 25 μC per second. After what time the P.D. across the capacitor plates shall be 20 volts.

Solution

Total charge on the capacitor plates,

$Q = CV$

$= 2000\ \mu f \times 20$ volts

$= 4 \times 10^4\ \mu C$

Time needed for charging the capacitor to 20 volts is

$$t = \frac{4 \times 10^4}{25} = 1600\ S$$

❑ ❑ ❑

3

Inductors

Q. 1. What is an Inductor ?

Ans. *Inductor :* Inductor is a device that stores energy in the form of magnetic field. A straight wire bent into the form of a coil possesses inductance which is known as self inductance of the coil.

Q. 2. How does an inductor act ?

Ans. *Function :* An inductor offers resistance to the flow of a.c. through it. A pure inductance doesn't offer any resistance to the flow of d.c. through it.

Q. 3. Discuss on various types of inductors.

Ans. *Types of inductors :* Different types of inductors are :

(a) An air core inductor

(b) Iron core inductor

(c) A special type of inductor called toroid

(a) An air core inductor. In most electronic applications of inductance where space is limited, wire is wound into either single-layer or multi-layer coils with air. Such an inductor is known as air core inductor.

(b) Iron core inductor. When an inductor has core made of powdered-iron compound (Ferrite) or laminated (thin iron sheet) cores, it is called an iron core inductor. Ferrite materials like zinc or manganese-iron oxides have magnetic permeability but are insulators. Therefore a coil with a ferrite core makes an inductor of inductance many times more than that of an air core inductor. In many applications, coils are wound around a ferrite cylinder that can be screwed into or out of the core space of the coil. This results in a controlled variations of inductance. The inductor has maxm. inductance when the iron core (slug) is screwed into the coil and the inductance is minimum when the slug is backed out.

(c) A special type of inductor or toroid. A special type of inductor called toroid which consists of a doughnut-shaper ferrite core, either single-layer-wound as shown in the figure below or multi-layer-wound. Such an inductor has high values of inductance with less wire and therefore little resistance in the coil. Moreover all the lines of force remain in the core. As a result of which, no shielding is required to prevent its field from surrounding circuits or electrical appliances. Even two toroids very close to each other will have no interaction between them.

Q. 4. Give the symbols of inductors.

Ans. *Inductor symbols :*

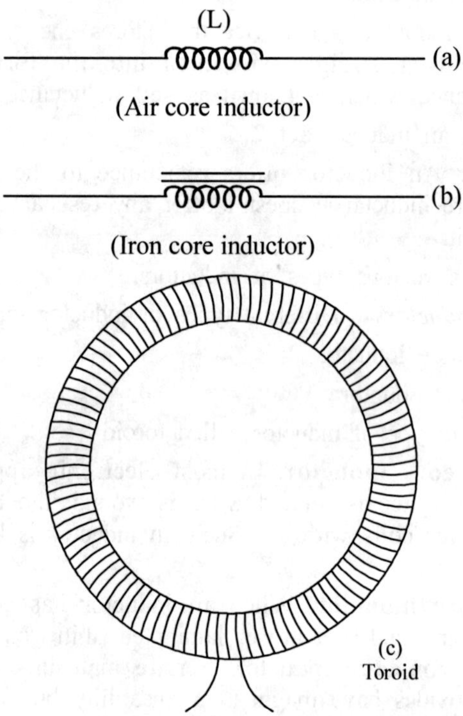

Fig. 3.1. (a) Air core inductor, (b) Iron core inductor, (c) A special type of inductor or toroid.

Q. 5. What are variable inductors ?

Ans. *Variable inductors :* There are some inductors, mainly iron core inductors whose inductance is required to be varied in accordance with the need of the applications. In such inductors, coils are wound around a ferrite cylinder which is screwed into or out of the core space of the coil. Consequently, there is a controlled variations of inductance. In such cases, the inductance may be maximum or minimum depending upon the position of the "slug".

Q. 6. Deduce the law of combination of inductors in series and in parallel.

Ans. *Combination of inductors :* Inductors can be combined in two ways.

(a) In series
(b) In parallel

Self-inductances combine like the series and parallel combinations of resistances.

(a) Inductors in series. When inductances are placed in series. The total induced emf is the sum of the separate emf's. Hence, if 'i' is the current flowing through the inductances and 'L' is the resultant or equivalent inductance of the series, then

$$L\frac{di}{dt} = L_1\frac{di}{dt} + L_2\frac{di}{dt} + \ldots\ldots + L_n\frac{di}{dt}$$

Since,

$$i_1 = i_2 = \ldots\ldots = i_n = i$$

So, $$L = \sum_{1}^{n} L_j \qquad \ldots(29)$$

Where $j = 1, 2, 3.......n$

Expanding equn. (29) we have,

$$L = L_1 + L_2 + L_3 \ldots\ldots L_n \qquad \ldots(30)$$

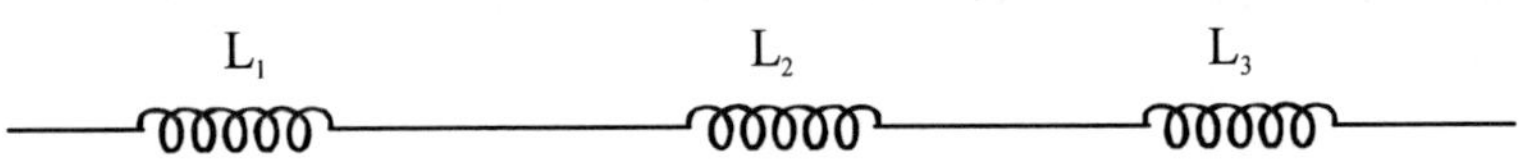

Fig. 3.2. Inductors in series

(b) Inductors in parallel : When inductors are joined in parallel the induced emf across each inductor is the same as the induced emf across the combination. But the currents are different in different inductors. Let the current in each inductor be i_1, i_2.... i_n respectively and the current across the combination is i. Then $i = i_1 + i_2 + i_3 + \ldots\ldots i_n$ = total current

Thus, $L\frac{di}{dt} = L_1\frac{di_1}{dt} = L_2\frac{di_2}{dt} = ... = L_n\frac{di_n}{dt}$ ...(31)

From the above equation we get

$$\frac{di_1}{dt} = \frac{L}{L_1}\frac{di}{dt}$$

$$\frac{di_2}{dt} = \frac{L}{L_2}\frac{di}{dt}$$

$$- \quad - \quad - \quad -$$

$$\frac{di_n}{dt} = \frac{L}{L_n}\frac{di}{dt}$$

Adding the above relations we get,

$$\left(\frac{L}{L_1}+\frac{L}{L_2}+\frac{L}{L_3}+....+\frac{L}{L_n}\right)\frac{di}{dt} = \frac{di_1}{dt}+\frac{di_2}{dt}+\frac{di_3}{dt}+....\frac{di_n}{dt}$$

But $\frac{di}{dt} = \frac{di_1}{dt}+\frac{di_2}{dt}+.....\frac{di_n}{dt}$

So, we have, $\frac{L}{L_1}+\frac{L}{L_2}+\frac{L}{L_3}+.....+\frac{L}{L_n} = 1$

or, $\frac{1}{L} = \frac{1}{L_1}+\frac{1}{L_2}+\frac{1}{L_3}+....+\frac{1}{L_n}$ ...(32)

or, $\frac{1}{L} = \sum_{j=1}^{n}\frac{1}{L_j}$ ...(33)

Where, $j = 1, 2, 3, 4 .... n.$

Therefore, the sum of the reciprocals of individual inductance is equal to the reciprocal of the combined inductance of all inductors joined in parallel.

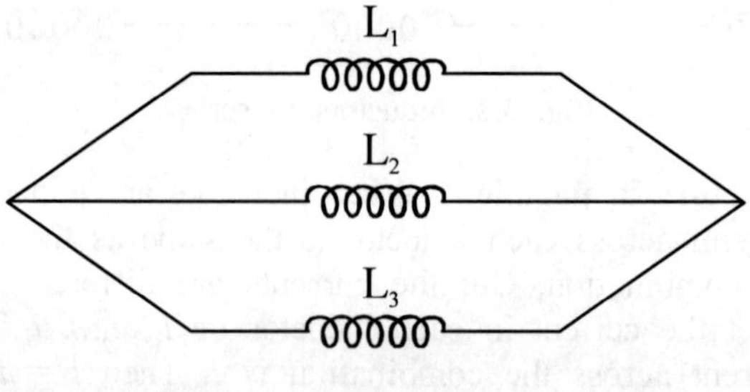

Fig. 3.3. Inductors in parallel

Q. 7. Define Inductance. Mention its units.

Ans. *Inductance :* A coil which produces induction effect in itself by a current flowing through it has co-efficient of self-induction or inductance L.

Definition of L : Self-inductance L of an inductor is defined as the magnetic flux linked with it when a unit current flows through it. Mathematically, it is given by

$$L = \frac{\phi}{i} \qquad ...(34)$$

When $i = 1, \ \phi = L$ ...(35)

where ϕ is the flux linked with the coil when the current is unity.

Units of inductance

MKS — Henry

$$1 \text{ Henry} = \frac{1 \text{ volt}}{1 \text{ amp./sec.}}$$

CGS — Emu of inductance

$$1 \text{ Emu of inductance} = \frac{1 \text{ emu of potential difference}}{1 \text{ emu of current / sec.}}$$

Relation between Henry and emu of inductance

1 Henry = 10^9 emu of inductance.

Q. 8. Derive an expression for the energy stored in an inductor.

Ans. *Energy stored in an inductor :* When current through an inductor changes, the flux linked with inductor also changes. The rate of change of flux induces an emf in the inductor which opposes the inducing current. To increase the current in the inductor energy must be spent externally. This energy gets stored in the inductor.

From Faraday's law of electro magnetic induction, we have,

$$e = -\frac{d\phi}{dt}$$

But $\phi = Li$

Hence
$$e = -\frac{d(Li)}{dt}$$

$$= -L\frac{di}{dt}$$

Electrical work done by the field = $ei\,dt$

$$= L\frac{di}{dt}(idt)$$

$$dw = Lidi$$

Total work done by a current i_0 in the inductor is given by

$$\int_0^w dw = \int_0^{i_0} Lidi$$

$$w = L\int_0^{i_0} idi$$

$$= L\left(\frac{i^2}{2}\right)_0^{i_0}$$

$$w = \frac{1}{2}Li_0^2 \qquad ...(36)$$

where 'L' is the self-inductance of the inductor and w is the total energy stored in the inductor.

Q. 9. What do you mean by d.c. and a.c. resistance of an inductor?

Ans. *A.C. and D.C. resistance* : As the inductor has negligible resistance it doesn't oppose the flow of d.c. But when alternating current passes through an inductor, induced emf is set up, which opposes the a.c. This opposition to a.c. is measured by inductive reactance X_L or a.c. resistance of an inductor.

$$X_L = \omega L = 2\pi fL \qquad ...(37)$$

where 'L' is the self-inductance of the coil measured in Henry and 'f' is the frequency of applied a.c. measured in Hertz.

Q. 10. How does inductance vary with emf ?

Ans. *Variation of L with emf* : In an inductive circuit, the current is given by,

$$i = i_0 \sin\left(\omega t - \frac{\pi}{2}\right)$$

where $i_0 = \dfrac{E_0}{\omega L}$

or, $L = \dfrac{E_0}{\omega i_0}$

where E_0 is the peak value of alternating emf applied across L and i_0 is the peak value of alternating current flowing through L. ω is the angular frequency given by

$$\omega = 2\pi f$$

where f is the frequency of the applied a.c. emf.

Thus inductance of the coil varies with the emf.

Q. 11. Write down the uses of an inductor.

Ans. *Application of inductor :* An inductor is generally used as a choke. An air core inductor maintains constant inductance even at a very high current while an iron core looses its ability to smooth out variations beyond that current. To prevent this, a small air gap may be left in the iron core to introduce sufficient reluctance in the magnetic circuit.

Q. 12. What is a choke coil? Mention various types of chokes.

Ans. *Types of chokes :* An inductor is known as a choke coil since it chokes out variation of amplitudes. There are different types of choke coils depending upon the reactances. These are :

(a) Radio frequency choke (RFC)
(b) Audio frequency choke (AFC)
(c) Power frequency choke (PFC)

(a) Radio frequency choke (RFC)

For radio frequency applications (20 KH_z to 500 MH_z), an air-core coil with low value of inductance is used as RFC.

(b) Audio frequency choke (AFC)

In case of lower frequency circuits, coils having greater inductance are required. For this reason iron core choke coils with higher inductance are used as AFC. Here the frequency is limited to 20 KH_Z.

A laminated iron core choke :

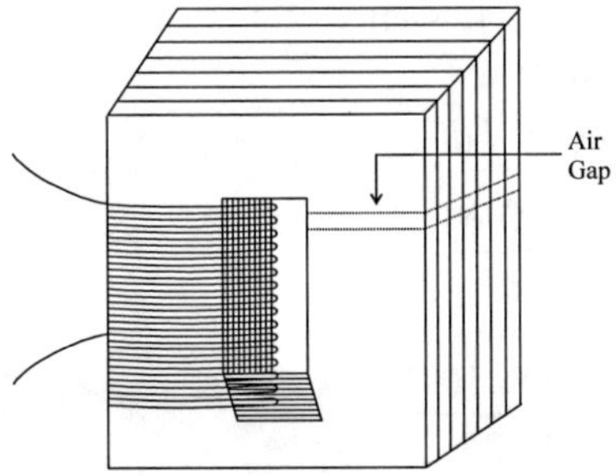

Fig. 3.4

(c) Power frequency choke (PFC)

The power frequency is about 50 Hz. Therefore inductors having still higher inductances are required for chokes.

EXERCISES

Numerical problems with solutions

1. Find out the self-inductance of a coil of 100 turns when a current of 5 amp passing through it produces a magnetic flux of 0.5 × 10^{-14} weber.

Solution

$N = 100$ turns

$I = 5$ amp

$\phi = 0.5 \times 10^{-4}$ weber

Here, $N\phi = LI$

or, $$L = \frac{N\phi}{I}$$

$$= \frac{100 \times 0.5 \times 10^{-4}}{5}$$

$= 1000 \times 10^{-5}$ henry

$= 1000 \times 10^{-6}$ henry

$= 1000\ \mu H$

2. Find out the co-efficient of self-inductance of a coil of 500 turns when a change of current of 1.25 amperes produces a magnetic flux of one microweber.

Solution

$N = 500$ turns

$dI = 1.25$ amps

$d\phi = 1\ \mu$ weber $= 1 \times 10^{-6}$ weber

$L = ?$

Here, $Nd\phi = LdI$

or, $$L = \frac{Nd\phi}{dI}$$

$$= \frac{500 \times 10^{-6}}{1.25}$$

$= 400 \times 10^{-6}$ henry

$$= 0.4 \times 10^{-3} \text{ henry}$$
$$= 0.4 \text{ millihenry}$$

3. The current in a circuit drops from 10A to $2A$ in $0.1S$. An average emf of 32 V is induced in the circuit while this happens. Find the inductance of the circuit.

Solution

$$I_i = 10A, \quad I_2 = 2A$$
$$dI = I_2 - I_1 = 2 - 10 = -8A$$

Induced emf. $= 32$ V, $t = 0.1\ S$

$$e = -L\frac{di}{dt}$$

$$32 = -L\left(\frac{-8}{0.1}\right)$$

$$= L\ (80)$$

$$L = \frac{32}{80} = \frac{2}{5} = 0.4 \text{ henry}$$

4. Two coils are coupled together. When current in one of the coils decays from $1A$ to Zero in $0.12S$, emf of 1V is induced across the second. Find the co-efficient of mutual induction.

Solution

$$dI = -1A$$
$$dt = 0.12\ S$$
$$e = 1V, \ M = ?$$

Here, $e = -M\dfrac{dI}{dt}$

$$1 = -M\left(-\frac{1}{0.12}\right)$$

$$= \frac{M}{.12}$$

or, $M = 0.12$ henry

5. A circuit in which there is a current of $5A$ is changed so that the current falls to Zero in 0.1 S. If an average emf of 200 V is induced, what is the self-inductance of the circuit ?

Solution

$$dI = -5\ A$$
$$dt = 0.1\ S$$
$$e = 200\ \text{V}$$
$$L = ?$$

Here, $e = -L\dfrac{dI}{dt}$

$$200 = -L\left(\frac{-5}{0.1}\right)$$
$$= L\ (50)$$

or, $L = \dfrac{200}{50} = 4$ henry

6. The current in a coil of 325 turns is changed from zero to 6.32 A, thereby producing a flux of 8.46×10^{-4} wb. What is the self-inductance of this coil ?

Solution

$$N = 325 \text{ turns}$$
$$I = 6.32 \text{ Amps}$$
$$\phi = 8.46 \times 10^{-4} \text{ wb}$$
$$L = ?$$

Here, $N\phi = LI$

or, $L = \dfrac{N\phi}{I}$

$$= \frac{335 \times 8.46 \times 10^{-4}}{6.32} \text{ henry}$$
$$= 435 \times 10^{-4} \text{ henry}$$
$$= 43.5 \times 10^{-3} \text{ henry}$$
$$= 43.5 \text{ millihenry}$$
$$= 43.5 \text{ mH}$$

7. An air-core solenoid is 50 cm long and has 120 turns per centimeter and a diameter of 12 cm. What is the self-inductance of the solenoid ?

Solution

$$l = 50 \text{ cm}$$

$$n = 120 \text{ turns/cm}, \; N = 50 \times 120 \text{ turns}$$

$$d = 12 \text{ cm}, \; r = \frac{d}{2} = 6 \text{ cm}$$

$$L = ?, \; \mu_0 = 4\pi \times 10^{-7} \text{ wb/A.m}$$
$$= 4\pi \times 10^{-9} \text{ wb/A.cm}$$

Here, $$L = \frac{\mu_0 N^2 A}{L} = \frac{4\pi \times 10^{-9} \times (6000)^2 \pi(6)^2}{50}$$

$$= \frac{4\pi^2 \times 36 \times 36 \times 10^6 \times 10^{-9}}{50} = \frac{4 \times 9.87 \times 1296 \times 10^{-4}}{5}$$

$$L = 10233 \times 10^{-4} \text{ henry}$$
$$= 1023.3 \times 10^{-3} \text{ henry}$$
$$= 1023.3 \text{ millihenry}$$
$$= 1023.3 \text{ mH}$$

8. Two coils A and B are placed near each other. A current change of $6A$ in A produces a flux change of 12×10^{-4} Wb in B, which has 2000 turns. What is the mutual inductance of the system.

Solution

$$dI = 6A, \; d\phi = 12 \times 10^{-4} \text{ wb}$$
$$N = 2000 \text{ turns}, \; M = ?$$

Here, $Nd\phi = MdI$

or, $$M = \frac{Nd\phi}{dI} = \frac{2 \times 10^3 \times 12 \times 10^{-4}}{6}$$

$$= 400 \times 10^{-3} \text{ henry}$$
$$= 400 \text{ mH}$$

9. In an inductor, a current of $5A$ is maintained. If the self-inductance of the inductor is 100 mH, find the energy stored in the inductor.

Solution

$$I = 5A, \; L = 100 \text{ mH} = 100 \times 10^{-3} \text{ H}$$
$$E = ?$$

Here, $$E = \frac{1}{2} LI^2$$

$$= \frac{1}{2} \times 100 \times 10^{-3} \times (5)^2$$

$$= \frac{1}{2} \times 10^{-1} \times 25$$

$= 12.5 \times 10^{-1}$ Joules

$= 1.25$ Joules.

10. What will be the self-inductance of an inductor if it stores up 20 joules for a current 4 A in it ?

Solution

$E = 20$ Joules, $I = 4\ A$

$L = ?$

Here, $E = \frac{1}{2} LI^2$

or, $$L = \frac{2E}{I^2} = \frac{2 \times 20}{(4)^2} = \frac{40}{16} H$$

$$= \frac{5}{2} H = 2.5\ H$$

❑ ❑ ❑

4

Electromagnetic Induction

Q. 1. What do you mean by electromagnetic induction ? Explain.

Ans. When there is a relative motion between a coil and a magnet, there is change in magnetic flux which causes a current in the coil. This phenomenon is called electromagnetic induction. This was discovered by Faraday in 1831.

Definition of electromagnetic induction :

The process of producing electric current or emf by the relative mation between a coil and a magnet is known as electromagnetic induction. In this process the electric current is produced by the magnetic field. The current so produced is called induced current and the emf which causes induced current is called induced emf.

Experimental remonstration of the phenomenon of electromagnetic induction :

Michael Faraday of England and Joseph Henry of America independently demonstrated that the change of magnetic flux linked with a coil induces an emf in the coil. The induced emf is found to oppose the inducing emf by Lenz.

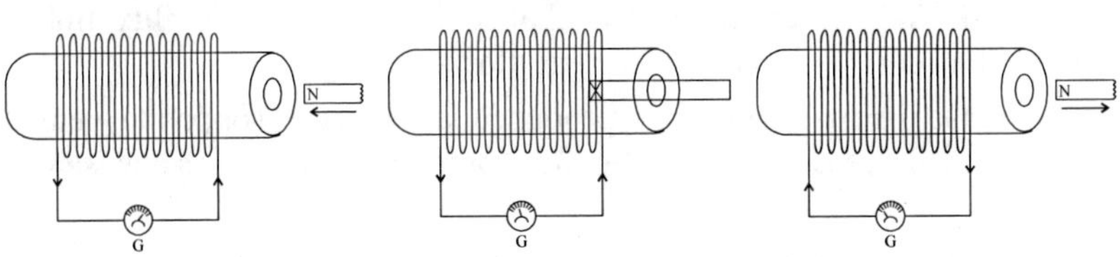

(a) Magnet moving towards the coil

(b) Magnet is stationary

(c) Magnet moving away from the coil

Fig. 4.1

The above figure 4.1 represents a coil of wire connected to a sensitive galvanometer 'G'. If the north pole of a bar magnet is moved towards the coil, there is deflection in the galvanometer indicating a momentary current in the coil in the anti clockwise direction [fig. 4.1(a)]. This current which is set up in the coil due to the change of flux linked with the coil is known as the induced current and the process of generating the induced emf is known as electromagnetic induction.

When the bar magnet remains at rest within the coil, no current is induced so there is no deflection by the galvanometer [fig. 4.2(b)]. When the magent is quickly removed from the coil, the galvanometer will indicate a current in the direction opposite to that observed in the 1st case. This current is in clockwise direction.

Conclusion. By conducting a series of such experiments, Faraday found that an emf is induced in a coil (conductor) when there is any change of magnetic flux linked with the conductor.

Note

(i) In the above experiments, magnet is moved towards or away from the coil and induction effect is noticed. The same effect can also be produced by moving the coil (conductor) towards a fixed magnet.

(ii) An emf may also be induced in a coil by the change in the magnetic field association with a change in current in a near by circuit. This is due to the fact that the change in current in a coil or in a circuit produces a magnetic field. In this magnetic field, if a conductor is moved, a current is set up in the conductor. This is nothing but the induced current.

Q. 2. State and explain Faraday's laws of electromagnetic induction.

Ans. *Faraday's laws of electromagnetic induction :*

Law-1. Whenever there is a change in the magnetic flux linked through the coil, an induced current flows through it.

Increase in magnetic flux produced an inverse (opposite sense) current and decrease in magnetic flux produces a direct (same sense) current.

Law-2. The magnitude of the induced emf 'E' is directly proportional to the number of turns (N) of the coil and the rate of change of flux ($d\phi/dt$) linked through the coil.

Mathematically, $E \propto N$

$$E \,\alpha\, \frac{d\phi}{dt}$$

$$\therefore \qquad E \propto N \frac{d\phi}{dt}$$

$$\text{or,} \qquad E = KN\frac{d\phi}{dt} = N\frac{d\phi}{dt} \qquad ...(38)$$

where K = constant of proportionality = 1, using suitable units. '$d\phi$' is the change in flux linked through the coil in time interval 'dt'. In M.K.S. system '$d\phi$' is expressed in webers, 'dt' in seconds and 'E' in volt.

Q. 3. State and explain Lenz's Law.

Ans. *Lenz's Law :* This law states that whenever an emf is induced, the induced current is in such a direction as to oppose (by its magnetic action) the change inducing the current.

Lenz's law and the principle of conservation of energy

Lenz's law is a particular example of the principle of conservation of energy. An induced current can produce heat to do chemical or mechanical work. The energy must come from the work done in inducing the current. When induction is due to motion of a magnet or a coil, work is done; therefore the motion must be resisted by a force. This opposing force comes from the action of the magnetic field of the induced current. When a change in current in a primary coil induces an emf in a neighbouring secondary coil, the current in the secondary will be in such a direction as to require the expenditure of additional energy in the primary to maintain the current.

Q. 4. How can Lenz's law be verified experimentally ? Describe.

Ans. *Experimental verification of Lenz's law :*

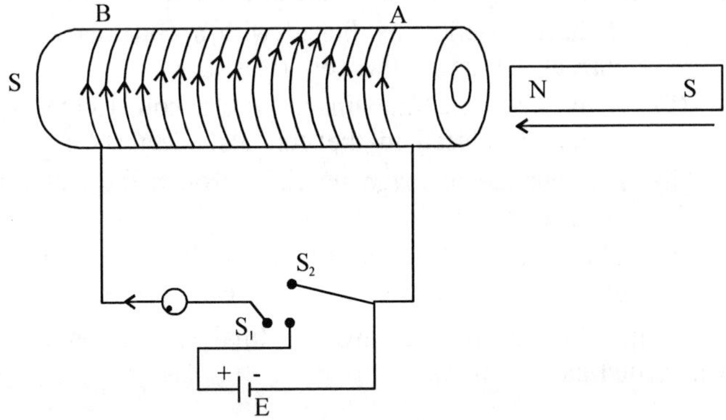

Fig. 4.2

Components

(i) Coil AB
(ii) Hallow wooden cylinder
(iii) Galvanometer (G)
(iv) Battery (E)
(v) Switches (S_1, S_2)

Procedure

The circuit is completed as shown in fig. 4.2. Gap 1 is closed by S_1 keeping gap 2 open. Consequently the source of emf E sends a current through the circuit in the direction as shown in the figure. The direction in which the pointer of galvanometer moves is noted. Let us say, it moves towards right. The direction of this current, when looked into face 'A' of the coil is anticlockwise. So we conclude that a current which appears anticlockwise from face 'A', gives a deflection toward right in galvanometer. Now gap 2 is closed by S_2 keeping gap 1 open, so that battery goes out of circuit. A magnet N-S towards the end 'A' of the coil is moved. It will be observed that if north pole, 'N' of the magnet approaches the coil, the current set up in the coil produces a deflection towards right indicating that the direction of induced current must be anticlockwise as seen through the face 'A'. According to the rule of magnetic effects of currents, face 'A' should develop a north polarity due to magnetic field of induced current. The approaching north pole of the magnet is opposed by this north polarity *i.e.*, the cause which produces induced emf. Hence Lenz's Law is verified.

Q. 5. What are the consequences of Lenz's law?

Ans. *Consequences of Lenz's law :* The following results are observed from the electromagnetic induction and Lenz's law :

(i) If the magnetic flux linked with a coil increases, the induced current set up in the coil rises.
(ii) If the magnetic flux linked with a coil decreases, the induced current produced in the coil also decreases.
(iii) Any change of magnetic flux through the coil causes induced current in the coil.
(iv) The induced current or the induced emf is always in a direction opposite to the very cause from which it is due.

In the light of Lenz's law, the final expression for induced emf is modified as :

$$E = -N\frac{d\phi}{dt} \qquad ...(39)$$

Q. 6. What do you mean by self-induction ?

Ans. *Self-induction :* Self-induction is the property of an electric circuit due to which any change in the magnetic flux linked through it, induces an emf in it. The emf induced is called self-induced emf or back emf.

Q. 7. Define self-inductance of a coil. Mention its MKSa Unit.

Ans. *Self-inductance :* The self-inductance of a coil is equal to the number of flux linkages per unit current.

The magnetic flux linked with the circuit is directly proportional to the flow of current I, *i.e.*, $\phi \propto I$

or, $\phi = LI$

where L the constant of proportionality is called the coefficient of self-induction or self-inductance.

When, $I = 1$, then $L = \phi$ ...(40)

Hence self-inductance of a coil is equal to the magnetic flux linked with it due to flow of unit current through it.

Now from the Faraday's law of electromagnetic induction the induced emf (e) is given by

$$e = -\frac{d\phi}{dt} = -\frac{d}{dt}(LI) = -L\frac{dI}{dt}$$

i.e., $$L = -\frac{e}{\left(\frac{dI}{dt}\right)} \quad \text{...(41)}$$

When, $$\frac{dI}{dt} = 1, \quad L = -e \quad \text{...(42)}$$

The –ve sign comes from the Lenz's law. Hence from equn. (42), L is numerically equal to 'e'.

Thus, self-inductance of the coil is defined as the ratio of the emf of self-induction to the rate of change of current in the coil.

Units of self-inductance :

MKS_a or SI unit of self-inductance is called Henry.

$$\text{One Henry} = \frac{1 \text{ volt}}{1 \text{ Amp / sec.}}$$

Hence, one Henry is the self-inductance of a coil in which an emf of 1 volt is induced when the current is changing at the rate of 1 Amp/Sec.

Q. 8. Find an expression for the self-inductance of an inductor.

Ans. *Self-inductance of an inductor :* The value of 'L' for a straight wire is negligibly small. But if the same wire is made into the form of a coil, its inductance increases. The value of 'L' of a coil depends upon :

(i) permeability of the material used in the core of the coil.
(ii) number of turns in the coil.
(iii) gap between the two adjacent turns of the coil.
(iv) the area of each turn of the inductor.

From the above we have

$$L = \frac{\mu N^2 A}{l} \qquad ...(43)$$

where μ = permeability of the magnetic core on which one coil is wound to prepare the inductor.

N = Number of turns.
A = The area of each turn.
l = Total length of the inductor.

Q. 9. What is mutual induction ?

Ans. *Mutual induction :* When there is a current in one of two adjacent circuits, some of the magnetic flux associated with the first or primary circuit 'P' links the near by or secondary circuit 'S' (Fig. 4.3). If the current changes in 'P' there results a change of flux in 'S'. Hence an emf is induced in the secondary circuit. This phenomenon is called mutual induction.

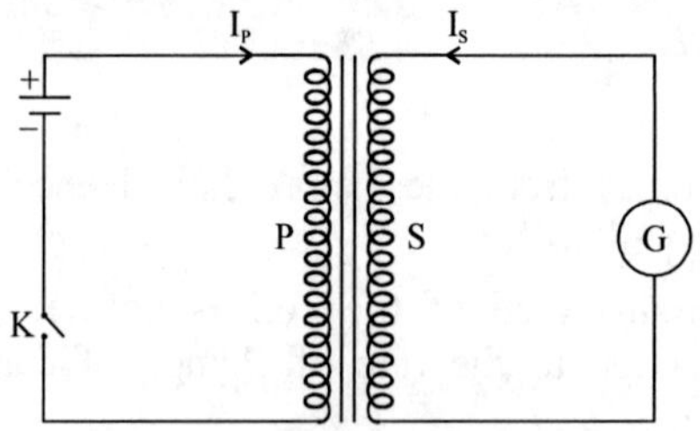

Fig. 4.3

Mutual induction of two circuits is the property by virtue of which a current changing in the first results in the induction of an emf in the second.

Q. 10. Define mutual inductance. Mention its units.

Ans. *Mutual inductance :* Let ϕ be the magnetic flux linked with the coil '*S*' due to a current '*I*' flowing through '*P*'.

$$\phi \propto I$$

$$\text{or,}\quad \phi = MI \qquad \text{...(44)}$$

where M is called the co-efficient of mutual induction of two circuits.
If $I = 1$, then $\phi = M$ ...(45)

Therefore, co-efficient of mutual induction of two circuits or mutual inductance is defined as the magnetic flux linked with the second circuit due to a unit current flowing through the first.

Differentiating equn. (44),

$$\frac{d\phi}{dt} = M\frac{dI}{dt}$$

According to Faraday's law of electromagnetic induction

$$e = -\frac{d\phi}{dt} \qquad \text{(if N is one)}$$

where 'e' is the emf induced in the second circuit

$$\text{Now,}\quad e = -\frac{d(MI)}{dt}$$

$$e = -\frac{MdI}{dt}$$

$$\Rightarrow \quad M = -\frac{e}{\frac{dI}{dt}}$$

$$\text{when}\quad \frac{dI}{dt} = 1, \quad M = -e \qquad \text{...(46)}$$

The –ve sign comes from the Lenz's law. Hence from equn. (46), M is numerically equal to e.

Thus, co-efficient of mutual induction of two circuits or mutual inductance is defined as the emf induced in the second circuit due to current changing at a unit rate in the first.

Units

(i) *SI system* : Unit of co-efficient of mutual induction or mutual inductance in SI unit is 'Henry'.

$$1 \text{ Henry} = \frac{1 \text{ volt}}{1 \text{ amp / sec.}}$$

Coefficient of mutual induction of two circuits is said to be 1 Henry if an emf of 1 volt is induced in the secondary circuit due to a current changing at the rate of 1 amp/sec in the first.

(ii) *CGS system* : Unit of mutual inductance in CGS system is emu of inductance.

Co-efficient of mutual induction of two circuits is equal to 1 emu of inductance, if an emf of 1 emu is induced in the second due to a current changing at the rate of 1 emu in the first.

Q. 11. Discuss on the mutual inductance of two circuits or coils.

Ans. *Mutual inductance of two circuits :* When two circuits or coils are closed to each other and current flowing in one circuit or coil, the mutual inductance 'M' between two circuits or coils is given by,

$$M = \frac{\mu N_1 N_2 A}{l} \quad ...(47)$$

where μ is the permeability of the cores on which the coils are wound and also the permeability of the medium separating the two coils.

N_1 = number of turns in primary
N_2 = number of turns in the secondary
A = area of the coils of primary and secondary
l = the length of the primary coil.

Thus the mutual inductance between the two coils or circuits depends upon :

(i) the permeability of the cores on which the coils are wound and also the permeability of the medium separating the two coils.
(ii) the number of turns of coils in primary and secondary.
(iii) the area of the coils primary and secondary.
(iv) the orientation of the two coils.

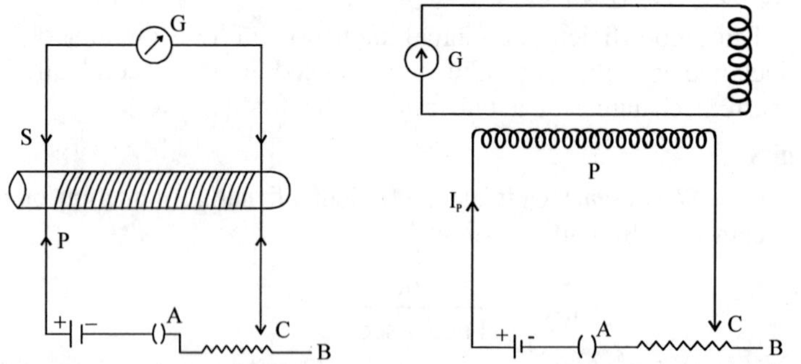

(a) Mutual induction in parallel (b) Mutual induction in ⊥er

Fig. 4.4

The value of 'M' is maximum when the two coils or circuits are parallel and wound over a single iron core. The value of 'M' is minimum when one coil is ⊥er to another. The value of 'M' varies by changing the orientation of one coil with respect to the other when all other factors remain the same.

EXERCISES

Numerical Problems with Solutions

1. Find out the flux linking through a single turn of a coil placed in an uniform magnetic field of induction 8×10^{-5} weber/m^2. The coil is rectangular of length 50 cm and breadth 20 cm and is so placed that the coil normal makes an angle of 60° to magnetic field direction.

Solution

$$B = 8 \times 10^{-5} \text{ weber/m}^2$$

$$A = 50 \times 10^{-2} \times 20 \times 10^{-2} \text{ m}^2$$

$$= 1000 \times 10^{-4} \text{ m}^2 = 10^{-1} \text{ m}^2$$

$$\alpha = 60°, \quad \phi = ?$$

Here, $\phi = BA \text{ Cos } \alpha$

$$= 8 \times 10^{-5} \times 10^{-1} \text{ Cos } 60°$$

$$= 8 \times 10^{-6} \times 1/2 = 4 \times 10^{-6} \text{ weber}$$

2. The north pole of a barmagnet is moved away from a coil of 400 turns, so that the flux linked through the coil changes from 4×10^{-6} weber to 2×10^{-6} weber in 1 millisecond. What is the average emf in the coil ?

Solution

$$N = 400 \text{ turns}$$

$$\frac{d\phi}{dt} = \frac{4 \times 10^{-6} - 2 \times 10^{-6}}{1 \times 10^{-3}} \quad \text{weber/second}$$

$$E = ?$$

Here, $E = -N \dfrac{d\phi}{dt}$

$$= -400 \times \frac{2 \times 10^{-6}}{10^{-3}}$$

$$= -400 \times 2 \times 10^{-3}$$

$$= -800 \times 10^{-3}$$

$$= -0.8 \text{ Volt.}$$

3. A pair of adjacent coils has a mutual inductance of 1 henry. The current in primary changes from 0 to 10 amp in 0.04 second. What is the average emf induced in the secondary ? If the secondary has 400 turns, what is the change in flux linked with the secondary during this time ?

Solution

M = 1 henry

$$\frac{dI}{dt} = \frac{10}{0.04} \text{ amp/Sec} = \frac{10\times 10^2}{4} = 250 \text{ amp/Sec.}$$

E_s = ? N_s = 400 turns, $d\phi$ = ?

Here, $$E_s = -M\frac{dI}{dt} = -1 \times 250 = -250 \text{ volts}$$

Again, $$E_s = -N_s\frac{d\phi}{dt} \text{ or } d\phi = -\frac{E_s dt}{N_s}$$

$$d\phi = -\frac{(-250)\times 0.04}{400} = \frac{250\times 4\times 10^{-2}}{400}$$

$$= \frac{25\times 4\times 10^{-1}\times 10^{-2}}{4}$$

$= 25 \times 10^{-3}$ weber

= 0.025 weber

4. A coil has an inductance of 10 millihenry. Find the emf induced in the coil when the current flowing through it changes at the rate of 120 amp per second.

Solution

L = 10×10^{-3} henry

$$\frac{dI}{dt} = 120 \quad \text{amp/Second}$$

E = ?

Here, $$E = -L\frac{dI}{dt} = -10\times 10^{-3}\times 120$$

$= -12 \times 10^{-1}$ volts

$= -1.2$ volts.

5. The plane of a coil is at right angles to a magnetic field of induction 10^{-12} weber/m^2. If the flux linkage is 1 weber, what is the area of the coil?

Solution

$B = 10^{-2}$ weber/m^2

$\phi = 1$ weber

$A = ?$

Here, $\phi = BA$

or, $$A = \frac{\phi}{B} = \frac{1 \text{ weber}}{10^{-2} \text{ weber / m}^2}$$

$= 10^2$ m^2

$= 100$ m^2

6. Find the flux linking through a coil of single turn of area 100 cm^2 if it is placed in a magnetic field of induction 5×10^{-2} weber/m^2, such that the coil normal makes an angle of 60° to field direction.

Solution

$A = 100$ cm^2 $= 100 \times (10^{-2}\text{m})^2$

$= 10^2 \times 10^{-4}$ m^2 $= 10^{-2}$ m^2

$B = 5 \times 10^{-2}$ weber/m^2

$\alpha = 60°$, $\phi = ?$

Here, $\phi = BA \text{ Cos } \alpha$

$= 5 \times 10^{-2} \times 10^{-2} \times \text{Cos } 60°$

$$= 5 \times 10^{-4} \times \frac{1}{2} = 2.5 \times 10^{-4} \text{ weber}$$

7. A coil of 500 turns is threaded by a changing magnetic flux, changing at the rate of 0.4×10^{-2} weber/sec. Find the average emf induced in the coil.

Solution

$N = 500$ turns

$$\frac{d\phi}{dt} = 0.4 \times 10^{-2} \text{ weber/Sec}$$

$E = ?$

Here, $$E = -N\frac{d\phi}{dt}$$

$= -\ 500 \times 0.4 \times 10^{-2}$ volts

$= -\ 2$ volts

8. A magnetic field of 2×10^{-3} T produces a flux of 2.4×10^{-5} Wb in an iron bar of cross-section 0.2 cm². Calculate its permeability.

Solution

$B = 2 \times 10^{-3}$

$\phi = 2.4 \times 10^{-5}$ Wb

$A = 0.2\ \text{cm}^2 = 0.2 \times (10^{-2}\text{m})^2 = 0.2 \times 10^{-4}\ \text{m}^2$

$\mu = ?$

Here, $\mu = \dfrac{\phi}{\phi_0}$

Where, $\phi_0 = BA = 2 \times 10^{-3} \times 0.2 \times 10^{-4}$ Wb

$= 0.4 \times 10^{-7}$ Wb

$= 4 \times 10^{-8}$ Wb

So, $\mu = \dfrac{2.4 \times 10^{-5}}{4 \times 10^{-8}}$

$= 0.6 \times 10^{3}$

$= 600$

9. A coil has inductance of 5 mH. Find the emf induced in it when the current is changed in it at the rate of 20 A/Sec.

Solution

$L = 5\ \text{mH} = 5 \times 10^{-3}$ H

$\dfrac{dI}{dt} = 20$ Amp/sec.

$E = ?$

Here, $E = -L\dfrac{dI}{dt}$

$= -\ 5 \times 10^{-3} \times (20)$

$= -\ 100 \times 10^{-3} = -\ 10^{-1}$ volt

$= -\ 0.1$ volt

10. The coil of 600 turns is threaded by a flux of 8×10^{-5} Wb. If the flux is reduced to 3×10^{-5} Wb in 0.015 S, what is the average induced emf ?

Solution

$$N = 600 \text{ turns}$$

$$\frac{d\phi}{dt} = \frac{8 \times 10^{-5} - 3 \times 10^{-5}}{0.015} \text{ Wb/S}$$

$$= \frac{5 \times 10^{-5}}{15 \times 10^{-3}} \text{ Wb/S}$$

$$= \frac{1}{3} \times 10^{-2}$$

$$\overline{E} = ?$$

Here, $\overline{E} = -N\dfrac{d\phi}{dt}$

$$= -600 \times \frac{1}{3} \times 10^{-2}$$

$$= -2 \text{ volts}$$

11. A wire "cuts" a flux of 0.03 Wb in 0.02 *S*. What average emf is induced in the wire?

Solution

$$d\phi = 0.3 \text{ Wb}, \quad dt = 0.02\ S$$

$$\overline{E} = ?$$

Here, $\overline{E} = -\dfrac{d\phi}{dt} = -\dfrac{0.3}{0.02} = -\dfrac{30}{2}$

$$= -15 \text{ volts}$$

❑ ❑ ❑

5

A.C. Fundamentals

Q. 1. What do you mean by alternating emf and alternating current.

Ans. *Alternating emf* : The emf produced by a coil rotating in a uniform magnetic field is called an alternating emf. This emf is so named because it alternates between the +ve and –ve half cycle after each half time period. This emf alternates reversing direction twice in each rotation of the coil. This emf is periodic in nature. It is given by

$$E = E_0 \sin \omega t$$

where E_0 is the steady values of the emf and ω is the angular velocity of the coil.

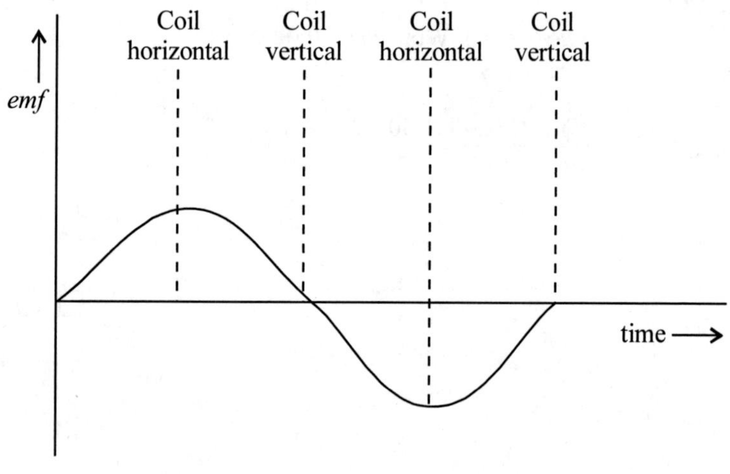

Fig. 5.1

Alternating current : It is the current which, when flowing through a conductor, changes continuously in magnitude and varies periodically in direction.

A graph between instantaneous value of a.c., 'I' and 't' is shown in fig. 5.2. The graph indicates that magnitude of current varies continuously in between '$+I_0$' and '$-I_0$'. Its direction also changes after equal intervals of time of ($T/2$) second where T is the time period of a.c. The curve, 'I' against 't' is a 'sine curve'.

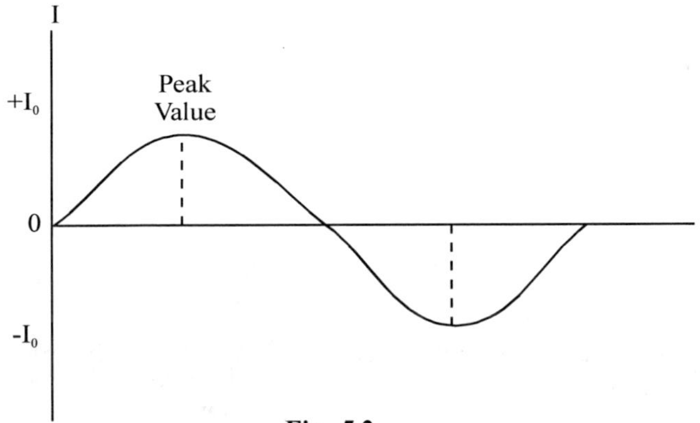

Fig. 5.2

Therefore instantaneous value 'I' of current is given by

$$I = I_0 \sin \omega t$$

where 'I_0' is called the peak value and 'ω' is the angular velocity of rotation of conductor.

$$\omega = \frac{2\pi}{T} = 2\pi f \quad \left(\therefore f = \frac{1}{T}\right)$$

'f' being the frequency of a.c.

$$I = I_0 \sin 2\pi f t.$$

Q. 2. Distinguish between a.c. and d.c.

Ans. *Distinction between a.c. and d.c. :*

Alternating current	Direct current
1. An alternating current is a current which continuously changes in magnitude and periodically reverses in direction.	1. A direct current changes continuously in magnitude but does not change in direction.
2. It fluctuates between +ve & –ve half cycle.	2. It is unidirectional having only +ve or –ve half cycle.

3.

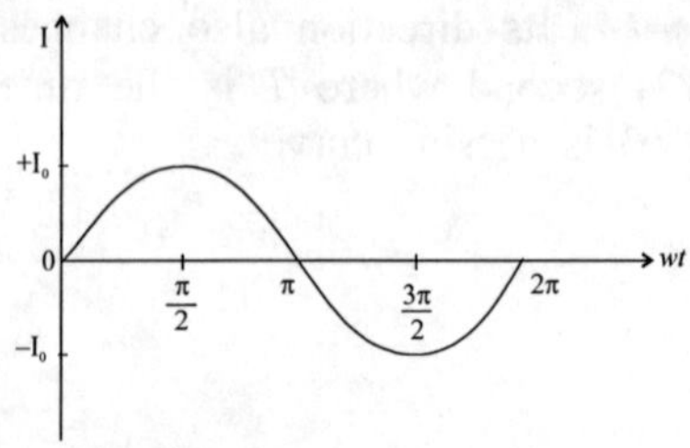

Fig. 5.3

3.

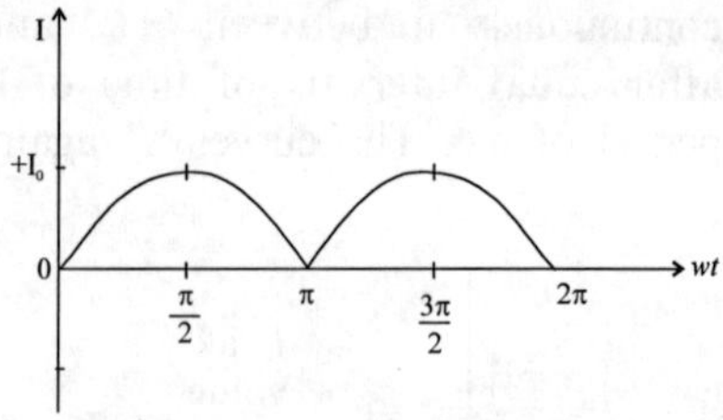

Fig. 5.4

4. Alternating current is given by $I = I_0 \sin\omega t$
5. Alternating current is more dangerous.

4. Direct current is given by $I_0 = E_0/R$
   where E_0 is the steady emf & R is the resistance in the circuit.
5. Direct current is less dangerous.

Q. 3. Find the average, mean and rms value of a.c.

Ans. *Average value of a.c. (for full cycle)* : It is the value of alternating current averaged over a complete cycle. It can be shown that the average value of a.c. is zero. Let 'T' be the time period of a.c. Alternating current is given by

$$I = I_0 \sin \omega t$$

$$I_{av} = \frac{\int_0^T I_0 \sin\omega t \, dt}{\int_0^T dt}$$

$$= \frac{1}{T} I_0 \int_0^T \sin \omega t \, dt$$

or, $$I_{av} = \frac{I_0}{T}\left[-\frac{\cos\omega t}{\omega}\right]_0^T$$

$$= \frac{I_0}{\omega T}\left[(-\cos\omega t)\right]_0^T$$

$$= \frac{I_0}{\omega T}\left[(-\cos\omega t) - (\cos 0)\right]$$

$$= \frac{I_0}{\omega T}(-\cos 2\pi + \cos 0) \quad (\because \omega T = 2\pi)$$

$$= \frac{I_0}{\omega T}(-1+1)$$

$$= \frac{I_0}{\omega T} \times 0$$

or, $I_{av} = 0$

Mean value of A.C. or D.C. value of A.C. (*for half cycle*) : Mean value of a.c. for half cycle is that value of steady current which sends the same charge through a circuit as is done by the a.c. in half cycle *i.e.*, in a time '$T/2$'.

Let 'dq' be the charge transferred in a small time 'dt'.

$$dq = I\,dt$$

Since, $I = I_0 \sin \omega t$

$$dq = I_0 \sin \omega t\,dt$$

If 'q' is the charge transferred in half the cycle of a.c.,

$$q = \int_0^q dq = \int_0^{T/2} I_0 \sin\omega t\,dt$$

or, $$q = I_0\left(-\frac{\cos\omega t}{\omega}\right)_0^{T/2}$$

$$= -\frac{I_0}{\omega}\left(\cos\frac{\omega T}{2} - \cos 0\right)$$

$$= -\frac{I_0}{\omega}(\cos\pi - \cos 0) \left(\because \frac{\omega T}{2} = \pi\right)$$

i.e., $$q = \frac{-I_0}{\omega}(-1-1)$$

or, $$q = \frac{2I_0}{\omega} \quad ...(48)$$

If 'I_{mean}' is the mean value of a.c.,

$$q = I_{mean} \times \frac{T}{2} \quad ...(49)$$

From equns. (48) and (49)

$$I_{mean} \times \frac{T}{2} = \frac{2I_0}{\omega}$$

$$\therefore \quad I_{mean} = \frac{4I_0}{\omega T}$$

But $\omega T = 2\pi$

$$\therefore I_{mean} = \frac{4I_0}{2\pi}$$

or, $$I_{mean} = \frac{2I_0}{\pi} \quad ...(50)$$

R.M.S. value of A.C. or virtual value of A.C. : R.M.S. value of A.C. is that value of steady current which when flowing through a resistance produces same heat as is done by a.c. in half of its cycle flowing through same resistance.

Let 'dH' be the heat produced by a.c. in a small time 'dt'.

$$dH = I^2 R \ dt$$

or, $dH = I_0^2 \ R \sin^2 \omega t \ dt$ (where $I = I_0 \sin \omega t$)

Heat produced in half the cycle of a.c. is

$$H = \int_0^H dH = \int_0^{T/2} I_0 R \sin^2 \omega t \, dt$$

or, $$H = I_0^2 \ R \int_0^{T/2} \left[\frac{1 - \cos\omega t}{2}\right] dt$$

$$= \frac{1}{2} I_0^2 \ R \left[\int_0^{T/2} dt - \int_0^{T/2} \cos \ 2\omega t \ dt \right]$$

or, $$H = \frac{I_0^2 \ R}{2} \left[t - \frac{\sin \ 2\omega t}{2\omega} \right]_0^{T/2}$$

or, $$H = \frac{I_0^2 \ R}{2} \left[\left(\frac{T}{2} - \frac{\sin \omega T}{2\omega} \right) - \left(0 - \frac{\sin \ 0}{2\omega} \right) \right]$$

or, $$H = \frac{I_0^2 R}{2} \times \frac{T}{2} \qquad ...(51)$$

If I_{rms} is the r.m.s. value of a.c.,

$$H = I^2{}_{rms} \times R \times \frac{T}{2} \qquad ...(52)$$

From equns. (51) & (52)

$$I^2{}_{rms} \times R \times \frac{T}{2} = \frac{I_0^2 R}{2} \times \frac{T}{2}$$

$$I^2{}_{rms} = \frac{I_0^2}{2}$$

or, $$I_{rms} = \frac{I_0}{\sqrt{2}} \qquad ...(53)$$

Q. 4. Find the rms value of a.c. emf.

Ans. *R.M.S. value of emf :* The a.c. emf is given by

$$E = E_0 \sin \omega t = E_0 \sin \alpha$$

The limiting value of 'α' for a complete cycle is o to 2π. So the mean square value of a.c. emf is given by

$$\overline{E}^2 = \frac{\int_0^{2\pi} E_0^2 \sin^2 \alpha \, d\,\alpha}{\int_0^{2\pi} d\alpha}$$

$$= \frac{E_0^2}{2\pi} \int_0^{2\pi} \sin^2 \alpha \, d\,\alpha$$

$$= \frac{E_0^2}{2\pi} \int_0^{2\pi} \left[\frac{1 - \cos 2\alpha}{2}\right] d\,\alpha$$

$$= \frac{E_0^2}{2\pi} \left[\frac{1}{2} \int_0^{2\pi} d\alpha - \frac{1}{2} \int_0^{2\pi} \cos 2\alpha \, d\alpha\right]$$

$$= \frac{E_0^2}{2\pi}\left[\frac{1}{2}\{\alpha\}_0^{2\pi} - \frac{1}{2}\left\{\frac{\sin\ 2\alpha}{2}\right\}_0^{2\pi}\right]$$

$$= \frac{E_0^2}{2\pi}\left[\frac{1}{2}2\pi - \frac{1}{4}\{\sin 4\pi - \sin 0\}\right]$$

$$= \frac{E_0^2}{2\pi}\left(\pi - \frac{1}{4}\times 0\right)$$

$$= \frac{E_0^2}{2\pi}.\pi$$

$$= \frac{E_0^2}{2}$$

i.e., $$\overline{E}^2 = \frac{E_0^2}{2}$$

or, $$\sqrt{\overline{E}^2} = \sqrt{\frac{E_0^2}{2}}$$

i.e., $$E_{rms} = \frac{E_0}{\sqrt{2}} \qquad ...(54)$$

Q. 5. Define virtual emf and virtual current.

Ans. *Virtual emf* : Virtual emf is defined as that potential difference which when maintained across a resistor produces the same heat as produced by a steady d.c. across the same resistor in the same time interval. The r.m.s. value of emf is the virtual emf.

Virtual current : Virtual current can be defined as that a.c. which produces the same heat as produced by a steady d.c. through the same resistance for the same time interval. The rms value of current is the virtual current.

Q. 6. What are different circuit elements ? Mention their functions.

Ans. *Circuit elements and their functions :* The circuit elements are :

(a) Resistance
(b) Inductance
(c) Capacitance

(a) Resistance. An electron current flows through a conductor, electrons move forward while they are colliding with each other also. Opposition offered by the conductor due to these collisions is called resistance. Schematic diagram of resistance is shown in fig. 5.5. It is denoted by R. It is equally effective in constant as well as variable currents.

Resistance of a conductor opposes the flow of current in the conductor.

Fig. 5.5

(b) Inductance. Let us consider a wire wound over a wooden cylinder having large number of turns. Let it have a negligible resistance. If a constant current flows through it, it offers no opposition ($\because$ R = 0). In case of variable currents an induced emf is set up in such a direction that it opposes the changing current. This opposition to the flow of electricity is called inductive reactance and is denoted by 'X_L'.

$$X_L = \omega L$$

Where $\omega = 2\pi f$, f being frequency of a.c. and

L = coefficient of self-induction of coil.

Schematic diagram of an inductor is shown in fig. 5.6.

Inductance in an electric circuit plays the role of friction in mechanics. Inductor is called as choking coil as it chokes a.c. The role of inductance in a.c. circuit is same as the role of resistances in d.c. circuit.

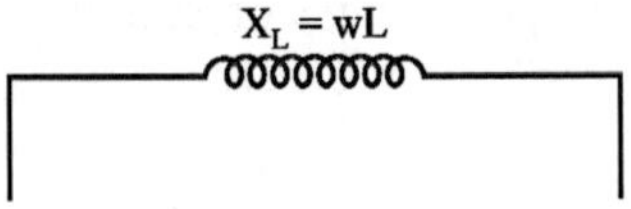

Fig. 5.6

Capacitance. When a condenser is inserted in a d.c. circuit, the plate gets charged up, and the electron current drops to zero. The capacitor (fig. 5.7), thereafter, acts as though it had an infinite resistance, when connected to an alternating source of emf, it acts differently. In reversing of emf reverses the flow of electrons to and from the plates of the capacitor and the alternating flow of charge constitutes an alternating current. The opposition offered by the capacitor to a.c. is called capacitative reactance. It is denoted by 'X_C',

$$X_C = \frac{1}{\omega C}$$

where C = Capacity of the condenser.

A capacitor allows a.c. to flow and blocks d.c.

$$X_c = \frac{1}{\omega C}$$

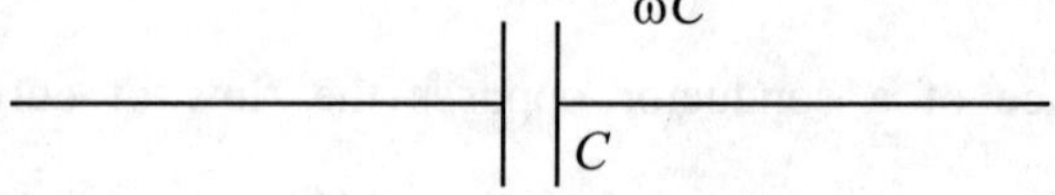

Fig. 5.7

Q. 7. Find current in an a.c. circuit containing pure resistance only. Show by graph the phase relationship between emf and current.

Ans. *Circuit containing pure resistance* : Let us consider a source of a.c. connected in a circuit comprising of pure resistance [fig. 5.8(a)]. If E is the instantaneous voltage of the source,

$$E = E_0 \sin \omega t \qquad ...(55)$$

where E_0 is the peak value of the emf. If 'I' is the instantaneous value of current, from ohm's law,

$$I = \frac{\text{Instantaneous value of emf}}{\text{resistance}}$$

$$= \frac{E_0 \sin \omega t}{R}$$

i.e., $I = I_0 \sin \omega t$ ...(56)

where $I = +\frac{E_0}{R}$, is the peak value of current.

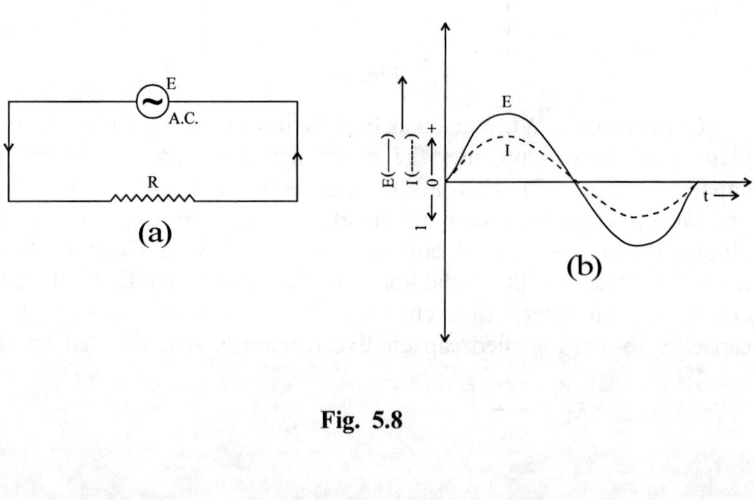

Fig. 5.8

Comparing expressions for 'E' & 'I' it is seen that there is no phase difference between emf and current in a circuit containing pure resistance *i.e.,* emf and current are in same phase in a pure resistive a.c. circuit.

A graph between time (along *x-axis*) and emf (—) and current (--) along *y-axis* is shown in fig. 5.8 (b).

Q. 8. Find current in an a.c. circuit containing pure inductance only. Show by graph the phase relationship between emf and current.

Ans. *Circuit containing inductance only :* Let a source of a.c. be applied to a circuit containing pure inductance [fig. 5.9(a)]. The instantaneous emf is

$$E = E_0 \sin \omega t \qquad \text{...(57)}$$

If E' is the emf induced in the coil,

$$E' = -L \frac{dI}{dt}$$

where 'L' is the co-efficient of self-inductance of the coil and 'I' is the instantaneous current.

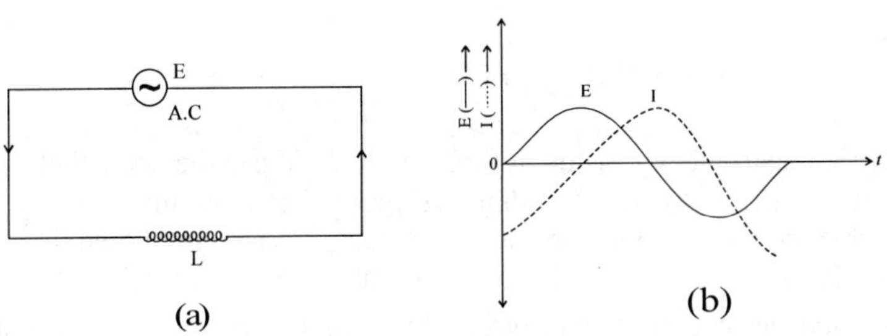

Fig. 5.9

At any instant the current can be in equilibrium if the applied voltage is equal and opposite to the induced voltage.

i.e., $\quad E = -E'$

or, $\quad E_0 \sin \omega t = -\left(-L\frac{dI}{dt}\right)$

or, $\quad L\frac{dI}{dt} = E_0 \sin \omega t$

$$\therefore \quad dI = \frac{E_0}{L} \sin \omega t \, dt$$

$\therefore$ Instantaneous current I is,

$$I = \int dI = \int \frac{E_0}{L} \sin \omega t \, dt$$

$$= \frac{E_0}{L} \int \sin \omega t \, dt$$

or, $$I = \frac{E_0}{\omega L}[-\cos \omega t] = \frac{E_0}{\omega L}[-\sin(\pi/2 - \omega t)]$$

$$= \frac{E_0}{X_L} \sin(\omega t - \pi/2) \quad [\because \sin(-\theta) = -\sin\theta]$$

where $X_L = \omega L$

Let $I_0 = \dfrac{E_0}{X_L}$

$$I = I_0 \sin\left(\omega t - \frac{\pi}{2}\right) \qquad ...(58)$$

By comparing expression for 'E' and 'I' it can be seen that emf leads the current by ($\pi/2$) radian. A graph between time (along *x-axis*) and emf (—) and current (...) (along *y-axis*) is shown in fig. 5.9(b). Hence, current lags behind the emf by an angle $\pi/2$.

Q. 9. Find current in an a.c. circuit containing pure capacitance only. Show by graph the phase relationship between emf and current.

Ans. *Circuit containing pure capacitance :* Let us consider a source of a.c. connected to a circuit containing capacitance only.

Instantaneous emf is,

$$E = E_0 \sin \omega t \qquad ...(59)$$

If q = Instantaneous charge on the condenser

$$E = \frac{q}{c} \qquad ...(60)$$

where c = capacity of the condenser

Substituting (60) in (59) we get,

$$\frac{q}{c} = E_0 \sin \omega t$$

or, $q = E_0 \; C \sin \omega t$

$\therefore$ Instantaneous current '*I*' is,

$$I = \frac{dq}{dt} = \frac{d}{dt} \quad (E_0 \; C \sin \omega t)$$

$$= E_0 \, C \frac{d}{dt} (\sin \omega t)$$

$$= E_0 \; C \cos \omega t \frac{d}{dt} (\omega t)$$

$$= \frac{E_0}{\frac{1}{\omega C}} \sin \left(\frac{\pi}{2} + \omega t \right)$$

$$= \frac{E_0}{X_C} \sin \left(\omega t + \frac{\pi}{2} \right)$$

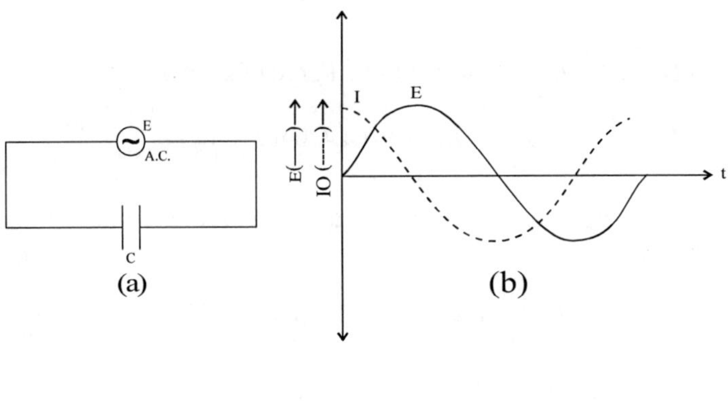

Fig. 5.10

Thus, in a circuit consisting of pure capacitance the emf lags behind the current by a phase of ($\pi/2$) radian.

A graph between time and '*E*' or '*I*' is shown in fig. 5.10(b).

Q. 10. Find current in an a.c. circuit containing inductance and resistance in series. Give phase-relationship between emf and current.

Ans. *A.C. circuit containing inductance and resistance in series :* Let us consider an a.c. circuit containing inductance and resistance in series. An alternating emf $E = E_0 \sin \omega t$ is applied to this circuit.

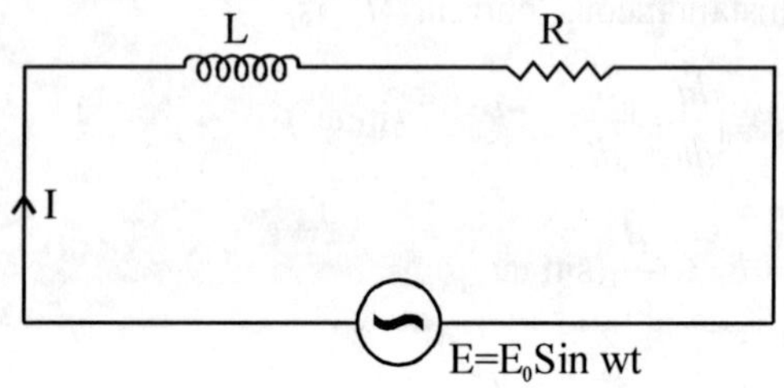

Fig. 5.11. Circuit diagram

By applying Kirchhoff's second Law, the equn. of the circuit can be written as

$$E_L + E_R = E \qquad ...(61)$$

where E_L is the potential difference across L

E_R is the potential difference across R

E is the applied a.c. emf.

Now, $\quad E_L = L \dfrac{dI}{dt}$ (from Faraday's Law)

$E_R = RI$ (from Ohm's Law)

$E = E_0 \sin \omega t$

Substituting these values in equn. (61) the equn. of the circuit, becomes

$$L \frac{dI}{dt} + RI = E_0 \sin \omega t \qquad ...(62)$$

Let the soln. of this equn. be

$$I = A \sin \omega t + B \cos \omega t \qquad ...(63)$$

Differentiating (63) with respect to 't' we have

$$\frac{dI}{dt} = A\omega \cos \omega t - B\omega \sin \omega t$$

Substituting the values of $\dfrac{dI}{dt}$ and I in equn. (62) we have,

$$L\omega (A \cos \omega t - B \sin \omega t) + R (A \sin \omega t + B \cos \omega t) = E_0 \sin \omega t \qquad ...(64)$$

This is true for all values of t.

When $t = 0$, $\sin \omega t = 0$ and $\cos \omega t = 1$

$$\therefore\ L\omega A + RB = 0 \qquad ...(65)$$

when $\omega t = \pi/2$, $\sin \omega t = 1$ and $\cos \omega t = 0$

$$\therefore\ -L\omega B + RA = E_0 \qquad ...(66)$$

From equn. (65)

$$RB = -L\omega A$$

$$\text{or,}\quad B = -\frac{\omega LA}{R} \qquad ...(67)$$

Substituting (67) in (66), we get,

$$\frac{\omega^2 L^2 A}{R} + RA = E_0$$

$$\text{or,}\quad A\left[\frac{\omega^2 L^2 + R^2}{R}\right] = E_0$$

$$\text{or,}\quad A = \left[\frac{R}{\omega^2 L^2 + R^2}\right] E_0$$

$$= \frac{RE_0}{R^2 + \omega^2 L^2} \qquad ...(68)$$

From equns. (68) and (67)

$$B = -\frac{\omega L}{R}\ \frac{RE_0}{R^2 + \omega^2 L^2} = -\frac{\omega LE_0}{R^2 + \omega^2 L^2} \qquad ...(69)$$

Substituting the values of A and B in equn. (63), we get,

$$I = \frac{RE_0}{R^2 + \omega^2 L^2}\sin \omega t + \left(\frac{-\omega LE_0}{R^2 + \omega^2 L^2}\right)\cos \omega t$$

$$= \frac{RE_0}{R^2 + \omega^2 L^2}\sin \omega t - \frac{\omega LE_0}{R^2 + \omega^2 L^2}\cos \omega t$$

$$= \frac{E_0}{\sqrt{R^2 + \omega^2 L^2}}(\cos\theta.\sin \omega t - \sin\theta \cos \omega t) \qquad ...(70)$$

where $\quad \cos\theta = \dfrac{R}{\sqrt{R^2 + \omega^2 L^2}}$

and $$\sin\theta = \frac{\omega L}{\sqrt{R^2 + \omega^2 L^2}}$$

$$\tan\theta = \frac{\sin\theta}{\cos\theta} = \frac{\omega L}{R} = \frac{X_L}{R}$$

i.e., $$\theta = \tan^{-1}\left(\frac{\omega L}{R}\right) = \tan^{-1}\left(\frac{X_L}{R}\right) \quad ...(71)$$

This angle 'θ' is known as the angle of lag or phase difference between emf and current.

Phase triangle :

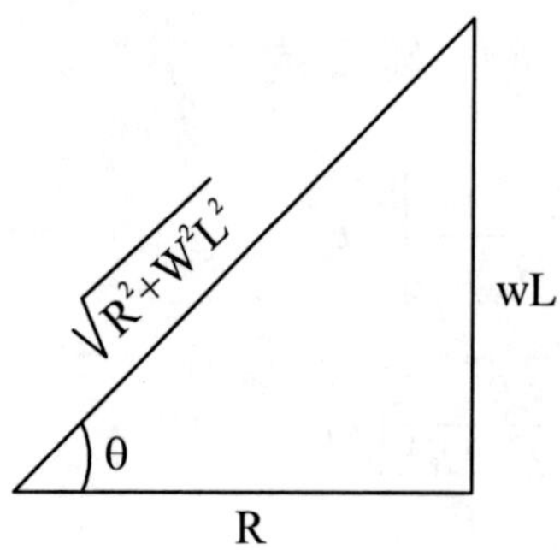

Fig. 5.11(a)

Now, equn. (70) becomes

$$I = \frac{E_0}{\sqrt{R^2 + \omega^2 L^2}} \sin(\omega t - \theta)$$

$$= I_0 \sin(\omega t - \theta) \quad ...(72)$$

where $I_0 = \frac{E_0}{\sqrt{R^2 + \omega^2 L^2}}$ = current amplitude $= \frac{E_0}{Z}$

where 'Z' is the impedance of the circuit. The current I as seen from equn. (72) is simple harmonic in nature. The current lags behind the emf by an angle 'θ'.

Amplitude of current :

The current amplitude is given by,

$$I_0 = \frac{E_0}{\sqrt{R^2 + \omega^2 L^2}} \quad \text{...(73)}$$

The current amplitude becomes maxm. when L = 0

so $$(I_0)_{\max} = \frac{E_0}{R} \quad \text{...(74)}$$

which is in accordance with ohm's law.

In this case, $\theta = 0$ *i.e.*, the current is in phase with the emf.

Time period :

The time period of oscillation of current is given by

$$T = \frac{2\pi}{\omega} \quad \text{...(75)}$$

Frequency :

The frequency of current oscillation is given by,

$$f = \frac{1}{T} = \frac{\omega}{2\pi} \quad \text{...(76)}$$

Impedance :

$$Z = \sqrt{R^2 + \omega^2 L^2} = \sqrt{R^2 + X_L^2}$$

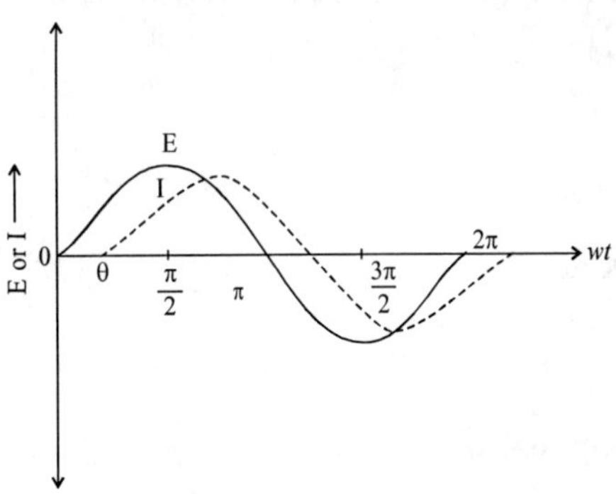

Fig. 5.12. Phase diagram

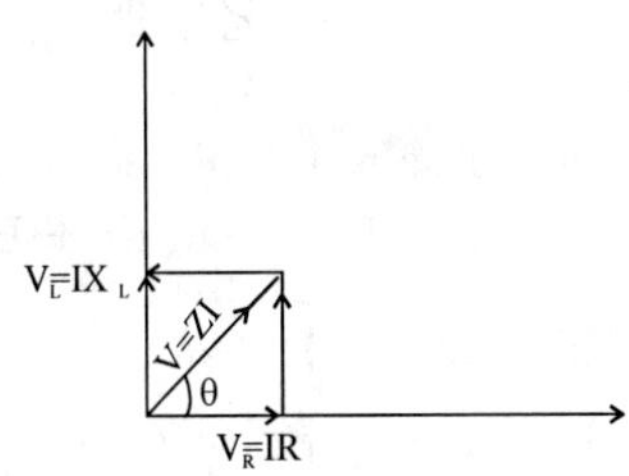

Fig. 5.13. Vector representation

Q. 11. Find current in an a.c. circuit containing resistance and capacitance in series. Give phase-relationship between emf and current.

Ans. *A.C. circuit containing resistance and capacitance in series :* Let's consider an a.c. circuit containing a capacitance '*C*' and a resistance R in series. The a.c. emf $E = E_0 \sin \omega t$ is applied to this circuit. The circuit connections are given by,

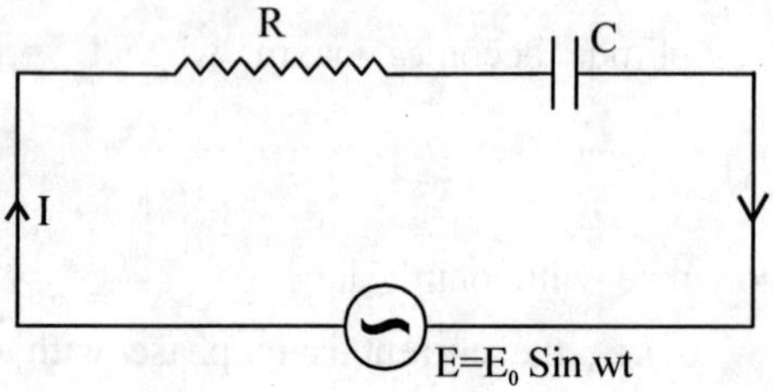

Fig. 5.14. Circuit diagram

By Kirchhoff's second law the equn. of this circuit becomes

$$E_R + E_c = E \qquad ...(77)$$

where E_R is the potential difference across R.

E_c is the potential difference across C.

E is the applied a.c. emf.

Now $E_R = RI$

$$E_c = \frac{Q}{C}$$

$$E = E_0 \sin \omega t$$

Substituting these values in equn. (77) the equn. of the circuit becomes

$$RI + \frac{Q}{C} = E_0 \sin \omega t \qquad ...(78)$$

Let the soln. of this equn. be

$$I = \text{A} \sin \omega t + \text{B} \cos \omega t \qquad ...(79)$$

Then
$$Q = \int I \, dt$$

$$= \int [A \sin \omega t + \text{B} \cos \omega t] \, dt$$

$$= \int A \sin \omega t \, dt + \int B \cos \omega t \, dt$$

$$= A \int \sin \omega t \, dt + \text{B} \int \cos \omega t \, dt$$

$$= -\frac{A}{\omega} \cos \omega t + \frac{B}{\omega} \sin \omega t \qquad ...(80)$$

Substituting (79) and (80) in (78) we get

$$R\left(A\sin\omega t + B\cos\omega t\right) + \frac{1}{C}\left(-\frac{A}{\omega}\cos\ \omega t + \frac{B}{\omega}\sin\ \omega t\right) = E_0\sin\omega t$$

when $\omega t = 0$, $\sin\ \omega t = 0$ and $\cos\ \omega t = 1$

$$\therefore\ -\frac{1}{C\omega}A + RB = 0 \qquad ...(81)$$

when $\omega t = \pi/2$, $\sin\ \omega t = 1$ and $\cos\ \omega t = 0$

$$\therefore\ RA + \frac{1}{C\omega}B = E_0 \qquad ...(82)$$

From (81) we have

$$RB = \frac{A}{C\omega}$$

or, $$B = \frac{A}{\omega RC} \qquad ...(83)$$

Substituting (83) in (82) we get

$$RA + \frac{1}{\omega C}\frac{A}{\omega RC} = E_0$$

or, $$A\left(R + \frac{1}{\omega^2 C^2 R}\right) = E_o$$

or, $$A\left(R^2 + \frac{1}{\omega^2 C^2}\right) = E_0 R$$

or, $$A = \frac{RE_0}{R^2 + \frac{1}{\omega^2 C^2}} \qquad ...(84)$$

From (84) and (83) we have

$$B = \frac{E_0 \frac{1}{C\omega}}{R^2 + \frac{1}{\omega^2 C^2}} \qquad ...(85)$$

From (84), (85) and (79) we have

$$I = \frac{E_0}{\sqrt{R^2 + \frac{1}{\omega^2 c^2}}}\left\{\frac{R}{\sqrt{R^2 + \frac{1}{\omega^2 c^2}}}\sin\ \omega t + \frac{\frac{1}{\omega c}}{\sqrt{R^2 + \frac{1}{\omega^2 c^2}}}\cos\ \omega t\right\}$$

$$= \frac{I_0}{\sqrt{R^2 + \frac{1}{\omega^2 c^2}}} \{\cos\theta \sin \omega t + \sin\theta \cos \omega t\}$$

$$= \frac{E_0}{\sqrt{R^2 + \frac{1}{\omega^2 c^2}}} \sin (\omega t + \theta) = I_0 \sin(\omega t + \theta) \qquad ...(86)$$

where $\tan\theta = \frac{X_c}{R} = \frac{\frac{1}{\omega c}}{R} = \frac{1}{\omega RC}$

$$\therefore \theta = \tan^{-1}\left(\frac{X_c}{R}\right)$$

$$\left[\begin{array}{l} \because \cos\theta = \dfrac{R}{\sqrt{R^2 + \frac{1}{\omega^2 C^2}}} \\ \sin\theta = \dfrac{\frac{1}{\omega C}}{\sqrt{R^2 + \frac{1}{\omega^2 C^2}}} \end{array}\right]$$

and $Z = \sqrt{R^2 + \frac{1}{\omega^2 c^2}} = \sqrt{R^2 + X_c^2}$

Hence the current leads in phase the applied emf by an angle

$$\tan^{-1}\left(\frac{1}{\omega RC}\right)$$

Phase relationship between emf and current

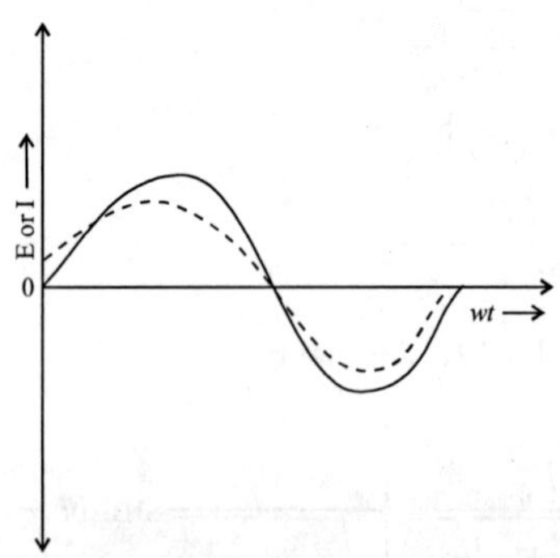

Fig. 5.15

Vector diagram

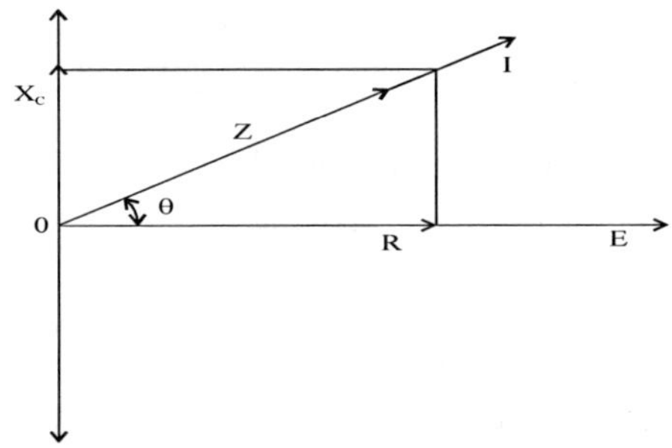

Fig. 5.16

Impedance :

The impedance of the circuit is given by

$$Z = \sqrt{R^2 + \frac{1}{\omega^2 C^2}} = \sqrt{R^2 + X_c^2} \qquad \text{...(87)}$$

The impedance in a.c. circuit plays the role of resistance in d.c. circuit.

Q. 12. Find current in an a.c. circuit containing inductance and capacitance in series. Give the phase-relationship between emf and current.

Ans. *A.C. circuit containing inductance and capacitance* :

Let the circuit contain an inductance L and a capacitance C and let the applied emf be $E = E_0 \sin \omega t$.

By Kirchhoff's second law the equn. of emf is

$$L \frac{dI}{dt} + \frac{Q}{C} = E = E_0 \sin \omega t \qquad \text{...(88)}$$

Let $I = A \sin \omega t + B \cos \omega t$ ...(89)

be a soln.

then $$\frac{dI}{dt} = A\omega \cos \omega t - B\omega \sin \omega t$$

$$Q = \int I \, dt = \int (A \sin \omega t + B \cos \omega t) \, dt$$

$$= -\frac{A}{\omega} \cos \omega t + \frac{B}{\omega} \sin \omega t$$

Substituting in equn. (88), we have

$$L(A\omega \cos \omega t - B\omega \sin \omega t) + \frac{1}{C}\left(-\frac{A}{\omega}\cos \omega t + \frac{B}{\omega}\sin \omega t\right) = E_0 \sin \omega t$$

This equn. is valid for all values of ωt

when $\omega t = 0$, $\sin \omega t = 0$ & $\cos \omega t = 1$

$$A\left(L\omega - \frac{1}{C\omega}\right) = 0$$

when $\omega t = \pi/2$, $\sin \omega t = 1$ & $\cos \omega t = 0$

$$-B\left(L\omega - \frac{1}{C\omega}\right) = E_0$$

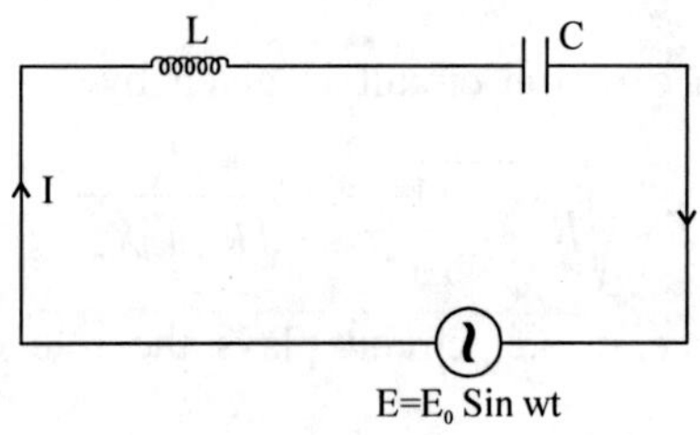

Fig. 5.17. Circuit diagram

$$\therefore A = 0, = -\frac{E_0}{L\omega - \frac{1}{C\omega}}$$

$$I = -\frac{E_0}{L\omega - \frac{1}{C\omega}} \cos \omega t$$

$$= \frac{E_0}{L\omega - \frac{1}{C\omega}} \sin\left(\omega t - \frac{\pi}{2}\right)$$

$$= \frac{E_0}{X_L - X_C} \sin\left(\omega t - \frac{\pi}{2}\right)$$

The current thus lags behind emf by $\pi/2$ and hence the circuit oscillates at this frequency, the power consumption being nil *i.e.*, the circuit is wattless. Again if $X_L = X_C$, the current I becomes very high.

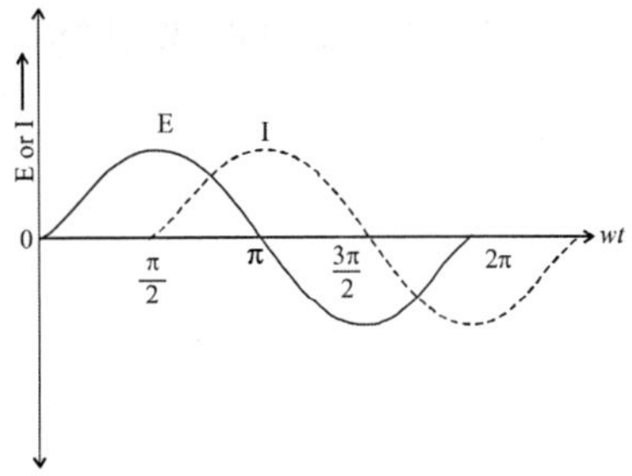

Fig. 5.18. Phase diagram

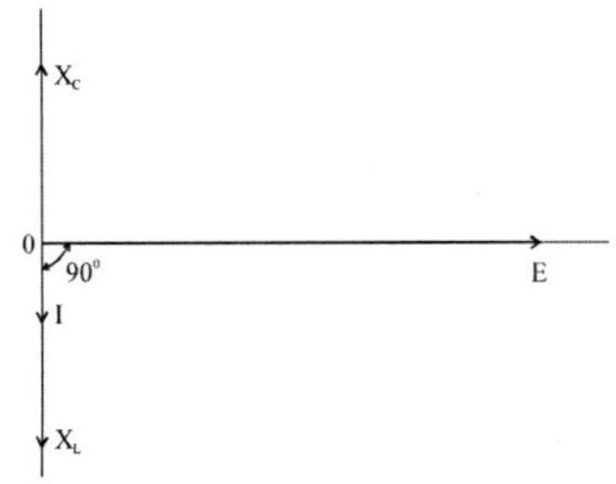

Fig. 5.19. Vector diagram

Q. 13. Find current in an a.c. circuit containing *L–C–R* in series. Give phase relationship between emf. and current.

Ans. *A.C. circuit containing inductance, capacitance, resistance:* Let us consider a circuit containing *L*, *C* & *R* in series. The a.c. emf applied to the circuit is $E = E_0 \sin \omega t$.

By Kirchhoff's second law the equn. of the circuit is given by,

$$E_L + E_R + E_C = E$$

where
$$E_L = L\frac{dI}{dt}$$

$$E_R = RI$$

$$E_C = \frac{Q}{C}$$

Then the equn. of the circuit becomes

$$L\frac{dI}{dt} + RI + \frac{Q}{C} = E_0 \sin \omega t \quad ...(90)$$

Let $$I = A \sin \omega t + B \cos \omega t \quad ...(91)$$

be a soln. to the above equn. Then

$$Q = \int I\, dt$$

$$= -\frac{A}{\omega}\cos \omega t + \frac{B}{\omega}\sin \omega t$$

$$\frac{dI}{dt} = A\omega \cos \omega t - B\omega \sin \omega t$$

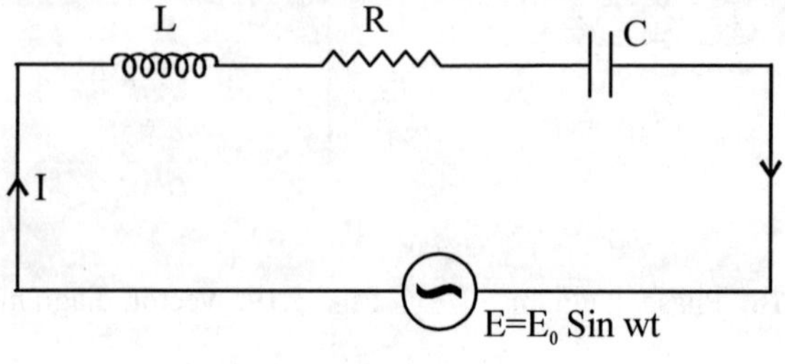

Fig. 5.20

Substituting in equn. (90) we get

$$L\,(A\omega \cos \omega t - B\omega \sin \omega t) + R\,(A \sin \omega t + B \cos \omega t)$$

$$+\frac{1}{C}\left(-\frac{A}{\omega}\cos \omega t + \frac{B}{\omega}\sin \omega t\right) = E_0 \sin \omega t \qquad ...(92)$$

This is valid for all values of ωt.
when $\omega t = 0$, $\sin \omega t = 0$ and $\cos \omega t = 1$

$$LA\omega + RB + 0 - \frac{A}{C\omega} = 0$$

$$A\left(L\omega - \frac{1}{C\omega}\right) + BR = 0 \qquad ...(93)$$

when $\omega t = \pi/2$, $\sin \omega t = 1$ and $\cos \omega t = 0$

$$AR - B\left(L\omega - \frac{1}{C\omega}\right) = E_0 \qquad ...(94)$$

Solving (93) and (94) for A and B, we get

$$A = \frac{E_0 R}{R^2 + \left(L\omega - \frac{1}{C\omega}\right)^2}$$

$$B = \frac{-E_0\left(L\omega - \frac{1}{C\omega}\right)}{R^2 + \left(L\omega - \frac{1}{C\omega}\right)^2}$$

$$\therefore \quad I = \frac{E_0 R}{R^2 + \left(L\omega - \frac{1}{C\omega}\right)^2} \sin \omega t - \frac{E_0\left(L\omega - \frac{1}{C\omega}\right)}{R^2 + \left(L\omega + \frac{1}{C\omega}\right)^2} \cos \omega t$$

$$= \frac{E_0}{\sqrt{R^2\left(L\omega - \frac{1}{C\omega}\right)^2}} \times \left[\frac{R}{\sqrt{R^2 + \left(L\omega - \frac{1}{C\omega}\right)^2}} \sin \omega t - \frac{\left(L\omega - \frac{1}{C\omega}\right)}{\sqrt{R^2\left(L\omega - \frac{1}{C\omega}\right)^2}} \cos \omega t\right]$$

$$= \frac{E_0}{\sqrt{R^2 + \left(\omega L - \frac{1}{C\omega}\right)^2}} (\cos\theta \sin \omega t - \sin\theta \cos \omega t)$$

$$= \frac{E_0}{\sqrt{R^2 + \left(L\omega - \frac{1}{C\omega}\right)^2}} \sin (\omega t - \theta) \qquad ...(95)$$

Where $\cos\theta = \dfrac{R}{\sqrt{R^2 + \left(\omega L - \frac{1}{\omega C}\right)^2}}$

$\sin\theta = \dfrac{\left(\omega L - \frac{1}{\omega C}\right)}{\sqrt{R^2 + \left(\omega L - \frac{1}{\omega C}\right)^2}}$

Hence $\tan\theta = \dfrac{\sin\theta}{\cos\theta}$

$= \dfrac{\left(\omega L - \frac{1}{\omega C}\right)}{R}$

or, $\theta = \tan^{-1}\left[\dfrac{\left(\omega L - \frac{1}{\omega C}\right)}{R}\right]$

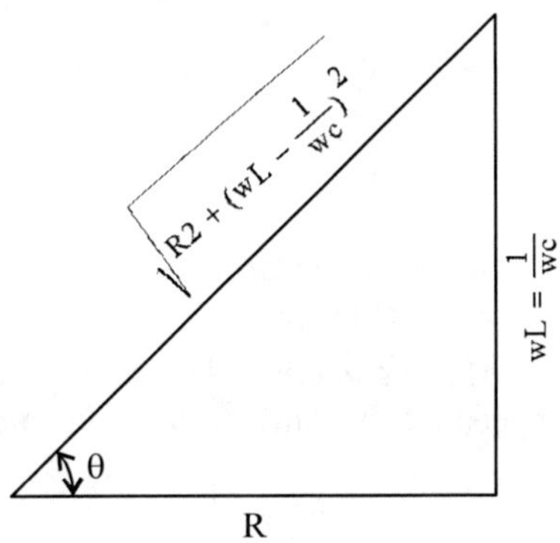

Fig. 5.21. Phase of triangle

θ is the phase difference between emf and current. Equn. (95) can also be written as

$$I = I_0 \sin (\omega t = - \theta) \qquad ...(96)$$

Thus the current lags behind the emf by an angle θ.

Now,

$$I_0 = \frac{E_0}{\sqrt{R^2 + \left(\omega L - \frac{1}{\omega c}\right)^2}} \quad = \text{ Current amplitude}$$

$$= \frac{E_0}{Z}$$

where Z is the impedance of the circuit. The impedance in a.c. circuit plays the role of resistance in d.c. circuit. If impedance is low the current increases and if the impedance is high the current decreases.

$$Z = \sqrt{R^2 + \left(\omega L - \frac{1}{\omega c}\right)^2} \qquad \text{...(97)}$$

$$= \sqrt{R^2 + (X_L - X_C)^2} \qquad \text{...(98)}$$

where, $X_L = \omega L$ = inductive reactance

$X_C = \dfrac{1}{\omega c}$ = capacitive reactance

So the impedance has two parts, one resistive part and the other one is reactive part.

Max^m current : The current in the circuit becomes maxm. when impedance becomes min^m. The impedance Z becomes minm. when

$$\omega L = \frac{1}{\omega c}$$

or, $$\omega^2 = \frac{1}{Lc}$$

or, $$\omega = \frac{1}{\sqrt{Lc}} \quad \text{or,} \quad 2\pi f = \frac{1}{\sqrt{Lc}}$$

or, $$f = \frac{1}{2\pi\sqrt{Lc}} = fr \qquad \text{...(99)}$$

This frequency is known as resonant frequency. In this case $\theta = 0$. Hence the emf and the current are in phase with each other.

The circuit behaves as a series resonant circuit.

Phase diagram (Series Circuit)

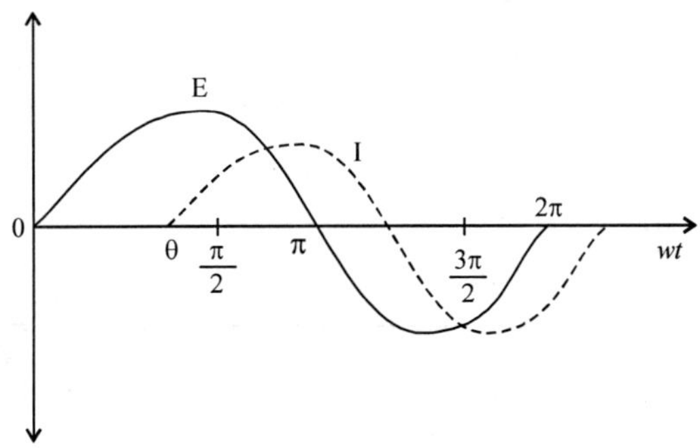

Fig. 5.22

Vector diagram (Series Circuit)

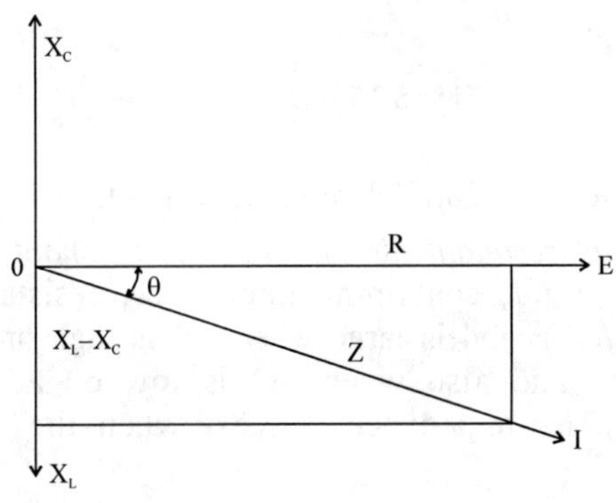

$$Z=\sqrt{R^2+(X_L-X_C)^2}$$

Fig. 5.23

Series resonant circuit

Phase diagram

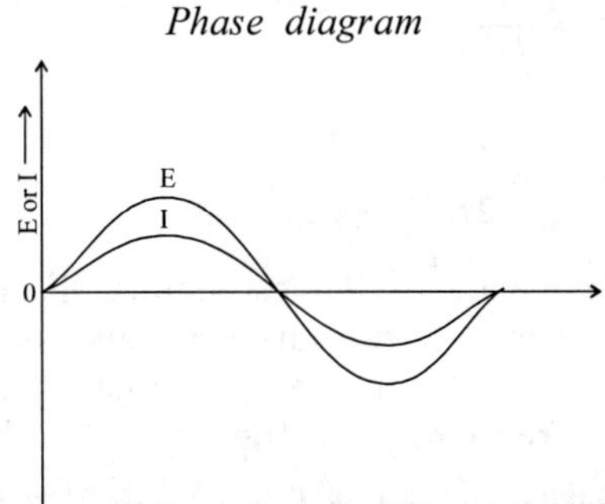

Fig. 5.24

Vector diagram

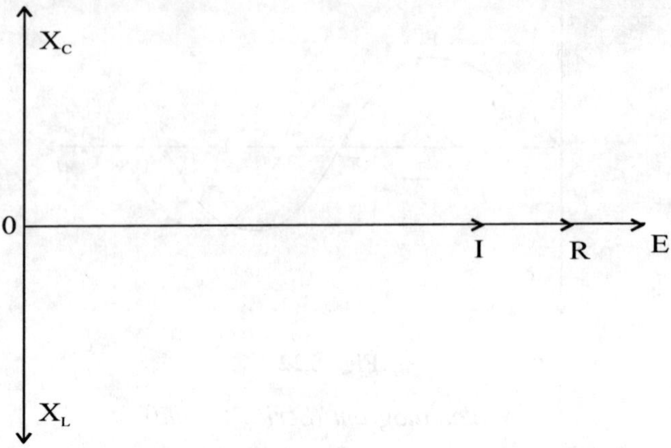

Fig. 5.25

Q. 14. Discuss on Series and Parallel resonant Circuits.

Ans. *Series and parallel resonant circuits : series resonant circuit :* The impedance of a circuit, containing inductance, resistance and capacity is $R^2 + (\omega L\text{-}1/\omega C)^2$ and is large when 'ω' is high on account of inductive resistance and also when 'ω' is low on account of capacitive reactance. The impedance is least when the resultant reactance is zero.

i.e., when $$\omega L = \frac{1}{\omega C}$$

or, $$\omega = \frac{1}{\sqrt{Lc}}$$

and the frequency $$f = \frac{\omega}{2\pi} = \frac{1}{2\pi\sqrt{Lc}}$$

But this is the frequency of the oscillatory discharge of a circuit containing *L* & *C*, when its resistance is low. The current in such a circuit is maxm., when the frequency of the applied emf coincides with the natural frequency of the circuit. This is called the phenomenon of resonance and the frequency $f = \frac{\omega}{2\pi} = \frac{1}{2\pi\sqrt{Lc}}$ c/sec. or Hertz (Hz)

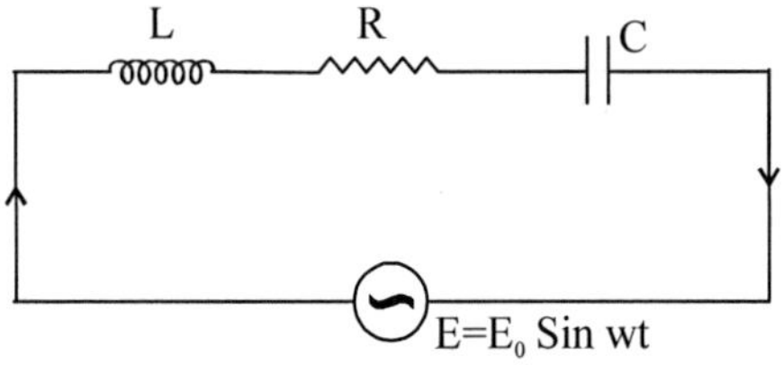

Fig. 5.26. Series resonant ct.

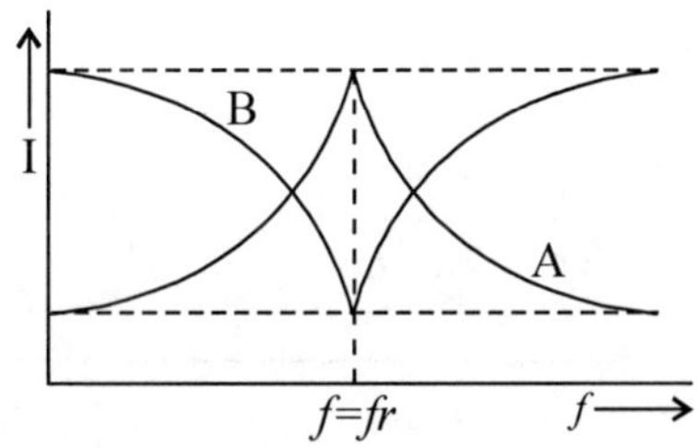

Fig. 5.27. A $\rightarrow$ Series resonant ct., B $\rightarrow$ Parallel resonant ct.

Parallel resonant circuit : Such a circuit is shown in fig. 5.28. The resistance '*R*' of the inductor *L* is very small as compared to its reactance. The current in *L* will, therefore, lag in phase by 90° behind the applied emf and is given by $I_L = \frac{E_0}{\omega L} \sin\left(\omega t - \frac{\pi}{2}\right)$. The current in the capacitance '*C*' will lead in phase by 90° the applied emf and is given by $I_c = \frac{E_0}{\frac{1}{\omega C}} \sin\left(\omega t + \frac{\pi}{2}\right)$. These two current being nearly 180° out of phase, may be deemed equivalent to an alternating current flowing round the *LC* Loop. The current in the main circuit *i.e.*, through the supply mains,

$$I = I_L + I_c$$

$$= \frac{E_0}{\omega L} \sin\left(\omega t - \frac{\pi}{2}\right) + E_0\, \omega C \sin\left(\omega t + \frac{\pi}{2}\right)$$

$$= E_0\left(\omega C - \frac{1}{\omega L}\right) \cos \omega t$$

and is zero when $\omega C = \frac{1}{\omega L}$

or, $\omega L = \frac{1}{\omega C}$

i.e., $X_L = X_C$

hence $\omega^2 = \frac{1}{Lc}$

or, $\omega = \frac{1}{\sqrt{Lc}}$

$$f = \frac{\omega}{2\pi} = \frac{1}{2\pi\sqrt{Lc}} \text{ Hertz} = fr$$

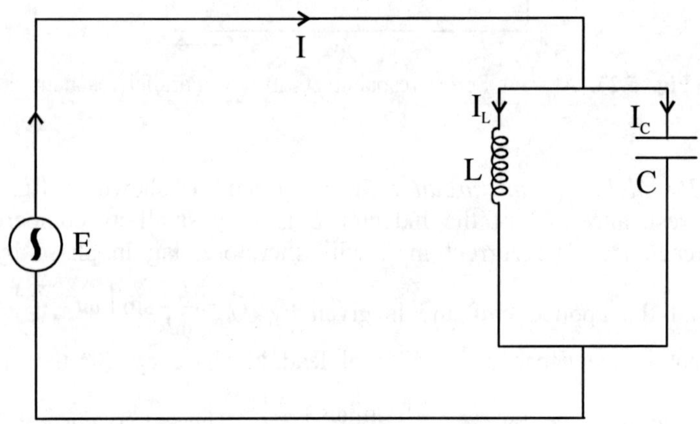

Fig. 5.28

Q. 15. Discuss on the use of resonance to reject unwanted signals.

Ans. *Use of phenomenon of resonance to reject unwanted signals :* For this frequency, the parallel resonant circuit does not allow any current to flow through the a.c. supply mains and works as a perfect choke for A.C. At the resonant frequency, the impedance of a series *Lc* circuit is least where as for a parallel *Lc* circuit it is max^m. Because, for a parallel resonant circuit,

$$I = E_0\left(\omega C - \frac{1}{\omega L}\right) \cos \omega t$$

$$= \frac{E_0}{\left(\frac{1}{\omega C - \frac{1}{\omega L}}\right)} \cos \omega t = \frac{E_0}{Z} \cos \omega t$$

where $$Z = \frac{1}{\omega C - \frac{1}{\omega L}} = \frac{1}{0} = \infty \quad \left(\because \omega C = \frac{1}{\omega L}\right)$$

Thus, impedance is max^m at resonant frequency and hence $I = 0$ at resonant frequency.

The series resonant circuits are used to produce max^m current whereas parallel resonant circuits are used to provide max^m voltage.

The supply current is least when the circuit offers max^m impedance, that is, when an emf of resonant frequency is applied, parallel resonance circuits are, therefore also called rejector or anti-resonance circuits and are employed in radio circuits as filter circuits.

Q. 16. Compare between series and parallel resonant circuits.

Ans. *Comparison between series and parallel resonant circuits :*

Series	*Parallel*
(i) It is an acceptor circuit.	(i) It is a rejecter circuit.
(ii) Resonant frequency, $f_r = \frac{1}{2\pi\sqrt{Lc}}$	(ii) Resonant frequency, $f_r = \frac{1}{2\pi\sqrt{Lc}}$
(iii) At resonant frequency, the impedance Z of the circuit is minimum.	(iii) At resonant frequency, the impedance Z of the circuit is max^m. (Admitance is minimum).
(iv) At resonant frequency, current value is max^m.	(iv) At resonance voltage across the circuit is max^m.
(v) At resonant frequency the inductive and capacitive reactances are equal in magnitude.	(v) At resonance, the inductive and capacitive susceptances are equal in magnitude. (Susceptance = 1/reactance)
(vi) At resonant frequency, the reactive component of voltage are equal in magnitude, but opposite in phase. Thus they cancel each other.	(vi) At resonance, the reactive component of current are equal in magnitude, but opposite in phase. Thus resultant is zero.

Q. 17. Discuss on *Q*-factor of a circuit.

Ans. *Q-factor :* The quality factor, *Q*, determines the band-width of a selective circuit. In other words, the sharpness of resonance is governed by the *Q* of the circuit.

From energy consideration, *Q* can be defined as

$$Q = 2\pi \frac{\text{Max}^m \text{ energy stored per cycle}}{\text{Energy dissipated per cycle}}$$

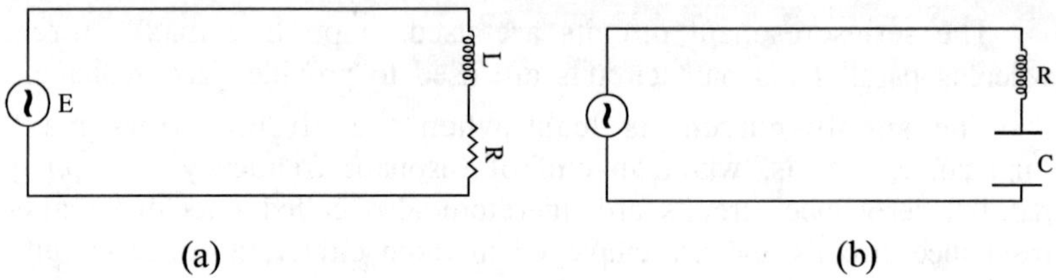

Fig. 5.29

For *L* & *R* in series,

for a coil or inductance,

$$Q_L = \frac{X_L}{R} = \frac{\omega L}{R} \quad \text{...(100)}$$

where *R* is the resistance of the coil.

For *C* & *R* in series,

for a capacitor, $Q_C = \dfrac{X_C}{R} = \dfrac{1}{\omega RC}$ ...(101)

where R is the effective resistance of the capacitor.

It can be seen from equn. (100), a coil with higher resistance has smaller *Q*. So, when a resistance is used in series, it produces lower *Q*. When a resistance is used in parallel, the effective '*Q*' of the circuit will vary directly with the value of the resistance. A high shunting resistance means a high *Q*; a low parallel resistance results in low *Q*.

The value of *Q* of resonant circuit, however, differs from that of a simple coil or capacitor. The *Q* of a series resonant circuit,

$$Q_s = \frac{\omega L}{R_s} = \frac{1}{R_s}\sqrt{L/C} \quad \text{...(102)}$$

where R_s is the total series resistance of the circuit. For a parallel resonant circuit, *C* & *R* are parallel,

$$Q_p = \omega RC \qquad ...(103)$$

$$Q_p = \frac{R}{\sqrt{L/C}} \qquad ...(104)$$

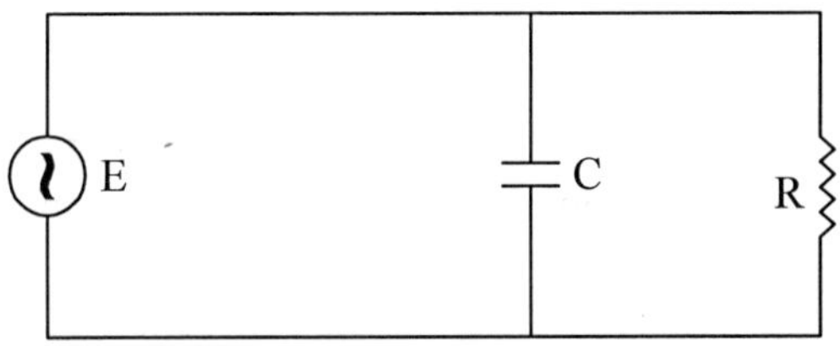

Fig. 5.30

Thus, for a high *Q*-series resonant circuit, *L/C* value should be high whereas for a high *Q*-parallel resonant circuit *L/C* value should be low.

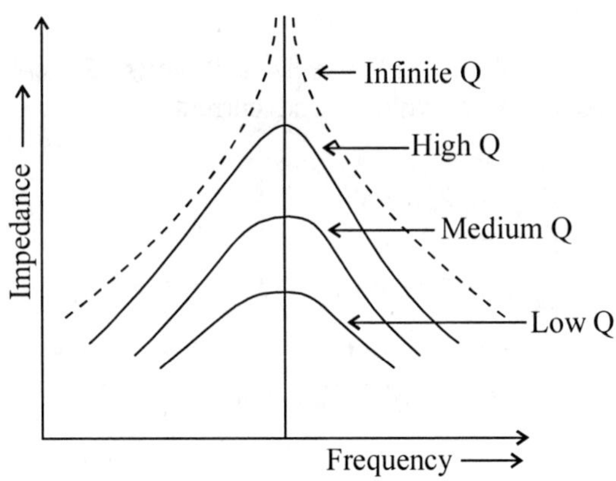

Fig. 5.31

In fig. 5.31 we have shown how the bandwidth changes for circuit with different *Q*. While designing tuned circuits, Quality factor is an important factor to be taken account of. Hence equn. (100) is usually used for *Q* of the turned circuit.

The bandwidth of a tuned circuit is also related to its *Q* by the relation,

$$Q = \frac{f_r}{\Delta f} \qquad ...(105)$$

where f_r is the resonant frequency and Δf is the bandwidth. Hence Q circuit results in narrow bandwidth.

So, $$\Delta f = \frac{f_r}{Q} \qquad ...(105a)$$

EXERCISES

Numerical Problems with Answers

1. What is the peak value of voltage when the rms value is 220 volts?

Solution

V_{rms} = 220 volts

V_{peak} = ?

Here, $V_{peak} = \sqrt{2}\ V_{rms}$

= 2 × 220 volts

= 1.414 × 220 volts

= 311.08 volts

2. The a.c. supply to a house is 220 volts, 50 c/s, 5 amp. Find the peak values of voltage and current.

Solution

V_{rms} = 220 volts

f = 50 c/s

I_{rms} = 5 amp, V_o = ? I_o = ?

Here, $V_o = \sqrt{2} \times V_{rms} = \sqrt{2} \times 220$ volts

= 1.414 × 220 volts = 311.08 volts

$I_o = \sqrt{2} \times I_{rms}$ = 1.414 × 5 amp

= 7.070 = 7.07 amp

3. What is the inductive reactance of a coil when the peak current through it is 7 amp and the peak applied emf across it is 311 volts.

Solution

I_o = 7 amp

V_o = 311 volts

X_L = ?

Here, $X_L = \frac{V_0}{I_0} = \frac{311}{7}$ ohms

$= 44.4$ ohms

4. What is the reactance of 0.5 H inductor at 500 Hz?

Solution

$L = 0.5$ H, $f = 500$ Hz, $X_L = ?$

Here, $X_L = 2\pi f L$

$= 2 \times 3.142 \times 500 \times 0.5$ ohms

$= 314.2 \times 5$ ohms

$= 1571.0$ ohms

$= 1571$ ohms

5. A what frequency a 7H inductor shall offer a reactance of 2200 ohms?

Solution

$L = 7\text{H}, X_L = 2200$ ohms

$f = ?$

Here, $X_L = 2\pi f L$

or, $f = \frac{X_L}{2\pi L}$

$$= \frac{2200}{2 \times 3.142 \times 7}$$

$$= \frac{100}{2} = 50 \text{ Hz}$$

6. What is the capacitive reactance of a 0.1 μf capacitor at (i) 500 H_z (ii) 500 KHz?

Solution

$C = 0.1\ \mu f = 0.1 \times 10^{-6} f$

$X_C = ?$

(i) $f = 500$ Hz

So, $X_C = \frac{1}{\omega C} = \frac{1}{2\pi f C}$

$$= \frac{1}{2 \times 3.142 \times 500 \times 1 \times 10^{-7}}$$

$$= \frac{10^5}{31.42}$$

$$= 3182.686 \text{ ohms}$$

(ii) $f = 500 \text{ KHz} = 500 \times 10^3 \text{ Hz}$

$$X_C = \frac{1}{2\pi fC} = \frac{1}{2 \times 3.142 \times 500 \times 10^3 \times 1 \times 10^{-7}}$$

$$= \frac{10}{3.142} = 3.182 \quad \text{ohms}$$

7. Calculate the capacitive reactance of a 7pf capacitor when it is connected to (a) audio frequency a.c. source of 50 Hz, (b) radio frequency a.c. source of 50 KHz, (c) d.c. source.

Solution

$C = 7\text{p}f = 7 \times 10^{-12} f, \; X_C = ?$

(a) $f = 50 \text{ Hz}$

Here, $X_C = \frac{1}{\omega C} = \frac{1}{2\pi fC} = \frac{1}{2 \times \frac{22}{7} \times 7 \times 10^{-12} \times 50}$

$$= \frac{10^{11}}{44 \times 5} = \frac{10^{11}}{220}$$

$$= \frac{10^{10}}{22} = \frac{100}{22} \times 10^8 \quad \text{ohms}$$

$= 4.5 \times 10^8$ ohms (very high value)

(b) $f = 50 \text{ KHz} = 50 \times 10^3 \text{ Hz}$

Here, $X_C = \frac{1}{2\pi fC} = \frac{1}{2 \times \frac{22}{7} \times 50 \times 10^3 \times 7 \times 10^{-12}}$

$$= \frac{10^7}{22}$$

$= \frac{100}{22} \times 10^5 = 4.5 \times 10^5$ ohms (low value)

(c) $f = 0$ for d.c.

Here, $X_C = \dfrac{1}{2\pi fC} = \dfrac{1}{0} = \infty$ (Very high reactance)

This shows that the capacitor offers very high reactance to low frequency a.c. and low reactance to high frequency a.c. It offers infinite reactance to steady d.c. A capacitor is therefore said to by pass a.c. and block d.c.

8. A series resonant circuit has $L = 200\ \mu H$, $C = 100$. Find the resonant frequency.

Solution

$$L = 200\ \mu H,\ C = 100\ pf,\ f_r = ?$$

Here, $$f_r = \frac{1}{2\pi\sqrt{LC}}$$

$$= \frac{1}{2 \times 3.142\sqrt{200 \times 10^{-6} \times 100 \times 10^{-12}}}\ \text{Hz}$$

$$= \frac{1}{2 \times 3.142 \times 10^{-7}\sqrt{2}}\ \text{Hz}$$

$$= \frac{10^7}{2 \times 3.142 \times 1.414}\ \text{Hz}$$

$$= \frac{10^7}{8.88} = \frac{100}{8.88} \times 10^5\ \text{Hz}$$

$$= 11.26 \times 10^5\ \text{Hz}$$

$$= 1126\ \text{KHz}$$

9. A parallel resonant circuit is to be designed for a frequency of 1 MHz. If the capacitor is of 100 pf, then find the value of the inductor.

Solution

$$f = 1\ \text{MHz} = 10^6\ \text{Hz}$$
$$C = 100\ pf = 100 \times 10^{-12}\ f = 10^{-10}\ f$$
$$L = ?$$

Here, $$f_r = \frac{1}{2\pi\sqrt{LC}}$$

or, $$f_r^2 = \frac{1}{4\pi^2 LC}$$

or, $$L = \frac{1}{4\pi^2 f_r^2 C}$$

$$= \frac{1}{4 \times 9.87 \times 10^6 \times 10^6 \times 10^{-10}} \text{ H}$$

$$= \frac{10^{-2}}{39.48} \text{ H}$$

$$= 0.0253 \times 10^{-2} \text{ H}$$

$$= 0.253 \times 10^{-3} \text{ H}$$

$$= 0.253 \text{ mH}$$

10. One LCR series circuit consist of L of 0.02 mH, R of 4 Ω and C of 10 pf. Calculate the resonant frequency and Q-factor of the circuit.

Solution

$$L = 0.02 \text{ mH} = 0.02 \times 10^{-3} \text{ H} = 2 \times 10^{-5} \text{ H}$$

$$R = 4\ \Omega,\ C = 10\ pf = 10^{-12} \times 10f$$

$$f_r = ?,\ Q = ?$$

Here, $$f_r = \frac{1}{2\pi\sqrt{LC}}$$

$$= \frac{1}{2 \times 3.142 \times \sqrt{2 \times 10^{-5} H \times 10^{-12} \times 10}}$$

$$= \frac{1}{2 \times 3.142 \times 10^{-8}\sqrt{2}}$$

$$= \frac{10^8}{2 \times 3.142 \times 1.414}$$

$$= \frac{10^8}{8.88} = \frac{100}{8.88} \times 10^6 \text{ Hz}$$

$$= 11.26 \times 10^6 \text{ Hz} = 11.26 \text{ MHz}$$

$$Q = \frac{\sqrt{L/C}}{R} = \sqrt{\frac{2 \times 10^{-5}}{10^{-11} \times 4 \times 4}} = \sqrt{\frac{10^6}{8}}$$

$$= \frac{10^3}{2\sqrt{2}} = \frac{10^3}{2 \times 1.414} = \frac{10^3}{2.828} = 353$$

11. A resistance of 50 ohms and an inductance of 0.7 H is connected to a house mains (220 volt, 50 Hz). Find (a) the impedance of the circuit, (b) the current in the circuit, (c) the p.d. across each element, (d) the phase angle between current and voltage.

Solution

R = 50 ohms, L = 0.7 henry

V = 220 volts, f = 50 Hz

(a) $$Z = \sqrt{R^2 + X_L^2} = \sqrt{R^2 + (\omega L)^2}$$

$$= \sqrt{R^2 + \omega^2 L^2}$$

$$= \sqrt{R^2 + (2\pi f)^2 L^2}$$

$$= \sqrt{R^2 + 4\pi^2 f^2 L^2}$$

$$= \sqrt{(50)^2 + 4 \times 9.87 \times (50)^2 \times (0.7)^2}$$

$$= \sqrt{2500 + 4 \times 9.87 \times 2500 \times 0.49}$$

$$= \sqrt{2500 + (9.87 \times 0.49 \times 10^4)}$$

$$= \sqrt{2500 + (987 \times 49)}$$

$$= \sqrt{2500 + 48363}$$

$$= \sqrt{50863} = 225.5 \text{ ohms}$$

(b) $$I = \frac{V}{Z} = \frac{220}{225.5} \text{ amp}$$

$$= 0.975 \text{ amp}$$

(c) $V_R = IR = 0.975 \times 50 = 48.75$ volts

$VL = I.X_L = \text{I. } 2\pi f L = 0.975 \times 2 \times 3.142 \times 50 \times 0.7$ volts

$= 0.975 \times 3.142 \times 70$ volts

$= 214.44$ volts

(d) $$\tan\theta = \frac{X_L}{R} = \frac{\omega L}{R} = \frac{2\pi f L}{R}$$

$$= \frac{2 \times 3.142 \times 50 \times 0.7}{50}$$

$= 1.4 \times 3.142$

$= 4.3988$

$\theta = \tan^{-1} (4.3988) = 77.19°$

12. A current of $4A$ flows in one R-L series circuit when a 220 volt 50 Hz a.c. source is joined across it. The voltage across the resistor and the inductor is found to be the same. Find (a) voltage across each element (b) the resistance (c) the reactance and (d) the inductance.

Solution

$I = 4\ A$, $V = 220$ volts, $f = 50$ Hz

$V_R = ?$, $V_L = ?$, $R = ?$, $X_L = ?$

$L = ?$, Given, $V_R = V_L$.

(a) We have $V = \sqrt{V_R^2 + V_L^2}$

$$= \sqrt{2V_R^2} \quad (\because V_R = V_L)$$

$$= \sqrt{2}\, V_R$$

$$\therefore\ V_R = \frac{V}{\sqrt{2}} = \frac{220}{\sqrt{2}} = 110\sqrt{2}$$

$$V_L = V_R = 110\sqrt{2} \text{ volts.}$$

(b) $R = \frac{V_R}{I} = \frac{110\sqrt{2}}{4} = 27.5 \times 1.414$

$= 38.88$ ohms

(c) $X_L = \frac{V_L}{I} = \frac{110\sqrt{2}}{4} = 38.88$ ohms

(d) $L = \frac{X_L}{2\pi f} = \frac{38.88}{2 \times 3.142 \times 50}$

$= \frac{38.88}{314.2} = 0.1237$ H

$= 123.7 \times 10^{-3}$ H

$= 123.7$ mH

13. Compute the capacitance of a capacitor to run a 100 V, 20 watt lamp joined in series with an a.c. mains of 220 V and 50 Hz, so that the bulb glows with the desired brilliancy.

Solution

V_R = 100 V, P = 20 watt
V_C = ?, V = 220 volts, f = 50 Hz
C = ?

Here, the lamp to glow with its desired brilliancy, the p.d. across the lamp must be 100 volts and the rated current of $\frac{20}{100} = \frac{1}{5} = 0.2\ A$ has to flow through it.

If 'C' be the capacitance of the capacitor joined in series with a.c. mains, then the lamp and the capacitor form one $R - C$ series circuit.

The applied voltage V is given by

$$V = \sqrt{V_R^2 + V_C^2}$$

or, $$V_C = \sqrt{V^2 - V_R^2} = \sqrt{(200)^2 - (100)^2}$$

$$= \sqrt{48400 - 10000} = \sqrt{38400}$$

$$= 19.59 \times 10 = 195.9 \text{ volts.}$$

But, $$V_C = IX_C$$

or, $$X_C = \frac{V_C}{I} = \frac{195.9}{0.2} = \frac{1959}{2}$$

$$= 979.5 \text{ ohms}$$

or, $$\frac{1}{2\pi fC} = 979.5$$

or, $$C = \frac{1}{2 \times 3.142 \times 50 \times 979.5} \text{ farad}$$

$$= \frac{1}{3.142 \times 100 \times 979.5} \text{ farad}$$

$$= (1/314.2) \times 979.5 \text{ farad}$$

$$= 3.249 \times 10^{-6} \text{ farad}$$

$$= 3.249\ \mu f$$

14. A resistance of 400 ohms, an inductance of 500 mH and a capacitance of 30 μf are joined in series across a 220 V, 50 Hz mains. Calculate (a) the reactance, (b) the impedance, (c) the current in the circuit, (d) the power factor of the circuit, and (e) angle of lead or lag.

Solution

$R = 400$ ohms, $L = 500$ mH $= 500 \times 10^{-3}$ H,

$C = 30\ \mu f = 30 \times 10^{-6}\ f$

$V = 220$ V, $f = 50$ Hz

(a) $X_L = \omega L = 2\pi f L = 2 \times 3.142 \times 50 \times 5 \times 10^{-1}$

$= 50 \times 3.142$ ohms

$= 157.1 = 157.1$ ohms

(b) $$X_C = \frac{1}{2\pi fC} = \frac{1}{2 \times 3.142 \times 50 \times 30 \times 10^{-6}}$$

$$= \frac{10^6}{3.142 \times 3 \times 10^3} = \frac{10^3}{9.426}$$

$$= \frac{1000}{9.426} = 106 \text{ ohms}$$

Impedance $$Z = \sqrt{R^2 + (X_L - X_C)^2}$$

$$= \sqrt{(4 \times 10^2)^2 + (157.1 - 106)^2}$$

$$= \sqrt{16 + 10^4 + (51.1)^2}$$

$$= \sqrt{160000 + 2611} = \sqrt{162611}$$

$= 403.25$ ohms

(c) $$I = \frac{V}{Z} = \frac{220}{403.25} = 0.545 \text{ amp}$$

(d) Power factor $$= \frac{R}{Z} = \frac{400}{403.25} = 0.9919$$

(e) Also power factor $\cos\theta = 0.9919$

$\therefore\ \theta = \cos^{-1}(0.9919)$

$= 7.29°$

Hence, the angle θ is the angle of lag because $X_C < X_L$.

15. What is the minimum power-dissipation rating of a resistor of 2×10^4 ohms to be joined across a potential of 400 V?

Solution

$$R = 2 \times 10^4 \text{ Ohms}$$
$$V = 4 \times 10^2 \text{ Volts}$$
$$P = ?$$

Here, $P = \dfrac{V^2}{R} = \dfrac{16 \times 10^4}{2 \times 10^4} = 8$ Watts.

16. Compute the voltage that can be withstood by a resistor rating at 50 W and 50 mA.

Solution

$$P = 50\ W,\ I = 50 \times 10^{-3}\ A$$

Here, $V = \dfrac{P}{I} = \dfrac{50 W}{50 \times 10^{-3} A} = 10^3$ Volts

17. An amplifier has voltage gain of 125. Express the gain in decibels.

Solution

Gain = 125

Decibel Gain = ?

Here db gain = $20 \log_{10} 125$

$= 42$ db

18. Find the gain in db in the following cases :
 (i) voltage gain of 100
 (ii) power gain of 20

Solution

(i) db gain = $20 \log_{10}$ (voltage gain)

$= 20 \log_{10} (100)$

$= 40$ db

(ii) db gain = $10 \log_{10}$ (power gain)

$= 10 \log_{10} (20)$

$= 10 \times 1.3010$

$= 13.01$ db

19. Express the following in terms of voltage and power gain.
 (i) – 30 db (ii) – 5 db

Solution

(i) Voltage gain $= \text{Antilog}\left(-\frac{30}{20}\right)$

$= \text{Antilog}\ (-\ 1.5)$

$= 1/31.62$

Power gain $= \text{Antilog}\left(-\frac{30}{10}\right)$

$= \text{Antilog}\ (-3)$

$= 10^{-3} = \frac{1}{10^3} = \frac{1}{1000}$

(ii) Voltage gain $= \text{Antilog}\left(-\frac{5}{20}\right)$

$= \text{Antilog}\left(-\frac{1}{4}\right)$

$= \text{Antilog}\ (-\ 0.25)$

$= \frac{1}{1.778}$

Power gain $= \text{Antilog}\left(-\frac{5}{10}\right)$

$= \text{Antilog}\left(-\frac{1}{2}\right)$

$= \text{Antilog}\ (-\ 0.5)$

$= \left(\frac{1}{3.162}\right)$

Voltage gain and power gain calculated in this problem actually shows the reduction in voltage or power level with respect to the original value.

20. A power transformer has 80 turns in primary and 400 turns in secondary. The alternating voltage developed across secondary is 1000 volts. Find the turns ratio and the alternating voltage applied across primary.

Solution

n_p = 80 turns
n_s = 400 turns
V_s = 1000 volts
n = ?, V_p = ?

Turns ratio $n = \frac{n_s}{n_p} = \frac{400}{80}$

$= 5$

Again, $\frac{V_s}{V_p} = \frac{n_s}{n_p}$

or, $V_p = V_s \times \frac{n_p}{n_s}$

$= 1000 \times \frac{1}{5}$

$= 200$ volts

21. The efficiency of a step up transformer is 90%. Calculate the output power and power loss when the applied voltage across primary is 220 volt and the primary coil has a resistance of 40 ohms.

Solution

η = 90% = 0.90
V_p = 220 volts
r_p = 40 ohms
V_s = ?, Power loss = ?

Primary Current $I_p = \frac{V_p}{r_p} = \frac{220}{40} = \frac{22}{4}$ amp

= 5.5 amp.

$\therefore$ Input power = $V_p I_p = 220 \times 5.5$ Watts

= 22 × 55 Watts
= 1210 Watts

$$\eta = \text{Efficiency} = 0.9 = \text{P output/P input}$$

$$\begin{aligned} \therefore \text{P output} &= 0.9 \times \text{P input} \\ &= 0.9 \times 1210 \\ &= 9 \times 121 = 1089 \text{ watts} \end{aligned}$$

$$\begin{aligned} \text{Hence power loss} &= \text{Input power} - \text{Output power} \\ &= 1210 - 1089 \\ &= 121 \text{ watts} \end{aligned}$$

❑ ❑ ❑

6

Thermionic Emission and Vacuum Tubes

Q. 1. What do you mean by work function of a metal ?

Ans. *Work function* : The amount of additional energy required to emit an electron from a metallic surface is known as work function of that metal.

The work function of pure metals varies roughly from 2 to 6 ev. It depends upon the nature of metal, its purity and the conditions of its surface.

Q. 2. What are different types of electron emission?

Ans. *Types of electron emission* : There are four types of electron emission, *viz* :

(i) Thermionic emission
(ii) Field emission
(iii) Photo electric emission
(iv) Secondary emission

(i) Thermionic emission. In this method, the metal is heated to sufficient temperature (about 2600°C) to enable the free electrons to leave the metal surface. The number of electrons emitted depends upon the temperature of the metal. The higher the temperature the greater is the emission of electrons. This type of emission is employed in vacuum tubes.

(ii) Field emission. In this method, a strong electric field (*i.e.*, a high +ve voltage) is applied at the metal surface, which pulls the free electrons out of metal because of the attraction of +ve field. The stronger the electric field, the greater is the electron emission.

(iii) Photo electric emission. In this method, the energy of light falling upon the metal surface is transferred to the free electrons within the metal to enable them to leave the surface. The greater the intensity of light beam falling on the metal surface, the greater is the photo electric emission.

(iv) Secondary emission. In this method, a high velocity beam of electrons strikes the metal surface and causes the free electrons of that metal to be knocked out from the surface.

Q. 3. Define one ev and find its value.

Ans. *Electron Volt :* One electron volt is the amount of energy acquired by an electron when it is accelerated through a potential deference of 1 volt. So, the ev is the unit of energy.

$$1 \text{ ev} = 1 \text{ volt} \times \text{charge of one electron}$$
$$= 1 \text{ volt} \times 1.602 \times 10^{-19} \text{ coul}$$
$$= 1.602 \times 10^{-19} \text{ Joule}$$

Hence, $1 \text{ ev} = 1.062 \times 10^{-19}$ Joule.

Q. 4. Explain the phenomenon of Thermionic emission. State Richardson-Dushman Law.

Ans. *Thermionic emission :* The process of electron emission from a metal surface by supplying thermal energy to it is known as thermionic emission.

At ordinary temperature, the energy possessed by free electrons in the metal is inadequate to cause them to escape from the surface. When heat is applied to the metal, some of heat energy is converted into kinetic energy, causing accelerated motion of free electrons. When the temperature rises sufficiently, these electrons acquire additional energy equal to the work function of the metal. Consequently, they overcome the restrain surface barrier and leave the metal surface.

Metals with lower work function will require less additional energy and therefore, will emit electrons at lower temperatures. The commonly used materials for electron emission are tungsten, thoriated tungsten and metallic oxides of Barium and Strontium. It may be added here that high temperatures are necessary to cause thermionic emission. For example, Pure Tungsten must be heated to about 2300°C to get electron emission. However, oxide coated emitters need only 750°C to cause thermionic emission.

Richardson-Dushman Law

The amount of thermionic emission increases rapidly as the emitter temperature is raised. The emission current density is given by Richardson-Dushman equation,

$$I_s = AT^2 e^{-b/T} \text{ amp/m}^2 \qquad ...(1)$$

where I_s = emission current density *i.e.*, current per square metre of the emitting surface.

T = absolute temperature of emitter in °k.

A = constant, depending upon the type of emitter and is measured in amp/m²/°k².

b = a constant for the emitter

e = natural log arithmic base.

The value of b is constant for a metal and is given by,

$$b = \frac{\phi e}{k}$$

where ϕ = work function of emitter.

e = electron charge = 1.602×10^{-19} coulomb

k = Bollzmann's constant = 1.38×10^{-23} g/°k

$$\therefore\ b = \frac{\phi \times 1.602 \times 10^{-19}}{1.38 \times 10^{-23}} = 11600\ \phi°\text{k}$$

Putting the value of b in the exp. (1), we get,

$$I_s = AT^2\, e^{-\frac{16000\theta}{T}} \qquad ...(2)$$

The following points may be noted from equn. (2).

(i) The emission is markedly affected by temperature changes. Doubling the temp. of an emitter may increase electron emission by more than 10^7 times. For instance, emission from pure tungsten metal is about 10^{-6} ampere per square cm. at 1300°C but rises to enormous value of about 100 amperes when temperature is raised to 2900°C.

(ii) Small change in the work function of the emitter can produce enormous effects on emission. Halving the work function has exactly the same effects as doubling the temperature.

Q. 5. What is a vacuum tube? what are different vacuum tubes?

Ans. *Vacuum Tube :* An electronic device, in which the flow of electrons is through a vacuum, is known as a vacuum tube.

A vacuum tube usually contains a cathode which is the electron emitter, an anode (also called plate) which is the electron collector and one or more electrodes (called grids) for controlling the flow of electrons between cathode and plate. These electrodes are housed in a highly evacuated glass envelope. The plate is held at +ve potential with respect to cathode, so that emitted electrons are attracted to plate to provide current in vacuum. The ability of vacuum tubes to conduct current in vacuum enables them to perform different functions.

Classification of vacuum tubes

There are several ways of classifying vacuum tubes. However, according to the number of electrodes, vacuum tubes are classified as under :

(i) vacuum diode
(ii) vacuum triode
(iii) vacuum tetrode
(iv) vacuum pentode

The diode, triode, tetrode and pentode contain two, three, four and five electrodes respectively.

Q. 6. What is the function of control grid? What are different types of grids used in vacuum tubes?

Ans. *Function of control grid :* The control grid is used in a triode valve to control the flow of electrons between cathode and plate.

Types of grids. There are three types of grids used in three different vacuum tubes. These are :

(i) control grid
(ii) screen grid
(iii) suppressor grid

The control grid is the only grid used in triode valve. It is placed in between cathode and the plate. It is made –ve w.r.t. plate and +ve w.r.t. the cathode. The surface charge effect is eliminated by the grid. Hence a vacuum tube performs for a longer period by the Introduction of the grid.

In a tetrode valve, two grids are used. These are control grid and screen grid. These two grids are placed inbetween the cathode and the plate. Nearer to the cathode, the control grid is kept and then the screen grid. Thus a tetrode valve works for a longer period than a triode valve.

Three grids are used in a pentode valve. These grids are control grid, screen grid, and suppressor grid. From cathode the order of the grids are control grid, screen grid and suppressor grid. Thus a pentode valve works for a longer period than the tetrode valve.

Q. 7. Describe the construction and function of a vacuum diode.

And. *Construction :* A vacuum diode consists of two electrodes, a cathode and an anode (plate). These two electrodes are enclosed in a highly evacuated glass envelope. The cathode is in the form of nickel cylinder coated with Barium and Strontium oxides. The cathode is heated indirectly by a heater circuit in order to provide electron emission. The anode is generally a hollow cylinder made of nickel or molybdenum and surrounds the cathode. Fig. 6.1(a) shows the construction of vacuum diode whereas Fig. 6.1(b) shows its symbol. Note the symbol of diode where plate is represented by straight line, cathode by a straight line with sides folded down and heater by inverted *V*.

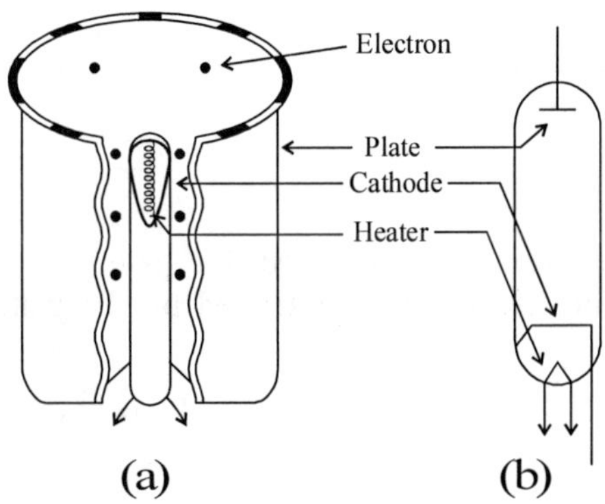

Fig. 6.1

Function

When the cathode is heated indirectly by the heater, it emits electrons. The behaviour of these emitted electrons depends upon the anode-cathode potential. If the anode is at zero potential with respect to cathode Fig. 6.2(a), then the emitted electrons can't be attracted by the plate. Therefore, there will be no current in the circuit. In this case, the emitted electrons are accumulated near the cathode and form a cloud of electrons. This is known as space charge.

When the plate is made +ve w.r.t. cathode [Fig. 6.2(b)] then the electrons from the space charge are attracted to the plate. These electrons flow from cathode to plate and this results in giving plate current. The plate circuit consists of wires through which the electrons attracted by the plate flow. A meter connected in series with the plate circuit can record the plate current. A battery is connected between plate and cathode to give plate-cathode potential. The electrons attracted by the plate flow through the external circuit and finally return to the cathode, thus filling up the supply of electrons lost by emission.

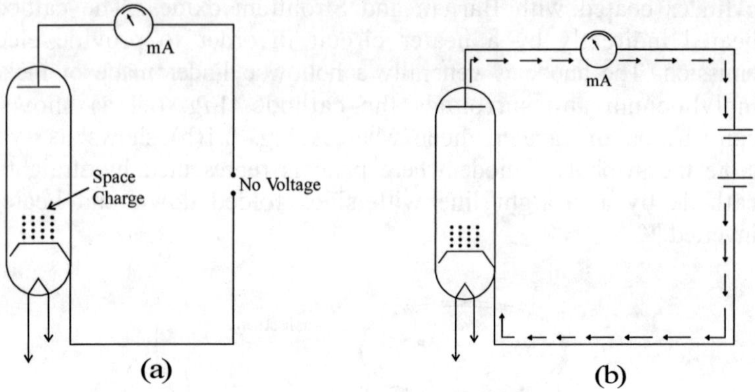

Fig. 6.2

If the plate potential is increased, the plate current also increases as more and more electrons are attracted by the plate.

Plate negative

If the plate is made –ve w.r.t. cathode as shown in fig. 6.3, the emitted electrons are repelled back and no current flows in the plate circuit.

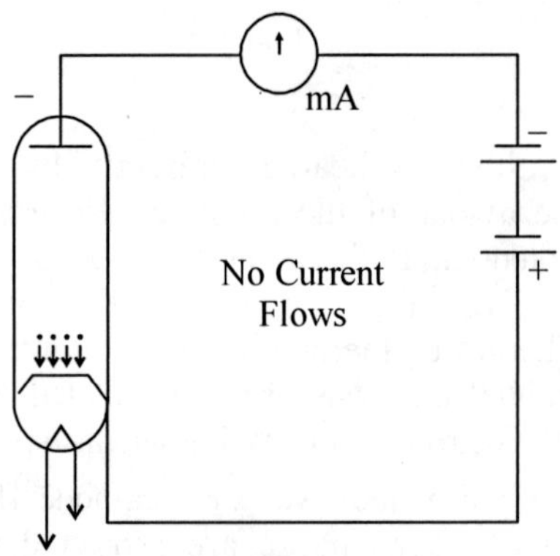

Fig. 6.3

Q. 8. How can you plot the plot characteristic curves of a vacuum diode? What is d.c. and a.c. plate resistance? How can they be determined from plate characteristics of a diode?

Ans. *Characteristics of vacuum diode :* The most important characteristic of a vacuum diode is the plate characteristic which gives the relation between plate voltage and plate current for a given cathode temperature. The circuit arrangement for determining the plate characteristics of an indirectly heated vacuum diode as shown in fig. 6.4 (a). The cathode temperature can be changed by varying R_1 connected in the heater circuit. The plate voltage can be varied to desired value by means of resistor R_2 arranged as potential divider in the plate circuit.

Keeping the cathode temperature constant, say at T_1, the plate voltage is varied in steps from zero by means of potential divider. Corresponding to each value of plate voltage, the value of plate current is noted. Then curve is drawn, taking plate voltage along X-axis and plate current along Y-axis. This gives the plate characteristic at cathode temperature T_1 as shown in fig. 6.4(b). Following similar procedure, plate characteristics at various cathode temperatures can be obtained:

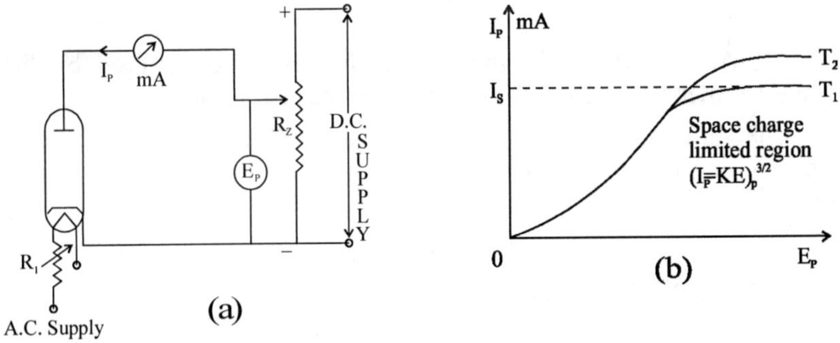

Fig. 6.4

Plate resistance of diode

The internal resistance offered by the diode to the plate current is known as its plate resistance.

(1) D.C. Plate resistance. The resistance offered by the diode to direct current is known as d.c. plate resistance.

The value can be calculated by finding the ratio of total d.c. plate voltage across diode to the resulting current. Fig. 6.5 shows the typical plate characteristic of a vacuum diode. At point P on the curve, the plate voltage is OA and the corresponding plate current is OB. The d.c. plate resistance R_p is given by :

$$R_p = \frac{OA}{OB} \qquad ...(1)$$

As the plate characteristic isn't a straight line, therefore, d.c. plate resistance isn't constant, but depends upon the operating point. Thus plate resistance at point X is different from that at point P. Hence d.c. plate resistance must be determined at the actual operating point.

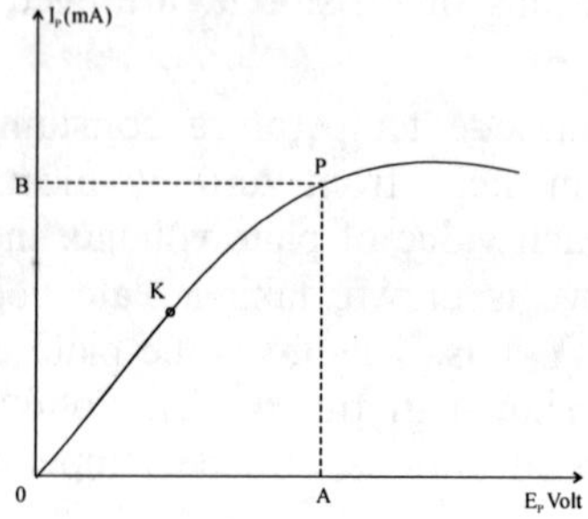

Fig. 6.5

(2) A.C. Plate resistance. It is the resistance offered by the diode to alternating current and may be defined as under.

The ratio of a small change on plate voltage across a diode to the resulting change in plate current is known as a.c. plate resistance *i.e.*,

$$\text{a.c. plate resistance, } r_p = \frac{\Delta E_p}{\Delta I_p} \quad \text{...(2)}$$

As tubes are gradually used with a.c. or varying potentials, therefore a.c. plate resistance is much more important than d.c. plate resistance. The a.c. plate resistance can be determined from plate characteristic by considering the small change of plate voltage half way on each side of the operating point. For example in fig. 6.6, the a.c. plate resistance at operating point P can be found by considering same equal changes OP plate voltage on either side of the operating point (*i.e.*, $AB = AC$).

Change in plate voltage = BC

Change in plate current = YZ

∴ a.c. plate resistance at P,

$$r_p = \frac{BC}{YZ} = \frac{(E_p)_2 - (E_p)_1}{(I_p)_2 - (I_p)_1}$$

$$= \frac{\Delta E_p}{\Delta I_p} \quad \text{...(3)}$$

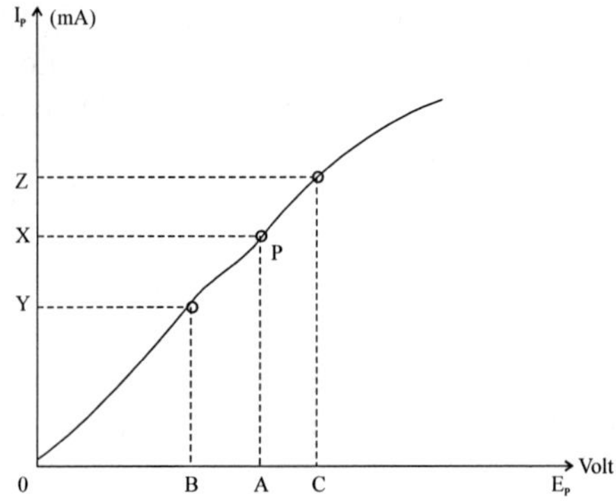

Fig. 6.6

Q. 9. State and explain Child's law.

Ans. *Child's law states that :* In space charge limited region, the plate current is directly proportional to the plate voltage raised to the power 3/2.

Mathematically, it is stated as

$$I_p = KE_p^{3/2}$$

where K is a constant whose value depends upon the shape of electrodes and geometry of tube. I_p is the plate current and E_p is the plate voltage.

Explanation. The low plate voltage region is known as space charge limited region. In this region, the plate current increases as the plate voltage is increased, because more +ve plate attracts electrons from this space charge at a greater rate. It is clear from Child's law that in space charge limited region, the plate current. 'I_p' is completely controlled by plate voltage 'E_p' and is independent of cathode temperature.

Q. 10. Discuss the nature of plate current of a vacuum diode beyond the space charge limited region.

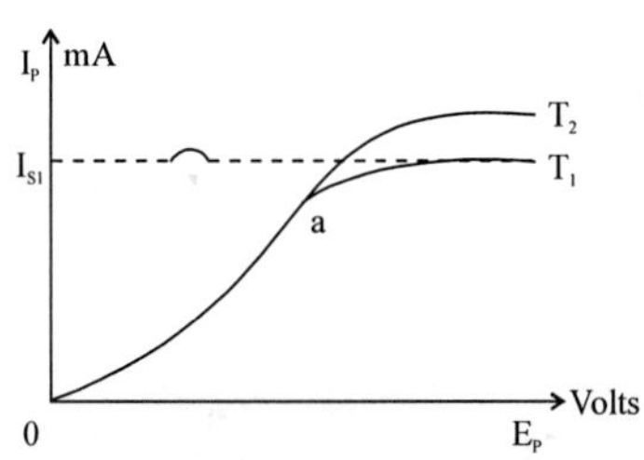

Fig. 6.7

Ans. As plate voltage is made progressively higher, greater portion of electrons from space charge is attracted to plate and eventually at some plate voltage, the space charge is completely eliminated. Under such conditions, the entire supply of emitted electrons (for a given cathode temperature) is attracted to the plate. Therefore, beyond this point plate current becomes constant and does not increase even if the plate voltage is increased. The max^{m} plate current is called saturation current. In fig. 6.7, I_{s_1}, is the saturation current at cathode temperature T_1. If the cathode temperature is raised (say from T_1 to T_2), the rate of emission is increased. Consequently, the saturation point is raised.

Q. 11. Describe the construction and function of a triode valve.

Ans. *Construction :* A triode valve has three electrodes, namely, cathode, plate and control grid. The cathode is located at the centre of the tube and is surrounded by control grid which is inturn surrounded by the plate. The cathode and plate have similar

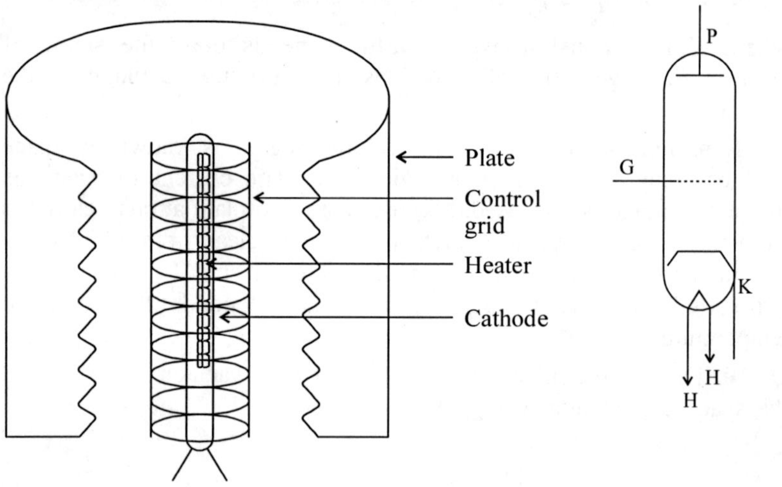

(a) (Cut away view of Triode)

(b) (Triode symbol)

Fig. 6.8

construction as for a diode. The control grid consists of a fine wire mesh placed very close to the cathode. The spacing between the turns of the mesh are wide enough so that the passage of electrons from cathode to plate is not obstructed by grid. The electrons attracted to the plate from cathode go through the openings in the grid. The cathode is indirectly heated by the filament (heater) to emit electrons. Fig. 6.8(a) shows the cut-away view of triode whereas fig. 6.8(b) shows its symbol. The dotted line between the plate and cathode in the symbol represents the control grid.

Function

Here we will discuss the action of each part of the triode valve.

Action of filament

The filament means the heater circuit. It is actually a low tension Ohm's law circuit where the resistor is heated by an emf series with it.

Action of cathode

As discussed above, the filament is surrounded by the cathode which is indirectly heated by the filament. The cathode emits electrons by way of heating.

Action of control grid

The control grid surrounds the cathode. The electrons emitted by cathode go to the plate through the openings in the grid. As the control grid is much closer to the cathode than the plate, therefore, a small voltage in the grid controls the flow of electrons from cathode to the plate more effectively than a high voltage on the plate. Fig. 6.9 shows the action of the control grid in a triode at different grid voltages, assuming the plate potential remains unchanged :

(i) when the grid is at zero potential with respect to cathode as shown in fig. 6.9(i), the triode valve just behaves like a diode. This is because the grid being at zero potential has no effect on the plate current.

(ii) If the grid is placed at low –ve potential w.r.t. cathode as shown in fig. 6.9 (ii), it has repelling effect on the electrons coming from the cathode. As a result of which fewer electrons reach the plate and hence the plate current is reduced.

(iii) If the –ve potential on the grid (called grid bias) is increased, the plate current decreases continuously. When the grid bias is large, all the electrons are repelled towards cathode, resulting zero plate current. At this cond^{n} the triode is said to be cut off [fig. 6.9 (iii)].

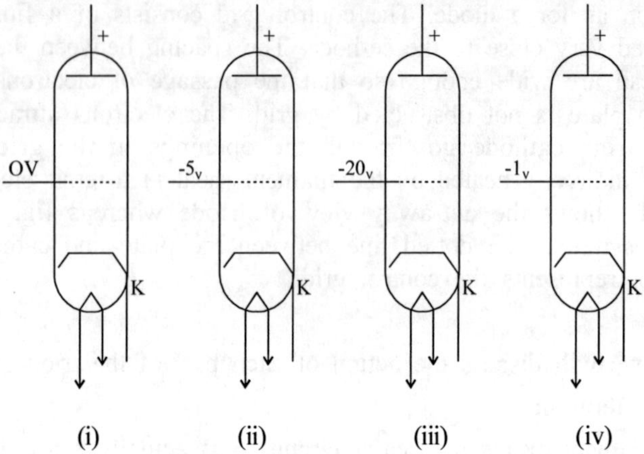

Fig. 6.9

(iv) If the control grid is made slightly +ve (say +1v) w.r.t. cathode as shown in fig. 6.9(iv), the electrons are accelerated towards the plate. This increases the plate current. Some of the electrons are also attracted to the grid to constitute grid current. The grid current isn't desirable because it causes power loss in the grid circuit. Therefore, grid is always kept at –ve potential w.r.t. cathode.

Conclusion

From the above discussion it is clear that a small change in grid potential brings about a large change in plate current. To affect the same change in plate current without changing the grid potential, a much larger plate potential is needed. Actually the current controlling property of the control grid is responsible for use of triode valves commonly as amplifiers.

Q. 12. What do you mean by triode characteristics? What are static and dynamic characteristics of a triode valve?

Ans. *Triode characteristics* : The graphical representations of relationship between plate current, plate voltage and grid voltage under normal operating conditions are known as triode characteristics. Assuming the cathode temperature to be constant, the plate current in a triode depends upon plate and grid potentials, *i.e.*,

$$I_p = f(E_p, E_g) \quad ...(1)$$

where I_p is the plate current, E_p is the plate potential and E_g is the grid potential.

Explanation. In the above equn. (1), there are three variables, namely, I_p, E_p and E_g. Keeping any one of the variables fixed, we can plot a graph between other two variables. Accordingly there will be three characteristic curves like :

(a) Plate characteristic curves, where $I_p \sim E_p$ when E_g is constant.

(b) Mutual characteristic curves, where $I_p \sim E_g$ when E_p is constant.

(c) Constant current characteristic curves, where $E_p \sim E_g$ when the plate current I_p is constant.

The set of curves showing the relationship between any two variables while the third one is kept constant, is known as triode characteristics or the characteristic curves of the triode valve.

The triode characteristics can be obtained under two sets of conditions like static cond^{n}s and dynamic cond^{n}s.

(i) Static conditions

Static conditions are achieved when various d.c. voltages are applied to the triode, electrodes with no load in the plate circuit and no signal at the input. Under such conditions, the plate potential remains constant or static and becomes independent of plate current. The curves obtained under static conditions are known as static characteristics.

Practical circuit of a triode valve

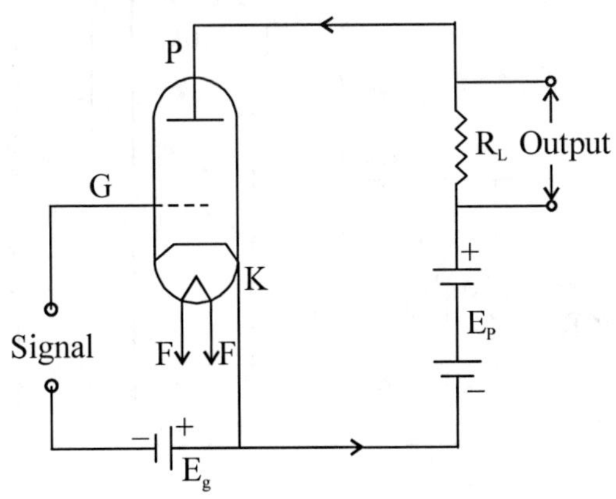

Fig. 6.10

(ii) Dynamic conditions

Dynamic conditions are achieved with signal applied in the grid circuit and load connected in the plate circuit. Under such conditions the plate current flowing through the load causes a voltage drop in it. Consequently, the plate potential varies with plate current. The curves obtained under this conditions are known as dynamic characteristics.

Conclusion

It should be noted here that dynamic characteristics represent actual operating cond^{n}s of a triode valve because a practical triode circuit has signal in the input and load in the plate circuit (output).

Q. 13. How can you plot the static characteristics of a triode valve? Give the circuit diagram you will use.

Ans. *Static characteristics of a triode :* Fig. 6.11 shows the experimental arrangement for drawing the static characteristic curves of a triode valve. Here, in this circuit no load is connected in the output circuit (plate circuit) and no signal is there in the grid circuit. So the plate potential can be kept constant for any desired value of the resistor R_1 arranged as potential devider. The grid can be given any +ve or –ve potential w.r.t. cathode by a reversing switch R_S.

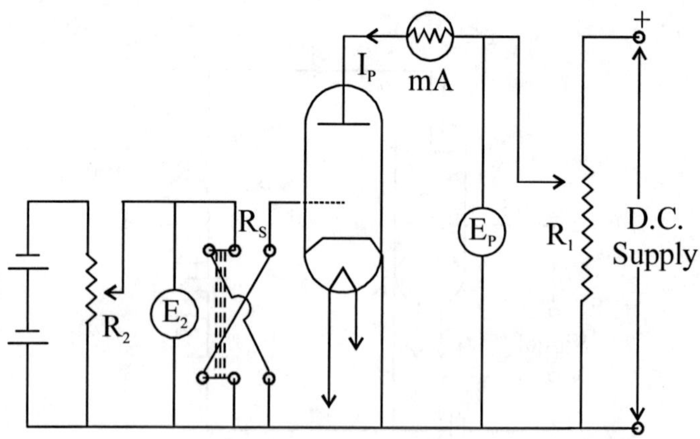

Fig. 6.11

1. Plate characteristics

It is the curve between plate voltage E_p and plate current I_p of a triode at constant grid voltage E_c.

For determining the static plate characteristics, refer to the circuit diagram shown in fig. 6.11. Firstly, the grid voltage is set at zero *i.e.*, $E_g = 0$. Keeping the grid voltage constant at this value, the plate voltage is changed in steps and the corresponding values of I_p are noted on the milliammeter connected in the plate circuit. The $E_p \sim I_p$ reading are plotted on the graph. This gives the plate characteristic at $E_c = 0$. The experiment is repeated with $E_c = -2$v and then $E_c = -4$v etc. Thus a family of plate characteristics is obtained at different grid voltages as shown in fig. 6.12. The following points may be noted from these characteristics :

(i) The characteristics are drawn for –ve values of grid voltage only as practically a triode is always operated at –ve grid voltage.

(ii) The characteristics are curved over the lower portion but fairly linear in the upper portion. It is because at low plate voltage, the plate current rises slowly, but at high plate voltages, it increases appreciably.

(iii) The curves are approximately equally spaced for equal differences of grid voltages.

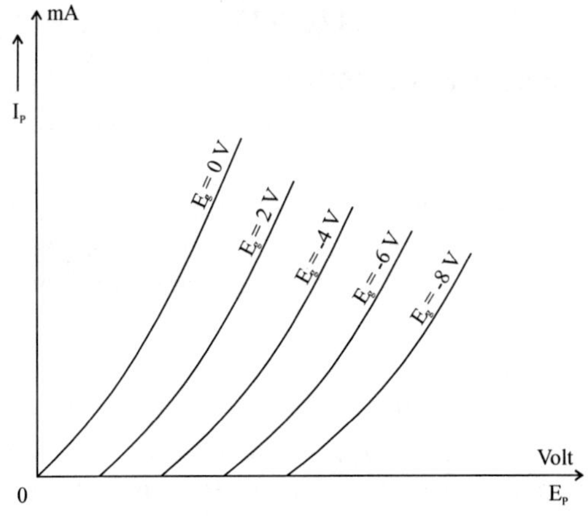

Fig. 6.12

2. Mutual characteristics

It is the curve between plate current I_p and grid voltage E_g at constant plate voltage.

For determining the static mutual characteristics of a triode, the circuit of fig 6.11 can be used. The slider on R_l is adjusted to give a constant value of plate voltage, say 200 volts. Keeping the plate voltage constant at this value, the grid voltage is changed from zero to –ve values in steps. The corresponding values of plate current I_p are noted on the milliammeter connected in the plate circuit. The $I_p \sim E_g$ readings are plotted on the graph. This gives the mutual characteristic at $E_p = 200$v. The experiment is repeated with $E_p = 150$v and $E_b = 100$v etc. Thus a family of mutual characteristics at different plate voltages is obtained as shown in fig. 6.13. It may be seen that mutual characteristics convey the same information as the plate characteristics. Often either of the two is used to study the performance of triode in a circuit.

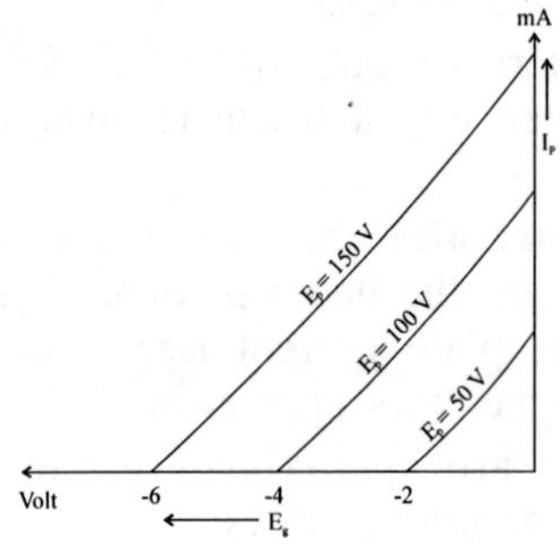

Fig. 6.13

(3) Constant current characteristic

It is the curve between plate voltage E_p and grid voltage E_g at constant plate current I_p.

For determining the constant current characteristics, the experimental arrangement of fig. 6.11 can be employed. However, this curve is relatively unimportant from the practical point of view and is seldom used.

Q. 14. Define plate resistance, transconductance and amplification factor of a triode and establish the relationship among them.

Ans. *A.C. Plate resistance* (r_p) : The ratio of small change in plate voltage (ΔE_p) to the resulting small change in plate current (ΔI_p) at constant grid voltage is known as a.c. plate resistance *i.e.*,

a.c. plate resistance, $r_p = \dfrac{\Delta E_p}{\Delta I_p}$ at constant E_g.

Transconductance (g_m)

The ratio of small change in plate current (ΔI_p) to the small change in grid voltage (ΔE_g) at constant plate voltage is known as transconductance or mutual conductance *i.e.*, mutual conductance,

$$g_m = \frac{\Delta T_p}{\Delta E_g} \text{ at constant } E_p$$

Amplification factor (μ)

The ratio of small change in plate voltage (ΔE_p) to a small change in grid voltage (ΔE_g) of a triode at constant plate current is known as amplification factor *i.e.*,

amplification factor, $\mu = \frac{\Delta E_p}{\Delta E_g}$ at constant I_p. Relationship between

μ, r_p and g_m :

We know, $\mu = \frac{\Delta E_p}{\Delta E_g}$

Multiplying and dividing the numerator and denominator on the R.H.S. by ΔI_p, we get,

$$\mu = \frac{\Delta E_p}{\Delta E_g} \times \frac{\Delta I_p}{\Delta I_p} = \left(\frac{\Delta E_p}{\Delta I_p}\right) \times \left(\frac{\Delta I_p}{\Delta E_g}\right)$$

$$\therefore \quad \mu = r_p \times g_m$$

i.e., amplification factor = plate resistance × mutual conductance. The amplification factor of a triode valve varies from 3 to 100.

Q. 15. What are the limitations of triode valve? Explain.

Ans. The triode valve has two serious drawbacks, namely :

(i) Inter electrode capacitance
(ii) Insufficient amplification

(i) Inter electrode capacitance

Capacitance exists between any two conducting surfaces separated by an insulating medium. As triode electrodes are made of metals and are separated from each other by vacuum, any two of them form a capacitor. Therefore, capacitance must exist between grid and cathode (C_{gk}), grid and plate (C_{gp}) and plate and cathode (C_{pk}). These capacitances are called inter electrode capacitances ranging from 2 to 12 pf (fig. 6.14).

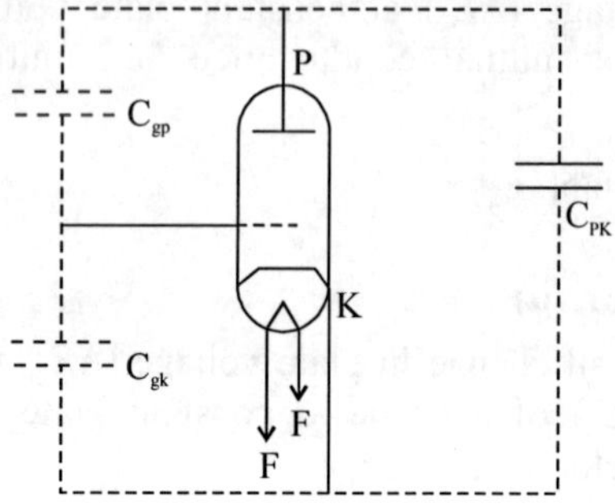

Fig. 6.14

At low frequencies the inter electrode capacitances have negligible effects. However, at high frequencies the inter electrode capacitance between plate and grid (C_{gp}) has serious effects. As the capacitive feedback is –ve, C_{gp} reduces the amplification of the triode valve at high frequencies. Thus, the triodes are generally used for amplifying low frequency signals (< 20KHz).

(ii) Insufficient amplification

The amplification factor of a triode is generally small, doesn't exceed 100. The amplification factor μ of a valve can be high if the effect of control grid on plate current is quite large as compared to that of plate. But this is not so in case of a triode due to insufficient shielding between plate and cathode. The control grid which is to provide the necessary shielding allows a part of electric field to penetrate through it.

However, if control grid is to be made more effective in controlling the plate current, the electric field from plate to cathode should be shielded as effectively as possible. This can be achieved by having a control grid with very finely spaced spiral. But there is a limit to it since in a very finely spaced spiral grid, the electrons will encounter difficulty in passing through the grid openings and consequently, the plate current will be reduced. This puts a limit to the effectiveness of control grid. Therefore, amplification factor of a triode is low.

Q. 16. What are the uses of triode valve?

Ans. Triodes can be used for amplifying a weak signal. Such a device which does this job is called an amplifier. So triodes can be used as amplifiers, called vacuum tube amplifiers.

The phenomenon of amplification is required in radio communication systems. Amplifiers are also used in industrial and commercial fields, *i.e.*, in television inter communication system, radio receivers.

In public address systems, amplifiers are also used inbetween microphone and loudspeaker. The microphone converts sound waves into electrical pulses (called the signal). Such a weak signal is amplified and then fed to the loudspeaker where the electrical pulses are again converted back to sound. Then the strength of the sound wave is increased.

Q. 17. Explain the construction and working of a tetrode valve.

Ans. *Construction of tetrode valve :* The tetrode is a four electrode valve. It consists of a plate, a cathode and two grids, called control grid and green grid.

The cathode is surrounded by the control grid 'G_1'. The control grid is surrounded by the screen grid 'G_2'. The plate 'P' surrounds the screen grid 'G_2'. In fig. 6.15(a) the position of the four electrodes of a tetrode valve is shown. In fig. 6.15(b) the electronic symbol of representing a tetrode is given.

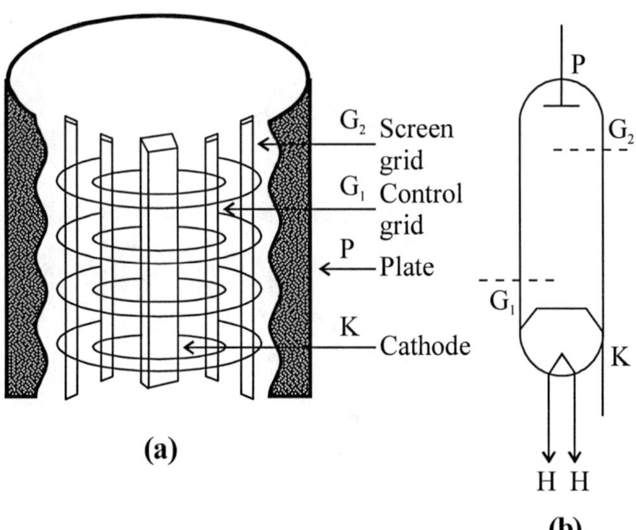

Fig. 6.15

Working

The working of tetrode is similar to triode with the additional action of screen grid. Like triode, the control grid in a tetrode is placed at a small –ve potential while the plate is at a fairly +ve potential with respect to cathode. The screen grid is also kept at a +ve potential w.r.t. cathode but somewhat lower than plate voltage.

When the cathode is indirectly heated by the filament, it emits electrons. These electrons are collected by the plate and move through the control grid and then through the screen grid before reaching the plate. Most of the emitted electrons reach the plate and form a plate current to flow. However, some electrons are attracted by the screen grid to constitute screen grid current. Therefore, in a tetrode the plate current and the screen grid current are produced. The screen grid current serves no useful purpose, it is only a small part of the total emission current. As the screen grid acts as an electrostatic shield between the plate and control grid, therefore change in plate voltage has little effect on the magnitude of the plate current. On the other hand, the control grid controls over the plate current. Thus, the addition of a screen grid makes the plate voltage less effective in controlling the plate current without effecting the influence of the control grid. Consequently, amplification factor is increased.

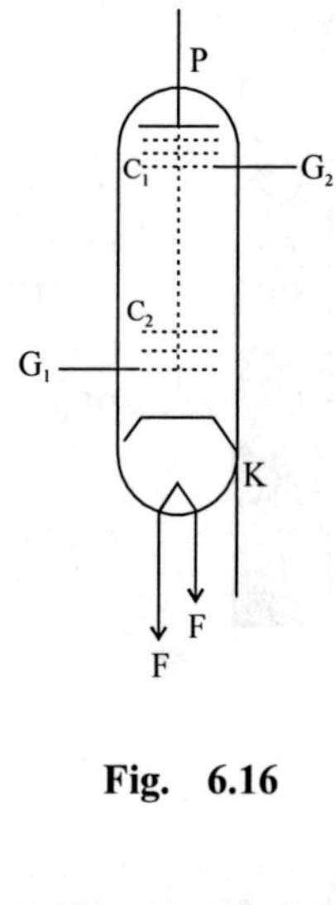

Fig. 6.16

Tetrode characteristics

The characteristic curves of a tetrode can be obtained in the same manner as were obtained for a triode. However more significant in this case is the plate characteristics. The plate characteristic if a tetrode gives the relationship between plate current (I_p) and the plate voltage (E_p) at constant control grid voltage (E_g) and at some fixed screen grid voltage. The circuit diagram for the determination of plate characteristics of tetrode is shown in fig. 6.17.

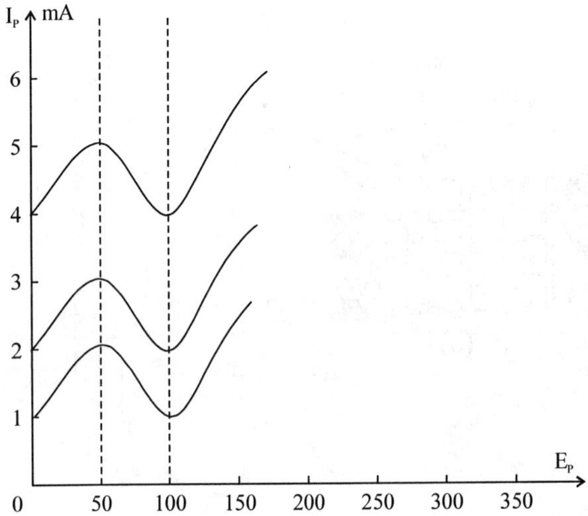

Fig. 6.17

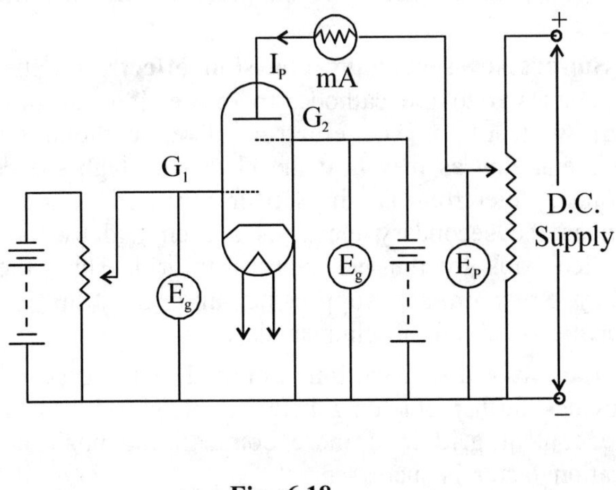

Fig. 6.18

Q. 18. Explain the construction and working of a pentode valve.

Ans. *Construction :* The pentode is a five electrode valve. It consists of a plate, a cathode, a control grid, a screen grid and a suppressor grid. The construction of pentode valve and its symbol are shown in fig. 6.19(a) and (b) respectively.

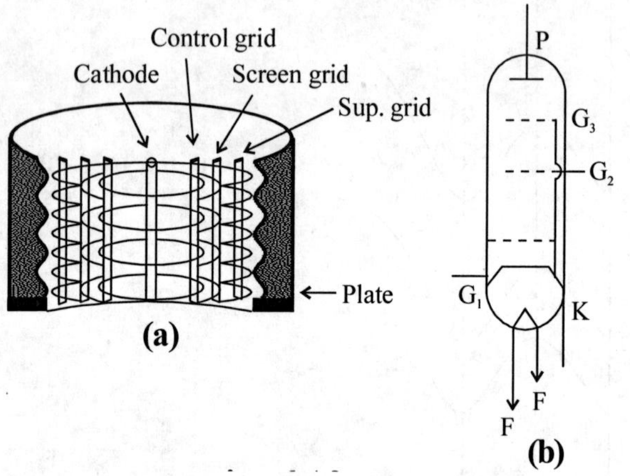

Fig. 6.19

Working

The working of pentode is similar to tetrode with the additional action of suppressor grid. The suppressor performs the following functions :

(i) Suppresses secondary emission effects. As the suppressor grid is connected to the cathode, therefore, it is at substantial –ve potential w.r.t. plate. The electrons from cathode still produce secondary emission as they heat the plate with high speeds. But, now the secondary electrons are no longer attracted to the screen grid. As soon as the secondary electrons are emitted by the plate, they are repelled back by the -ve suppressor grid. Thus, the effects of secondary emission are suppressed and the "dip" in the plate characteristic of tetrode is eliminated.

(ii) Increases amplification factor. The presence of suppressor grid provides further shielding between plate and control grid. This puts the control grid in a more commanding position and hence amplification factor is increased.

(iii) Reduces grid-plate capacitance (C_{gp}). The suppressor grid further reduces grid to plate capacitance. This completely eliminates the feedback action.

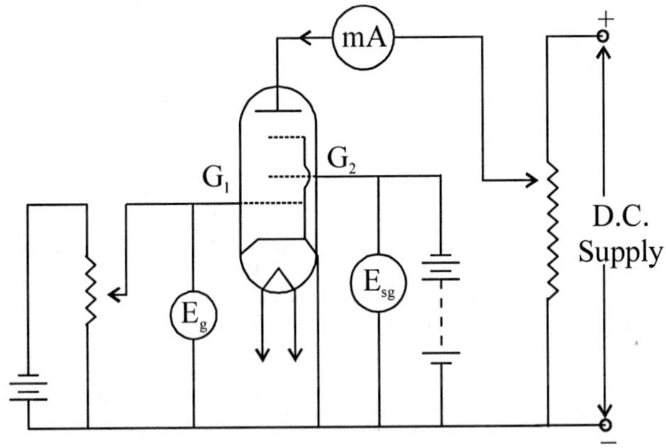

Fig. 6.20

Pentode characteristic curves

The plate characteristics of a pentode valve can be obtained by using the circuit shown in fig. 6.20. It is observed that :

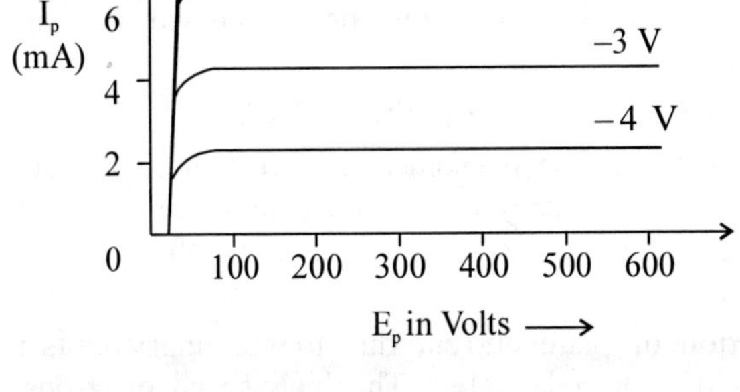

Fig. 6.21

(i) Over the major portion of the characteristics, plate current is largely independent of plate voltage changes. Therefore, pentode may be thought of as a constant current device. A pentode is generally operated in this region.

(ii) The dip found in the plate characteristics of tetrode is eliminated in the pentode plate characteristics.

(iii) The characteristics are non linear below the knee of the curve.

Q. 19. Give the comparative list of the constants of triode, tetrode and pentode.

S.N.	Particular	Triode	Tetrode	Pentode
1.	Amplification factor (μ)	10 to 100	Range around 500	1000 to 5000
2.	Plate resistance (r_p)	3000 Ω to 1000 KΩ	70 to 1000 KΩ	0.5 to 2MΩ
3.	Transconductance (gm)	About 2500 μ mho	About 1000 μ mho	1000 to 9000 mho

Q. 20. What do you mean by peak inverse voltage? (PIV)

Ans. *Peak Inverse Voltage :* In the normal situation, the plate of a tube is +ve w.r.t. cathode and hence current flows from cathode to plate. However, if small reverse voltage is applied to the tube (*i.e.*, plate is –ve and cathode is +ve), then no current will flow normally. When the reverse potential is increased, a stage may reach when the electric field becomes strong enough to tear electrons out of the cold plate. Consequently, current starts flowing in the reverse direction, *i.e.*, from plate to cathode. This situation is undesirable and may damage the tube along with the circuit elements.

The max^m reverse voltage that can be applied across the tube without conduction in the reverse direction is known as peak inverse voltage or PIV.

Q. 21. What do you mean by dynatron effect?

Ans. *Dynatron effect :* The phenomenon of reduction of plate current in a tetrode due to secondary emission is known as dynatron effect. As the plate acts here as a source of electrons, the region is called dynatron region.

The reduction of plate current due to secondary emission is a measure defect of a tetrode valve. The high speed electrons coming from cathode dash against the plate and emit secondary electrons from the plate. Since plate is +ve w.r.t. cathode, the electrons emitted from cathode are attracted towards plate and cause the emission of secondary electrons from the plate. In normal operation, with an a.c. signal applied to the grid circuit, the plate voltage of tetrode varies from time to time. As the screen grid voltage is likely to be greater than plate voltage during some part of a.c. signal, the secondary electrons are attracted by the screen grid. As a result of which, a current opposite to plate current starts flowing in the tetrode (fig. 6.22). This reduces the plate current. Due to reduction of plate current, "dip" is formed in $I_p \sim E_p$ characteristic of a tetrode. So tetrode operation is limited to a small portion of its plate characteristic curve. This disadvantage in the use of a tetrode is caused due to dynatron effect.

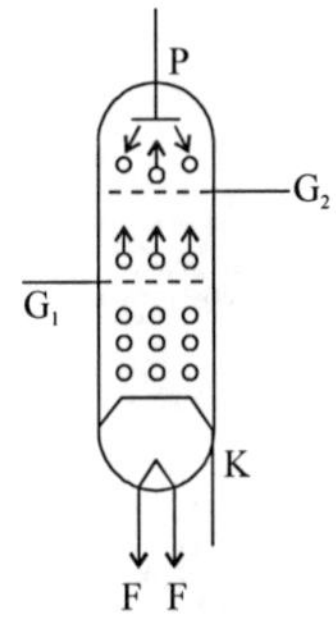

Fig. 6.22

In order to prevent secondary emission, a high supply voltage is necessary in the dynatron region. For this reason a tetrode is replaced by a pentode.

EXERCISES

Numerical Problems with Solutions

1. Temperature of a tungsten filament is increased from 2000°K to 2100°K. What is the ratio of emission current ?

Solution

$T_1 = 2000°K, \quad T_2 = 2100°K$

$b = 50{,}000°K, \quad I_2/I_1 = ?$

According to Richardson's law, we have

$$I_1 = AT_1^2 \, e^{-b/T_1}$$

$$I_2 = AT_2^2 \, e^{-b/T_2}$$

So, $$\frac{I_2}{I_1} = \frac{T_2^{\,2}\, e^{-b/T_2}}{T_1^{\,2}\, e^{-b/T_1}}$$

$$= \left(\frac{T_2}{T_1}\right)^2 \times e^{b\left(\frac{1}{T_1}-\frac{1}{T_2}\right)}$$

$$= \left(\frac{2100}{2000}\right)^2 \times e^{50,000\left(\frac{1}{2000}-\frac{1}{2100}\right)}$$

$$= \left(\frac{21}{20}\right)^2 \times e^{50,000\left(\frac{100}{2000\times 2100}\right)}$$

$$\frac{I_2}{I_1} = \frac{442}{400} \times e^{50\left(\frac{1}{2\times 21}\right)}$$

$$= \frac{4.42}{4} \times e^{50/42}$$

$$= 1.105 \times e^{25/21}$$

$$= 1.105 \times e^{1.19}$$

$$\log \frac{I_2}{I_1} = \log\ (1.105) + 1.19 \log e$$

$$= \log\ (1.105) + 1.19 \times 2.3026$$
$$= 0.212 + 2.74 = 2.95$$

or, $$2.3026\ \log_{10} \frac{I_2}{I_1} = 2.95$$

or, $$\log_{10} \frac{I_2}{I_1} = \frac{2.95}{2.3026} = 1.282$$

or, $$\frac{I_2}{I_1} = \text{Antilog } (1.282)$$

$$= 19.14$$

2. Under space charge limited condition the plate current of a diode is 375 milliamperes for a plate voltage of 200 volts. Find the plate current for plate voltage of 80 volts under the space charge limited condition.

Solution

$$I_{p_1} = 375 \text{ mA} = 375 \times 10^{-3} \text{ A}$$

$$V_{p_1} = 200 \text{ volts}$$

$$I_{p2} = ? \quad V_{p2} = 80 \text{ volts}$$

According to Child Langumuir's law, we have

$$I_{p1} \propto V_{p1}^{3/2}$$

and $$I_{p2} \propto V_{p2}^{3/2}$$

So, $$\frac{I_{p1}}{I_{p2}} = \frac{V_{p1}^{3/2}}{V_{p2}^{3/2}}$$

or, $$\frac{375 \times 10^{-3}}{I_{p2}} = \left(\frac{200}{80}\right)^{3/2}$$

or, $$\frac{375 \times 10^{-3}}{I_{p2}} = \left(\frac{5}{2}\right)^{3/2} = \frac{\sqrt{125}}{\sqrt{8}}$$

or, $$I_{p2} = \frac{375 \times \sqrt{8} \times 10^{-3}}{\sqrt{125}}$$

$$= \frac{375 \times 2\sqrt{2} \times 10^{-3}}{5\sqrt{5}}$$

$$= \frac{75 \times 2 \times 1.414 \times 10^{-3}}{2.236}$$

$$= \frac{150 \times 1.414 \times 10^{-3}}{2.236} = 94.8 \quad \text{mA}$$

3. The plate current in a vacuum diode is 10 mA at plate voltage of 100 volts when operating in the space charge limited region. What is the plate voltage required to triple the current still under space charge limited condition.

Solution

$$I_{p_1} = 10 \text{ mA}, \quad I_{p2} = 30 \text{ mA}$$

$$V_{p_1} = 100 \text{ volts}, \quad V_{p2} = ?$$

According to Child Langumuir's law,

$$I_{p1} \propto V_{p1}^{3/2}$$

$$I_{p2} \propto V_{p2}^{3/2}$$

So, $$\frac{I_{p1}}{I_{p2}} = \frac{V_{p_1}^{3/2}}{V_{p2}^{3/2}} = \left(\frac{V_{p1}}{V_{p2}}\right)^{3/2}$$

or, $$\frac{10}{30} = \left(\frac{100}{V_{p2}}\right)^{3/2}$$

or, $$\frac{1}{3} = \frac{10^3}{V_{p2}^{3/2}}$$

or, $$V_{p2}^{3/2} = 3 \times 10^3$$

Squaring both sides, we get

$$V_{p2}^3 = 9 \times 10^6$$

Taking log of both the sides, we have

$$\log V_{p2}^3 = \log\left(9 \times 10^6\right)$$

or, $$3 \log V_{p2} = \log 9 + 6 \log 10$$

or, $$\log V_{p2} \quad \frac{1}{3}\log(3^2) + \frac{6}{3}$$

$$= \frac{1}{3} \times 2 \log 3 + \frac{6}{3}$$

$$= \frac{2}{3} \times (.4771) + \frac{6}{3}$$

$$= 0.318 + \frac{6}{3} = 2.318$$

or, V_{p2} = Antilog (2.318)

= 208.0 = 208 volts

4. A tungsten wire of unknown composition emits 0.2 amp/cm^2 at a temperature of 2000°K. Find the work function of tungsten filament. Determine whether the tungsten is pure or contaminated with substance of lower work function. Given that A = 60.2 amp/cm^2/°K^2.

Solution

$$J_s = 0.2 \text{ amp/cm}^2$$
$$A = 6.2 \text{ amp/cm}^2/°\text{K}^2$$
$$T = 2000°\text{K}$$

If ϕ be the work function of the filament b = 11600 ϕ °K

Using Richardson-Dushman equation, emission current density is given by

$$J_s = AT^2 \, e^{-b/T} \text{ amp/cm}^2$$

$$\text{or,} \quad 0.2 = 60.2 \times (2000)^2 \, e^{-\frac{11600\phi}{2000}}$$

$$\text{or,} \quad e^{-5.8\phi} = \frac{0.2}{60.2 \times 4 \times 10^6}$$

$$= \frac{0.2 \times 10^{-6}}{240.8}$$

$$= 8.3 \times 10^{-10}$$

Taking log of both sides, we get

$$-5.8\phi \log_e e = \log_e 8.3 - 10 \log_e 10$$
$$\text{or,} \quad -5.8\phi = 2.3026 \times \log_{10} 8.3 - 10 \times 2.3026$$
$$= (2.3026 \times 0.919) - 23.026$$
$$= 2.116 - 23.026$$
$$= -20.9099$$

$$\text{or,} \quad \phi = \frac{20.9099}{5.8} = 3.6 \text{ eV}$$

Since the work function of pure tungsten is 4.52 eV, the sample must be contaminated. Thoriated tungsten has a work function ranging from 2.63 eV to 4.52 eV, depending upon the percentage of metallic thorium. Therefore, the sample is most likely to be thoriated tungsten.

5. The plate current in a diode is 5 mA at plate voltage of 50 volts under space charge limited condition. What will be the value of plate voltage to double the plate current under same condition.

Solution

$$I_{Pl} = 5 \text{ mA}, V_{Pl} = 50 \text{ V}$$
$$I_{P2} = 10 \text{ mA}, V_{P2} = ?$$

According to Child's law,

$$I_{p1} \propto V_{p1}^{3/2}$$

$$I_{p2} \propto V_{p2}^{3/2}$$

So, $$\frac{I_{p1}}{I_{p2}} = \frac{V_{p1}^{3/2}}{V_{p2}^{3/2}} = \left(\frac{V_{p1}}{V_{p2}}\right)^{3/2}$$

or, $$\frac{5}{10} = \left(\frac{50}{V_{p2}}\right)^{3/2}$$

or, $$\frac{1}{2} = \frac{50\sqrt{50}}{(V_{p2})^{3/2}}$$

or, $$V_{p2}^{3/2} = 2 \times 50 \times 5\sqrt{2} = 500\sqrt{2}$$

Squaring both sides, we get

$$V_{p2}^{3} = 250000 \times 2 = 25 \times 10^4 \times 2$$

or, $$V_{p2}^{3} = 50 \times 10^4 = 5 \times 10^5$$

or, $$\log_{10} (V_{p2}^3) = \log_{10} 5 + 5 \log_{10} 10$$

or, $$3 \log_{10} V_{p2} = \log_{10} 5 + 5$$

or, $$\log_{10} V_{p2} = \frac{1}{3} \times 0.6989 + \frac{5}{3}$$

$$= 0.2329 + 1.6666$$
$$= 1.8995$$

or, $$V_{p2} = \text{Antilog } (1.8995)$$
$$= 79.34 \text{ volts}$$

6. Find the mutual conductance of a triode if $\mu = 20$ and $r_p = 800\ \Omega$.

Solution

$$\mu = 20,\ r_p = 8000\ \Omega,\ g_m = ?$$

For a triode, $\mu = r_p \times g_m$

$$\text{or,}\quad g_m = \frac{\mu}{r_p} = \frac{20}{8000}\ \text{mho}$$

$$= \frac{1}{4} \times 10^{-2}\ \text{mho}$$

$$= 0.25 \times 10^{-2}\ \text{mho}$$

$$= 2500 \times 10^{-6}\ \text{mho}$$

$$= 2500\ \mu\ \text{mho.}$$

7. In a triode, change of grid voltage from –1.8 to –1.0 V produces a change in plate current from 5 to 9.8 mA at $V_p = 250$V. Calculate the mutual conductance of the tube.

Solution

$$\Delta V_g = (-1) - (-1.8)$$

$$= -1 + 1.8 = 0.8\ \text{V}$$

$$\Delta I_p = 9.8 - 5 = 4.8\ \text{mA}$$

$$= 4.8 \times 10^{-3}\ \text{A}$$

$$g_m = \left(\frac{\Delta I_p}{\Delta V_g}\right)_{V_p} = \frac{4.8 \times 10^{-3}}{0.8}\ \text{mho}$$

$$= 6 \times 10^{-3}\ \text{mho}$$

$$= 6000 \times 10^{-6}\ \text{mho}$$

$$= 6000\ \mu\ \text{mho}$$

8. A triode valve has a mutual conductance of 10 mA/V and a.c. plate resistance of 10,000 ohms. Calculate the amplification factor.

Solution

$$g_m = 10\ \text{mA/v} = 10 \times 10^{-3}\ \text{A/V}$$

$$r_p = 10{,}000\ \text{ohms} = 10^4\ \text{ohms}$$

$$\mu = ?$$

For a triode valve, $\mu = r_p \times g_m$

$$\text{So,}\quad \mu = 10^4 \times 10 \times 10^{-3}$$

$$= 10^5 \times 10^{-3}$$

$$= 10^2 = 100$$

9. A vacuum tube has amplification factor of 20. If the grid voltage is decreased by 0.2 V, by how much must the plate potential be raised to maintain the plate current constant?

Solution

$$\mu = 20, \quad \Delta V_g = 0.2 \text{ V}$$
$$\Delta V_p = ?$$

For a vacuum tube, $\mu = (\Delta V_p / \Delta V_g)_{I_p}$

So, $20 = \Delta V_p / 0.2$

or, $\Delta V_p = 20 \times 0.2 = 4$ volts.

10. The following readings were obtained from the linear portion of the static characteristics of a vacuum triode.

$$V_p = 100\text{V}, \quad 200\text{V}, \quad 200\text{V}$$
$$I_p = 10\text{mA}, \quad 15\text{mA}, \quad 7.5 \text{ mA}$$
$$V_g = -2\text{V}, \quad -2\text{V}, \quad -4 \text{ V}$$

Find the (i) a.c. plate resistance (ii) mutual conductance (iii) amplification factor.

Solution

(i) $$r_p = \left(\frac{\Delta V_p}{\Delta I_p}\right)_{V_g}$$

$$= \frac{200-100}{(15-10)\times 10^{-3}}$$

$$= \frac{100\times 10^3}{5} = 20\times 10^3 = 20{,}000 \text{ ohms}$$

(ii) $$g_m = \left(\frac{\Delta I_p}{\Delta V_g}\right)_{V_p}$$

$$= \frac{(15-7.5)\times 10^{-3}}{(-2+4)}$$

$$= \frac{7.5}{2}\times 10^{-3}$$

$$= 3.75\ 0 \times 10^{-3} \text{ mho}$$

(iii) $\mu = r_p \times g_m$

$$= 20 \times 10^3 \times \frac{7.5}{2} \times 10^{-3}$$

$$= \frac{150}{2} = 75$$

11. A triode has a mutual conductance of 2 mA/V and plate resistance of 20,000 ohms. Find the load resistance which must be introduced in the circuit to obtain a voltage gain of 20.

Solution

$g_m = 2\text{mA/V} = 2 \times 10^{-3}$ A/V

$r_p = 20 \times 10^3$ ohms

$R_L = ?, \ A_v = 20$

For a triode, we have

$\mu = r_p \times g_m$

$= 20 \times 10^3 \times 2 \times 10^{-3}$

$= 40$

$$A_v = \frac{\mu}{1 + \left(\frac{r_p}{R_L}\right)}$$

or, $$20 = \frac{40}{1 + \frac{20 \times 10^3}{R_L}}$$

or, $$1 + \frac{20 \times 10^3}{R_L} = 2$$

or, $20 \times 10^3 = (2 - 1)\ R_L = R_L$

or, $R_L = 20 \times 10^3$ ohms

❑ ❑ ❑

7

Semiconductors and Transistors

Q. 1. Explain the salient features of Bohr's atomic model.

Ans. *Introduction :* The study of atomic structure is of considerable importance for electronic engineering. Various scientists have given different models of an atom. These models are :

(a) Thomson's model
(b) Rutherford's model
(c) Bohr's model
(d) Sommerfeld's model

Out of the above four models, Bohr's atomic model is adequate to understand electronics.

Bohr's atomic model

In 1913, Neils Bohr, a Danish physicist gave clear explanation of atomic structure. According to Bohr :

(i) An atom consists of a +vely charged nucleus around which –vely charged electrons revolve in different circular orbits.
(ii) The electrons can revolve round the nucleus only in certain permitted orbits.
(iii) The electrons in each permitted orbits possess a fixed amount of energy. The larger the orbit (*i.e.*, larger radius), the greater is the energy of electrons.
(iv) If an electron is given additional energy in the form of heat or light etc., it is lifted to the higher orbit. The atom in this case is said to be excited.
(v) If an electron absorbs energy, it goes to the higher orbit. If an electron emits energy, it goes to the inner orbit of lower radius. This absorption or emission of energy is in the form of $h\nu$.

Explanation

Fig. 7.1 shows the structure of silicon atom. It has 14 electrons, 2 electrons revolve in the 1st orbit, 8 in the 2nd orbit and 4 in the 3rd orbit. The 1st, 2nd and 3rd orbits etc. are also known as K, L, M orbits respectively. As the electrons can revolve only in permitted orbits having radii r_1, r_2 and r_3 etc. any other radii between r_1 & r_2, r_2 & r_3, etc. are not allowed. Each orbit is associated with a fixed amount of energy. If an electron moving in the 1st orbit absorbs right amount of energy, it jumps to the 2nd orbit having larger radius. The reverse happens if the electron in the 2nd orbit radiates the acquired energy in the form of electromagnetic radiations.

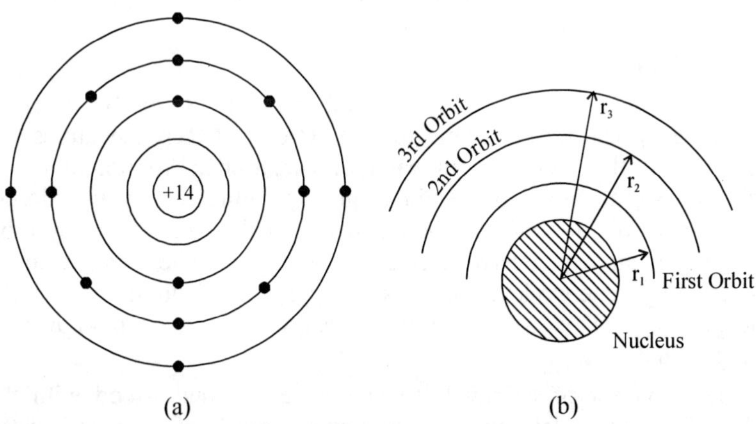

Fig. 7.1

Q. 2. What do you mean by energy levels?

Ans. From Bohr's theory of atomic structure it is clear that each orbit possesses certain fixed amount of energy. So the electrons

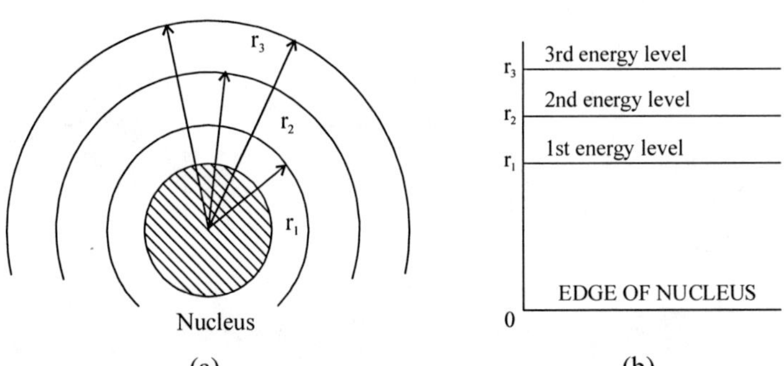

Fig. 7.2

moving in a particular orbit have the energy of that orbit. The larger the orbit, the greater is its energy. Thus, outer orbit electrons have more energy than the inner orbit electrons.

The energy of different orbits is different. The particular energy of a particular orbit is known as energy level. We can represent the energies of different orbits by the energy level diagram (fig. 7.2). The energy of 1st orbit is the 1st energy level, that of 2nd orbit is the 2nd energy level and so on. The larger the orbit of an electron the greater is its energy and higher is the energy level.

Q. 3. Explain the concept of energy bands in solids.

Ans. *Energy bands :* The range of energies possessed by an electron in a solid is known as energy band.

Explanation

In a solid, there are a large number of atoms, of the order of 10^{23} in a typical crystal. Due to interaction between the atoms the energies of the electrons in each of the atoms are bound to be affected. So, an atom in a solid is greatly influenced by the closely packed neighbouring atoms. The result is that the electron in any orbit of such an atom can have a range of energies rather than a single energy. Each energy level appears to spread continuously within the energy range. This will have the appearance of a band known as energy band.

The concept of energy band can be easily understood with the help of fig. 7.3 below. In this figure energy levels of a single isolated atom of silicon are shown. Each orbit of an atom has a single energy. Therefore, an electron can have only single energy corresponding to its orbit. However, when the atom is in a solid, the electron in any orbit can have a range of energies. Each energy level has a band of

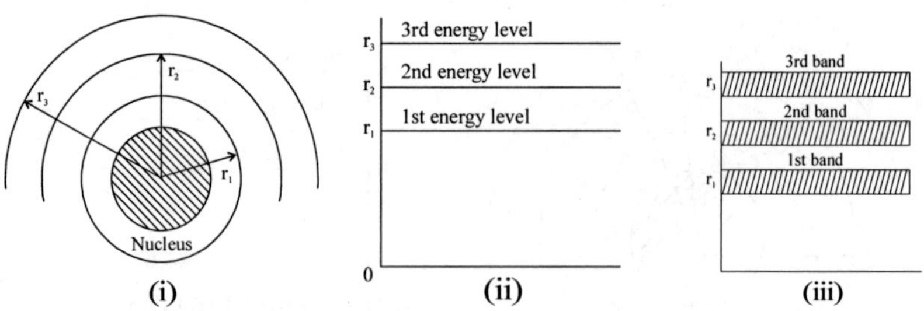

Fig. 7.3

energies. The electrons in the 1st orbit have 1st energy level which has in its turn 1st energy band. Similarly, the 2nd orbit electrons form 2nd energy band and so on.

Q. 4. Describe the valence band, conduction band and forbidden energy gap with the help of energy level diagram.

Ans. When an atom is in a solid, its energy levels like K, L, M etc. are converted into corresponding bands. Out of a number of energy bands in solids, the valence band, the conduction band and the forbidden energy gap are particularly important.

(i) Valence band

The range of energies possessed by valence electrons is known as valence band.

The electrons in the outermost orbit of an atom are known as valence electrons. In a normal atom valence band has the electrons of highest energy. This band may be completely or partially filled. For example, in case of inert gases, the valence band is completely filled whereas for other materials it is only partially filled. The partially filled band can accommodate more electrons.

(ii) Conduction band

In metals, the valence electrons are loosely attached to the nucleus. Even at ordinary temperature, some of the valence electrons may get detached to become free electrons. These free electrons are responsible for the conduction of electric current in a conductor. For this reason, they are known as conduction electrons.

The range of energies possessed by conduction band electrons is called simply conduction band.

All electrons in the conduction band are free electrons. If a substance has empty conduction band, it can't conduct electric current and such substances are known as insulators. The conduction band is partially filled for metallic conductors.

(iii) Forbidden energy gap

The separation between conduction band and valence band on the energy level diagram is known as forbidden energy gap.

No electron of a solid can stay in a forbidden energy gap. The greater the energy gap, more tightly the valence electrons are bound to the nucleus. An electron from valence band can go to the conduction band and for which external energy equal to the forbidden energy gap must be supplied.

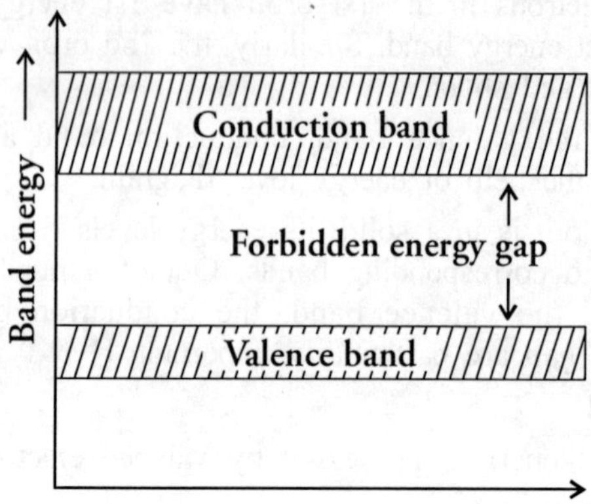

Fig. 7.4

Q. 5. Distinguish between conductors, semiconductors and insulators on the basis of energy band theory.

Ans. According to the property of conducting electric current, solids are divided into three classes, namely,

(a) Insulators
(b) Conductors
(c) Semiconductors

Such classification of solids can be well explained with the help of energy band theory. The valence and conduction bands are particularly important in giving the electrical behaviour of various solids.

(a) Insulators

Insulators are those substances which have no free electrons. So, they don't allow the passage of electric current through them. In case

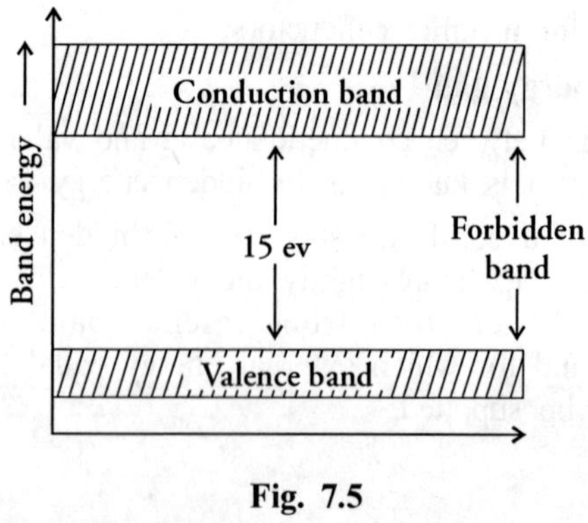

Fig. 7.5

of insulators (wood, glass, etc.) the valence band is completely filled while the conduction band is empty. Further, the forbidden energy gap between valence and conduction bands is very large (of the order of 15 ev). As a result of which, a very high electric field is required to push the valence electrons to the conduction band. Thus the electrical conductivity of insulators is extremely small and may be regarded as nil under ordinary conditions. However, when the temperature is raised) some of the valence electrons may acquire enough energy to go to the conduction band. Hence the resistance of an insulator decreases with the increase in temperature *i.e.*, an insulator has –ve temperature coefficient of resistance.

Some more examples of insulators :

(a) Mica, porcelain, Gutt apercha, ebonite, cotton, asbestos, bitumen, India rubber and various synthetic metals.

(b) Conductors

Conductors are those substances which have more number of free electrons. Such substances allow electric current to flow through them easily.

In the light of energy band theory, the valence and conduction bands overlap each other as shown in fig. 7.6. Due to this overlapping a slight potential difference across a conductor possesses the free electrons to constitute electric current. If the valence band is filled up from 10 to 90% by electrons, metals possesses conducting character. Thus the electrical behaviour of conductors can be explained by the use of energy bands.

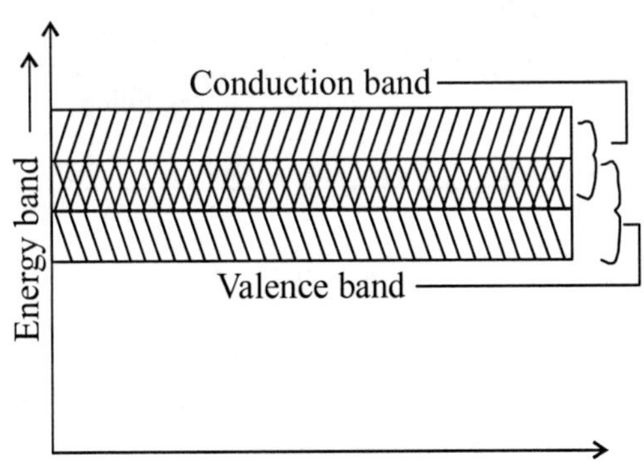

Fig. 7.6

(c) Semiconductors

Semiconductors are those substances whose electrical conductivity lies in-between conductors and insulators. Examples of semiconductors are germanium, silicon etc. In the light of energy band theory, the valence band is almost filled and conduction band is almost empty. Further the energy gap between valence and conduction bands is very small. Therefore comparatively smaller electric field is required to push the electrons from the valence band to the conduction band. In short, a semiconductor has almost full valency band, almost empty conduction band and small energy gap (of the order of 1 ev).

At low temperature the valence band is completely full and conduction band is completely empty. Therefore a semiconductor behaves as an insulator at low temperatures. But at room temperature, some electrons go to the conduction band and impart little conductivity to the semiconductor. If the temperature is increased, more number of valence electrons go to the conduction band and the conductivity increases. That means electrical conductivity of a semiconductor increases with the rise in temperature. Hence a semiconductor has –ve temperature coefficient of resistance.

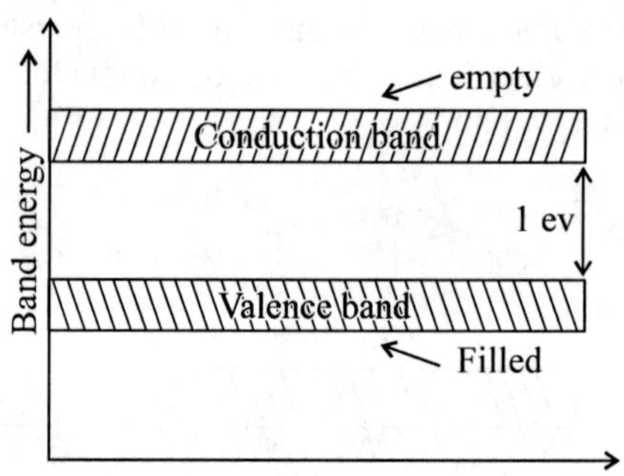

Fig. 7.7

Q. 6. What do you mean by a semiconductor? Discuss some important properties of semiconductors.

Ans. *Semiconductor :* A semiconductor is a substance having resistivity (10^{-4} to 0.5 ohm. m) in-between conductors and insulators. The examples are germanium, silicon, selenium, carbon etc.

Properties of semiconductors

(i) The resistivity of a semiconductor is less than an insulator but more than a conductor.

(ii) Semiconductors have –ve temperature coefficient of resistance *i.e.*, the resistance of a semiconductor decreases with the increase in temperature and *vice versa*. For example, Germanium is actually an insulator at low temperatures but it becomes a good conductor at high temperatures.

(iii) When a suitable metallic impurity (*i.e.*, Arsenic, gallium etc.) is added to a semiconductor its current conducting properties change appreciably. This property is the most important.

Q. 7. What do you mean by intrinsic semiconductor and extrinsic semiconductor?

Ans. *Intrinsic semiconductor :* A semiconductor in an extremely pure form is known as an intrinsic semiconductor.

In an intrinsic semiconductor, even at room temperature hole-electron pairs are created. When the electric field is applied across an intrinsic semiconductor, the current conduction takes place by two processes, namely, by free electrons and holes as shown if fig. 7.8. Hence the total current inside the semiconductor is the sum of electron-current and hole-current. It may be noted here that current in the external wires is only and wholly electronic *i.e.*, by the motion of electrons.

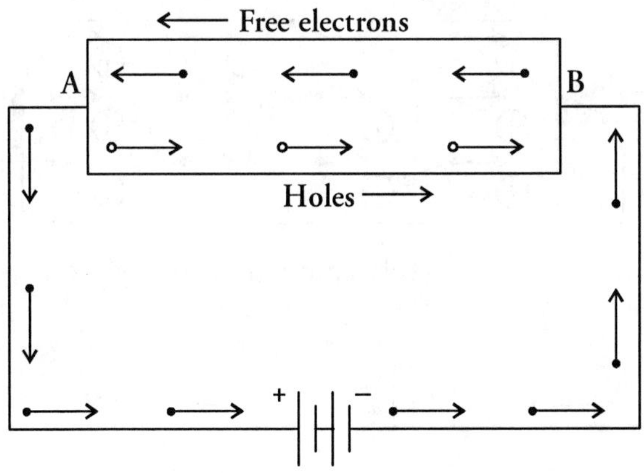

Fig. 7.8

Extrinsic semiconductor : The conducting properties of a pure semiconductor is increased by adding a small amount of suitable impurity to the semiconductor. It is then called impurity or extrinsic semiconductor. The process of adding impurities to a semiconductor is known as doping.

The purpose of adding impurity is to increase either the number of free electrons or holes in the semiconductor crystal. Depending upon the type of impurity added, extrinsic semiconductors are defined as P-type and N-type semiconductor.

Q. 8. What do you mean by doping?

Ans. The process of adding impurity to a semiconductor is known as doping. By this process the conducting properties of a semiconductor is increased.

Q. 9. Write short notes on (a) N-type semiconductor, (b) P-type semiconductor.

Ans. (a) *N-type semiconductor :* When a small amount of pentavalent impurity is added to a pure semiconductor, it is known as N-type semiconductor.

N-type semiconductor can be prepared by adding pentavalent impurities like arsenic or antimony to pure germanium crystal. Germanium is tetravalent. It is also a pure semiconductor. When arsenic is added to germanium crystal, a large number of free

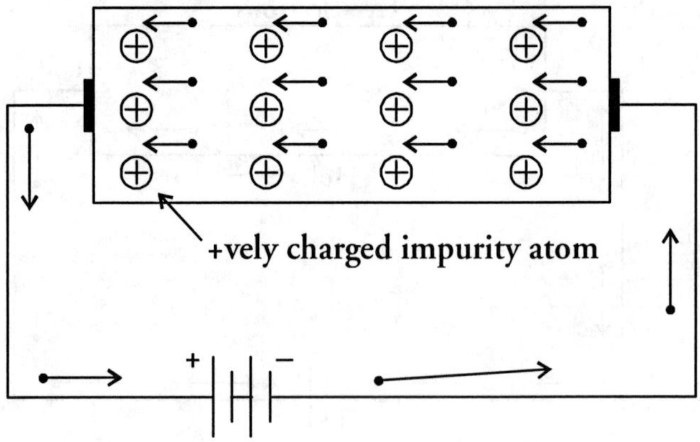

Fig. 7.9

electrons become available in the crystal. Germanium has 4 valency electrons where as arsenic has 5 such electrons. After covalent bonding between 4 germanium atoms, one electron of arsenic becomes excess.

(b) *P-type semiconductor :* When a small amount of trivalent impurity is added to a pure semiconductor, it is called P-type semiconductor.

The addition of trivalent impurity provides a large number of holes to the semiconductor. Trivalent impurities are gallium and indium. Such impurities are known as acceptor impurities because they create holes which can accept the electrons. When gallium is added to pure germanium crystal, holes are created in the crystal. This is because gallium has three valence electrons. Germanium has 4 valence electrons. The three valence electrons of gallium atom can form only 3 single covalent bonds with three germanium atoms as shown in fig. 7.10. In the fourth covalent bond, only germanium atom contributes one valency electron while gallium has no such electron to contribute. Therefore the fourth bond is incomplete, being short of one electron. This missing electron is called a hole. Thus one hole is created by adding one gallium atom. But a small amount of gallium provides millions of holes.

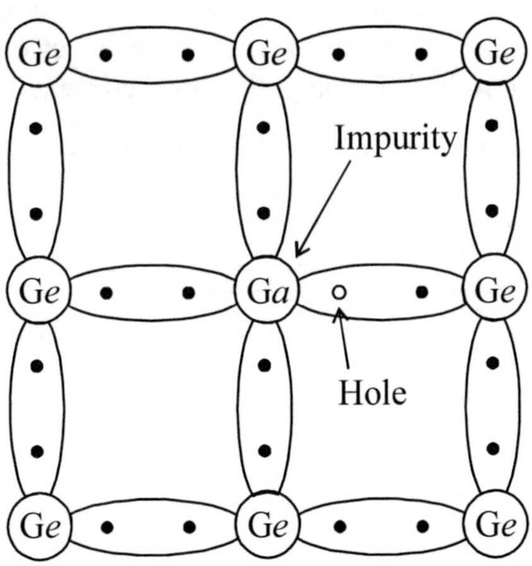

Fig. 7.10

Energy band diagram for P-type semiconductor : Fig. 7.11 shows the energy band diagram of the P-type semiconductor. Addition of trivalent impurity creates a large number of holes. However there are a few conduction band electrons due to thermal energy associated with room temperature. But the number of holes are more than the conduction band electrons. As a result of which holes predominate over free electrons and hence the name is P-type semiconductor or +ve type semiconductor.

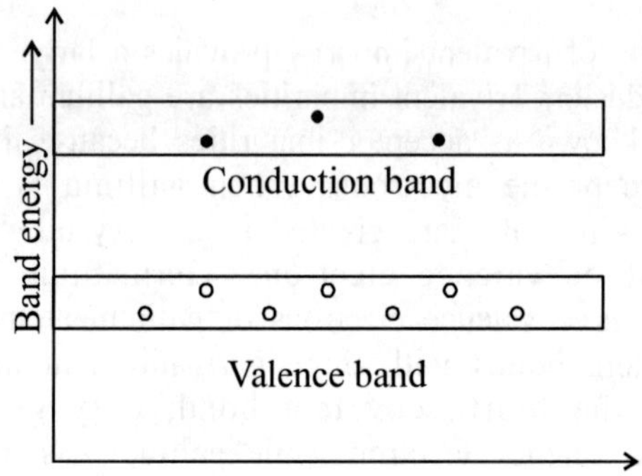

Fig. 7.11

Conductivity of P-type semiconductor

In P-type semiconductor the current conduction is due to holes or +ve charges. Therefore it is called P-type conductivity. When electric field is applied to the P-type semiconductor, the holes moves towards the -ve terminal forming the hole current.

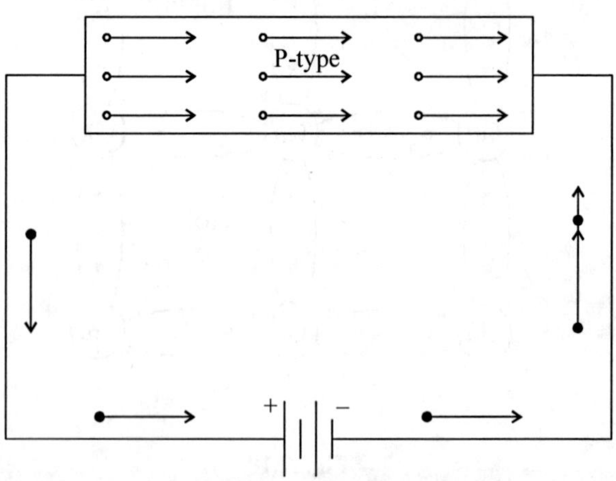

Fig. 7.12

Q. 10. What is a p-n junction? Describe the properties of p-n junction.

Ans. *p-n junction :* When a P-type semiconductor is suitably joined to n-type semiconductor, the contact surface is called p-n junction.

Properties of p-n junction

To explain the properties of a p-n junction let us consider p-type and n-type semiconductors as shown in fig. 7.13 below.

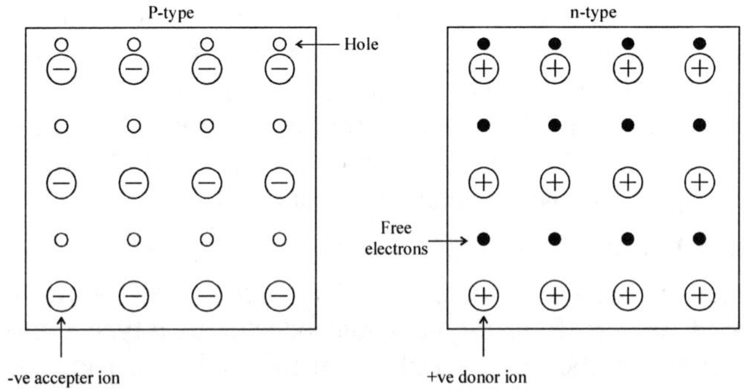

Fig. 7.13

The n-type material has more number of free electrons while the p-type material has a high concentration of holes. Therefore, when the two are joined together, there is a tendency for the free electrons to move towards p-type, material and holes to n-type material. This process is called diffusion of charged particles. As the free electrons

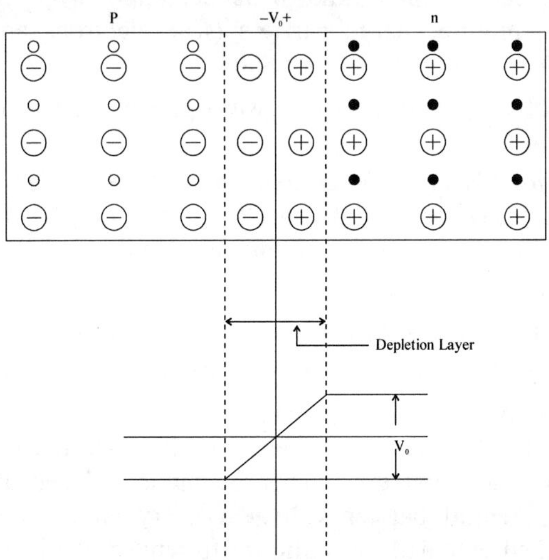

Fig. 7.14

move across the junction from n-type to p-type, +ve donor ions are uncovered *i.e.*, they are robbed of free electrons. Hence a +ve charge is built on the n-side of the junction. At the same time the free electrons cross the junction and uncover the –ve acceptor ions by filling in the holes. As a result of which, a net –ve charge is established on p-side of the junction. Diffusion ceases after sufficient number of donor and acceptor ions are uncovered. It is because now +ve charge on n-side repels holes to come to n-side and -ve charge on p-side repels free electrons to enter to p-side. Thus a barrier is set up against further movement of the charge carriers, *i.e.*, holes and electrons. This is called potential barrier or junction barrier V_o. The potential barrier so developed is of the order of 0.1 to 0.3 volts in fig. 7.14, the potential distribution is shown.

Potential barrier

Potential barrier is the potential set up to prevent the movement of charge carriers from p-type semiconductor to n-type semiconductor and *vice versa* when p material and n material joined together.

As the potential barrier is set up it gives rise to electric field. This field prevents the motion of the carriers from crossing the barrier region. It should be noted here that beyond this barrier on each side of the junction, the material is still neutral. Only inside the barrier there is +ve charge on n-side and -ve charge on p-side.

Depletion layer

The barrier region is known as depletion layer. It is named so because the mobile charge carriers (free electrons and holes) have been depleted (emptied) in this region.

Q. 11. What do you mean by forward biasing and reverse biasing in a p-n junction?

Ans. *Forward biasing :* When external voltage applied to the junction is in such a direction that it cancels the potential barrier and permits current to flow, it is known as forward bias.

Explanation

Forward biasing is done by connecting +ve terminal of the battery to p-type and -ve terminal to n-type material as shown in fig. 7.15.

The applied forward potential establishes an electric field which acts against the field due to potential barrier. This results in weakening the resultant field and the barrier height is reduced at the junction. Since the potential barrier voltage is very small, a small forward voltage applied externally is sufficient to remove the barrier completely. After removal of the barrier, resistance of the junction falls to zero

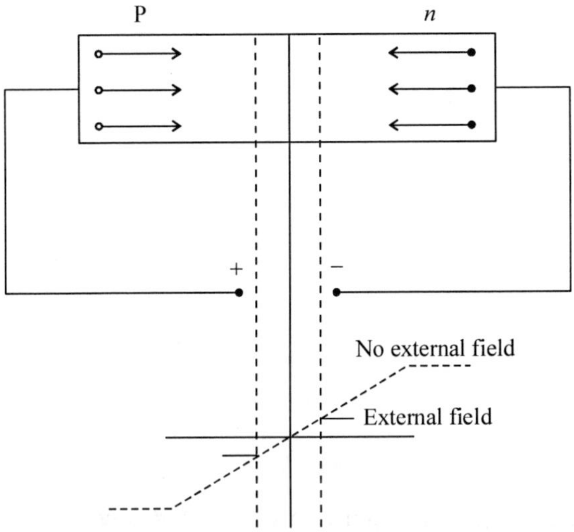

Fig. 7.15

and hence a low resistance path is established for the entire circuit. Therefore current flows in the circuit. This is the forward current. Forward biasing to p-n junction does the following :

(i) The potential barrier is reduced and at some forward voltage (0.1 to 0.3 volt) it is eliminated all together.

(ii) The junction offers few resistance (called forward resistance, R_f) to current flow.

(iii) Current flows in the circuit due to the establishment of low resistance path. The magnitude of current depends upon the applied forward voltage.

Reverse biasing

When the external voltage applied to the junction using such a direction that potential barrier is increased, it is called reverse biasing.

Explanation

Reverse biasing is done by connecting –ve terminal of the battery to p-type material and +ve terminal to n-type material as shown in fig. 7.16. It is clear that applied reverse voltage produces an electric field which acts in the same direction as the field due to potential barrier. Thus the resultant field at the junction becomes more and hence the barrier height is increased. The increase in potential barrier prevents the flow of charge carriers. Consequently, a high resistance path is formed for the entire circuit and the current conduction is ceased.

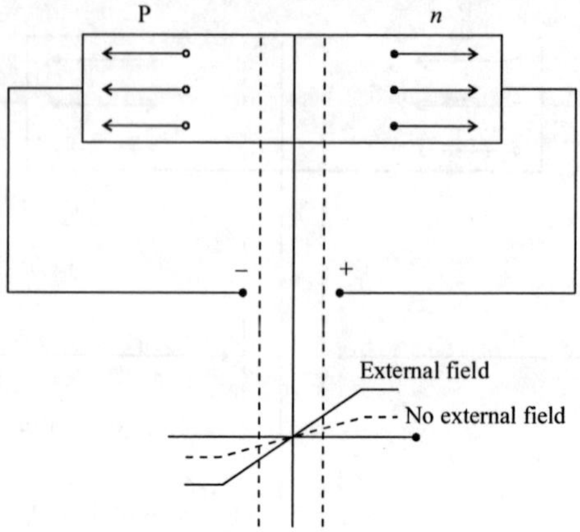

Fig. 7.16

Reverse biasing to p-n junction does the following :

(i) The potential barrier is increased.

(ii) The junction offers very high resistance to current flow.

(iii) No current flows in the circuit due to high resistance path.

Q. 12. With circuit diagram explain the V-I characteristic of a p-n junction.

Ans. Volt-ampere or V-I characteristic of a p-n junction is the curve between voltage across the junction and the circuit current. Usually, voltage is taken along the X-axis and current along the Y-axis. The circuit diagram for drawing the V-I characteristics for a p-n junction is shown in fig. 7.17.

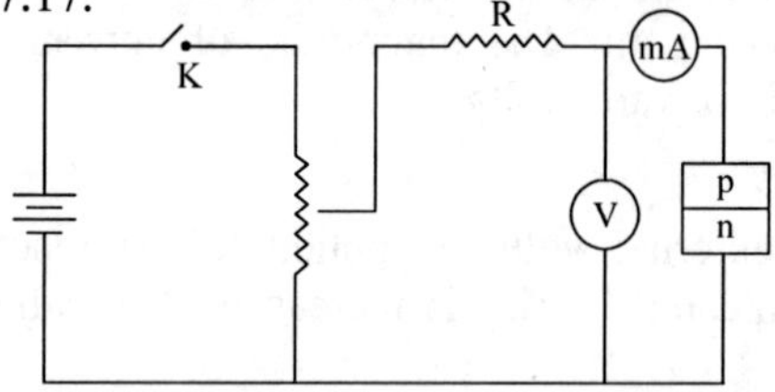

Fig. 7.17

The V-I characteristics can be studied under three heads, namely, zero external voltage, forward bias and reverse bias.

(i) Zero external voltage

When the circuit is open at K, no external voltage is applied. The potential barrier doesn't allow the motion of the charge carriers and hence no current flows in the circuit. This indicated by the point 'O' in the fig. 7.18.

(ii) Forward bias

When forward bias is given to p-n junction, the potential barrier is reduced. Forward biasing is done by connecting +ve terminal of the battery to p-type and -ve terminal of the battery to n-type semiconductor. If the forward voltage is sufficient to eliminate the potential barrier, the current flows in the circuit. For silicon this forward voltage is 0.7v and for germanium this is 0.3 v. The curve OB is obtained with forward bias (fig. 7.18). It is seen that at first (region OA), the current increases very slowly and the curve is non-linear. It is because the external applied voltage is used up in overcoming the potential barrier. However, once the external voltage exceeds the potential barrier voltage, the p-n junction behaves like an ordinary conductor. Therefore, the current rises very sharply with increase in external voltage (region AB on the curve). The curve is almost linear.

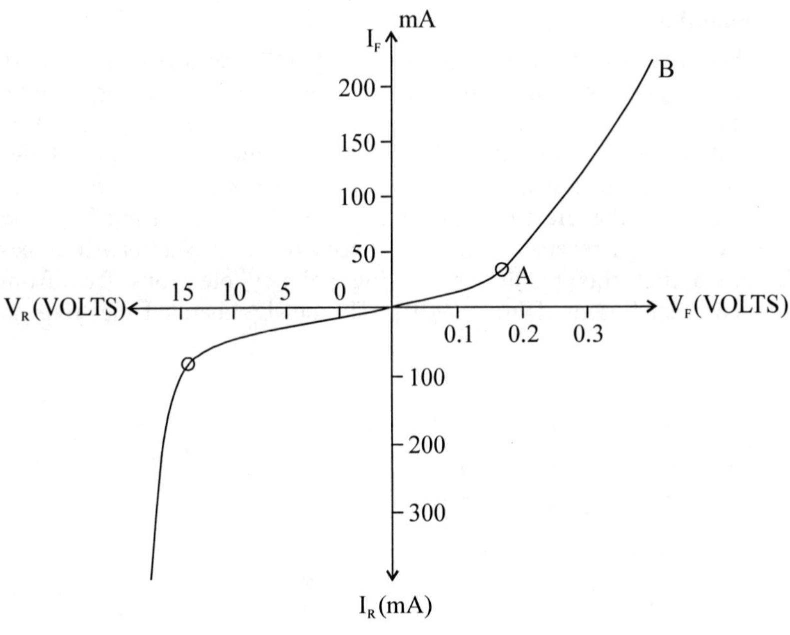

Fig. 7.18

(iii) Reverse bias

Reverse biasing is done by connecting –ve terminal of the battery to p-type and +ve terminal of the battery to n-type semiconductor. In this case potential barrier at the junction is increased. Therefore the junction resistance rises to a high value which ceases current flow through the circuit. But in practice, feeble current (of the order of μA) flows in the circuit. This is called reverse current. This reverse current is caused due to minority carriers in p and n-type semiconductor. Increasing reverse voltage, breakdown of the junction occurs and there is rise of reverse current due to sudden fall of the resistance of barrier region. This may damage the junction permanently.

Q. 13. Write short notes on the following :

(a) Breakdown voltage.

(b) Knee voltage.

(c) Limitations in the operating conditions of p-n junction.

(a) Breakdown voltage

It is the reverse voltage at which p-n junction breaks down with sudden rise in reverse current.

Explanation

When reverse voltage is small, a very little reverse current flows through a p-n junction. If the reverse voltage attends a high value, the junction may breakdown with sudden rise in reverse current. This can be followed from fig. 7.19. At room temperature, some hole-electron pairs are produced in the depletion layer [fig. 7.19(a)]. With reverse bias, the electrons move towards the +ve terminal of the battery. At high reverse voltage the speed of these electrons becomes high, so that this results in making valency electrons free from semiconductor atoms [fig. 7.19(b)]. The newly liberated electrons in

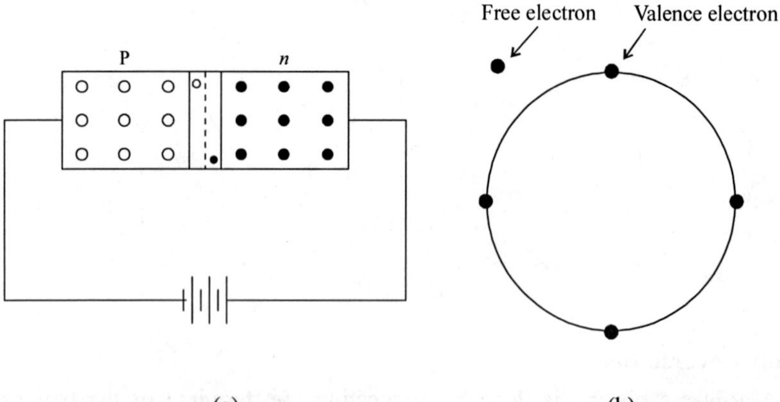

(a) (b)

Fig. 7.19

turn free other valency electrons. In this way large number of free electrons became available. Therefore, there is large reverse current in the p-n junction.

When the breakdown voltage is attained, high reverse current is produced to damage the junction. Precaution must be taken not to increase the reverse voltage across a p-n junction beyond the breakdown limit.

(b) Knee voltage

It is the forward voltage at which the current through the p-n junction starts to increase rapidly.

Explanation

If a p-n junction is forward biased, it conducts current very slowly till the potential barrier is eliminated. For silicon p-n junction the value of barrier voltage is 0.7V whereas its value is 0.3V for germanium junction. Therefore the knee voltage for silicon diode (p-n junction) is 0.7V and 0.3V for germanium diode. If the applied forward voltage becomes greater than the knee voltage, the current starts increasing rapidly. It should be remembered that the applied forward voltage to the p-n junction should be more than the knee voltage for producing useful current through the p-n junction. Fig. 7.20 shows the value of the knee voltage for silicon and germanium.

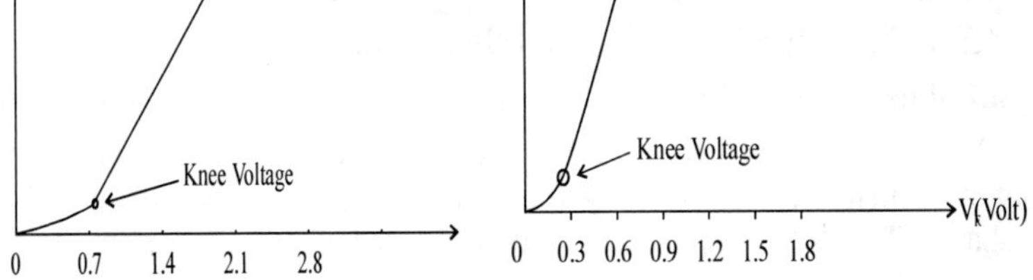

(a) Silicon (b) Germanium

Fig. 7.20

(c) Limitations in the operating conditions of p-n junction

Every p-n junction has limitations for the max^{m} forward current, peak inverse voltage and max^{m} power rating.

(i) Maximum forward current

The maximum value of the instantaneous forward current that a p-n junction can conduct without damage to the junction is called the max^m forward current. If the forward current exceeds this current the junction will be destroyed due to excess heat.

(ii) Peak inverse voltage (PIV)

The maximum reverse voltage that can be applied to the p-n junction without damage to the junction is known as *peak inverse voltage*. If the value of the reverse voltage exceeds PIV, the junction may be destroyed because of over heating. PIV has importance precifically in rectifier survice where a p-n junction changes alternating current to direct current by its rectifying action. In such applications, precaution should be taken not to increase the value of the reverse voltage beyond PIV of the p-n junction during -ve half-cycle of a.c.

(iii) Maximum power rating

It is the max^m power that can be dissipated at the junction without damaging it. The power dissipated at the junction is the product of junction current and the voltage across the junction. If the power rating at the junction exceeds the value of max^m power rating specified by the manufacturer, then the junction will be damaged. Therefore care is always taken to maintain the power rating within the limit specified.

Q. 14. What is a Zener diode? Describe the V-I characteristics of a Zener diode. How can you find Zener diode impedance from the V-I characteristics?

Ans. *Zener diode :* A properly doped crystal diode which has a sharp breakdown voltage is known as a Zener diode.

A Zener diode can be used as a voltage stabilizer.

Symbol of a Zener diode

Fig. 7.21 shows the symbol of a Zener diode. It may be seen that it is just like an ordinary diode except that the bar is turned into Z-shape. The following points may be noted about the Zener diode :

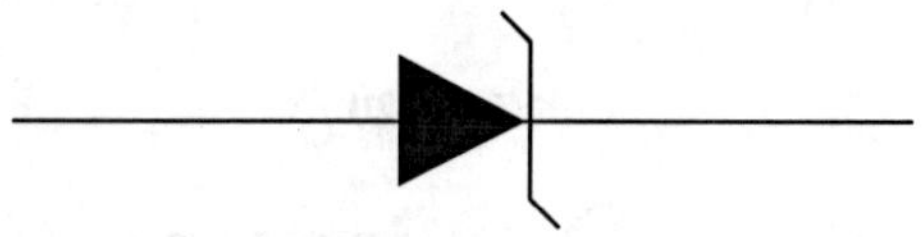

Fig. 7.21

(1) a Zener diode is like an ordinary diode except that it is property doped. So as to have a sharp breakdown voltage.
(2) A Zener diode is always reverse connected *i.e.*, it is always reverse biased.
(3) A Zener diode has sharp breakdown voltage, called Zener voltage v_z.
(4) When forward biased, its characteristics are just that of ordinary diode.
(5) The Zener diode is not immediately burnt just because it has entered the breakdown region. As long as the external circuit connected to the diode limits the diode current to less than burn out value, the diode will not burn out.

Equivalent circuit of Zener diode

When the reverse voltage across the Zener diode exceeds the breakdown voltage v_z, the current increases very sharply. In this region the curve is almost vertical. It means that voltage across Zener diode is constant at v_z even though the current through it changes. Therefore, in the breakdown region, an ideal Zener may be represented by a battery of voltage v_z as shown in fig. 7.22. In fig. 7.22(i), the Zener is reverse connected whereas in fig. 7.22(ii) shown its equivalent circuit.

In our further discussion we shall assume that in the breakdown region, the Zener diode can be replaced by a battery of voltage v_z. This simplified picture makes the analysis of Zener diode quite understandable.

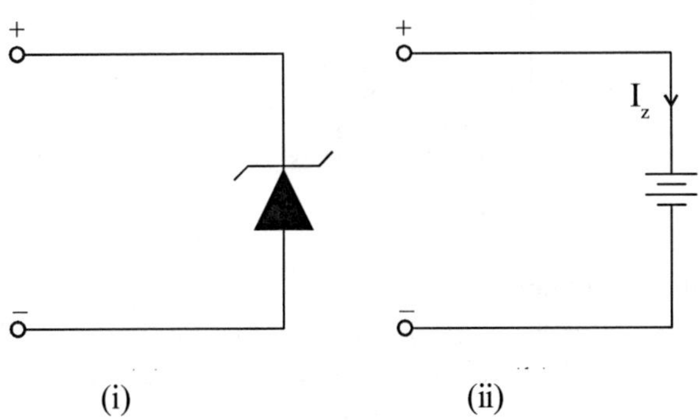

Fig. 7.22

V-I Characteristics of a Zener diode

When the reverse bias on a crystal diode increases and attains a critical value called breakdown voltage, then the reverse current increases sharply to a high value. If an ordinary crystal diode is properly doped to have a sharp breakdown voltage, a Zener diode is prepared. A satisfactory explanation of this breakdown of the p-n junction was first given by the American scientist C. Zener. Therefore, the breakdown voltage is sometimes called, Zener voltage and the sudden increase in current is known as Zener current.

The breakdown or Zener voltage depends upon the amount of doping. If the diode (p-n junction) is heavily doped, depletion layer will be thin and as a result of which the breakdown of the junction will occur at a lower reverse voltage. Under this condition, the crystal diode behaves like a Zener diode. On the other hand a lightly doped diode has a higher breakdown voltage. The V-I characteristics of a Zener diode is shown in fig. 7.23.

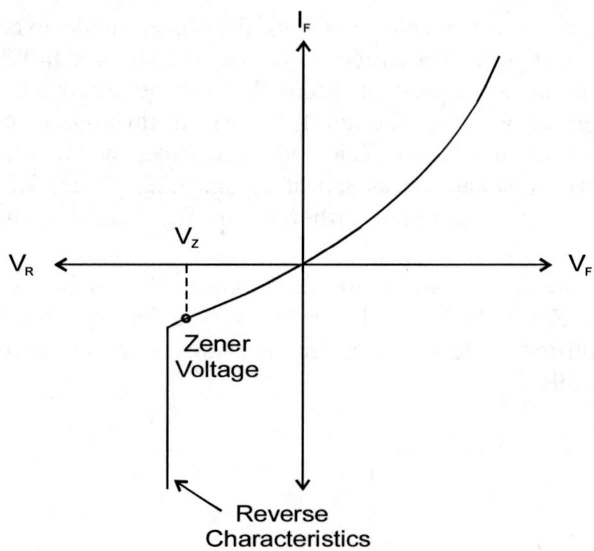

Fig. 7.23. V-I Characteristic of a Zener diode

Zener diode impedance

The Zener impedance is defined as the ratio of the change in voltage across the Zener diode to the change in Zener current in the Zener control region. Mathematically we can write,

$$Z = \frac{\Delta V}{\Delta I} \quad ...(1)$$

For a good Zener diode this value should be small. For an ideal Zener, the Zener impedance should be equal to zero. But, its value is generally specified by the manufacturers at two test points. One of them is near the Zener voltage and the second one is near the mid-range of the usable Zener current range.

Q. 15. Describe the action of a Zener diode as voltage stabiliser.

Ans. A Zener diode can maintain constant voltage across the load. The stabilising action of a Zener diode is utilized for regulating voltage from a source whose emf may vary over a wide range.

Zener diode as voltage stabiliser

The circuit diagram for a Zener diode as voltage stabiliser is shown in fig. 7.24 with its equivalent counterpart.

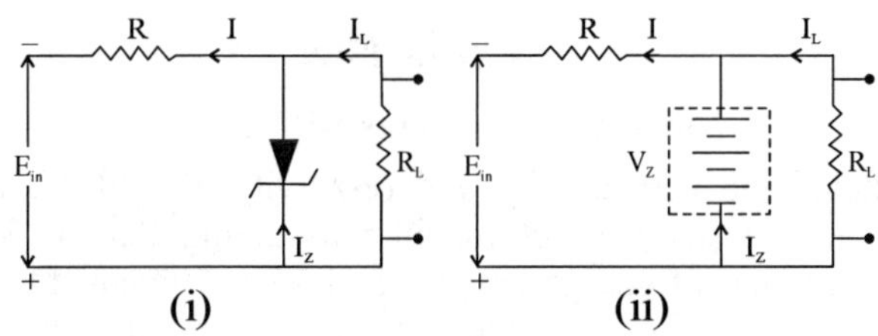

Fig. 7.24

The Zener diode of Zener voltage V_Z is reverse connected across the load R_L across which constant output is desired. The series resistance R absorves the output voltage variations so as to maintain constant voltage across the load. The Zener maintains a constant voltage V_z (equal to E_0) across the load so long as the input voltage doesn't decrease below V_z.

If the circuit is well designed, the load voltage E_0 remains essentially constant (equal to V_z) even though input voltage E_{in} and load resistance R_L may vary over a wide range. This can be explained as follows :

(i) Let us suppose that the input voltage increases since the Zener is in the breakdown region, the Zener diode acts as a battery of emf V_z. This is shown in [fig. 7.24(ii)]. It is clear that output voltage remains constant at V_z. The excess voltage is dropped across the series resistance R. Consequently, an increase in the value of total current I is caused. The Zener will conduct increase of current I while the load current remains constant. Hence, output voltage E_0 doesn't change in spite of the changes in the input voltage E_{in}.

(ii) Let us suppose that the input voltage is constant but the load resistance R_L decreases. This will cause an increase in load current. The extra current cannot come from the source because the voltage drop in R will not change as the Zener lies within its regulating range. The additional load current will come from a decrease in Zener current I_z. Thus the output voltage stays at constant value.

Voltage drop across $R = E_i - E_o$

Current through R, $I = I_z + I_L$

Applying Ohm's law, we have,

$$R = \frac{E_i - E_o}{I_z + I_L}$$

EXERCISES

Numerical Problems with Solutions

1. Compute the conductivity and resistivity at 300 °K of (a) pure silicon and (b) pure germanium assuming that at 300 °K the concentration of charge carriers is 1.75×10^{10}/cm^3 for silicon and 2.65×10^{13}/cm^3 for germanium. Assume electron mobility to be 1500 cm^2/volt-sec and hole mobility 500 cm^2/volt-sec in case of silicon. The corresponding values for germanium are assumed to be 3500 cm^2/volt-sec and 1500 cm^2/volt-sec respectively.

Solution

$q = e = 1.602 \times 10^{-19}$C

$p = n = n_l = 1.75 \times 10^{10}$/cm^3 for silicon

$p = n = n_l = 2.65 \times 10^{13}$/cm^3 for germanium.

$\mu_n = 1500$ cm^2/volt-sec for silicon

$\mu_p = 500$ cm^2/volt-sec for silicon

$\mu_n = 3500$ cm^2/volt-sec for germanium

$\mu_p = 1500$ cm^2/volt-sec for germanium.

$\sigma = ?\ \rho = ?$

Conductivity, $\sigma = q\,(p\mu_p + n\mu_n)$

$= qn_l\,(\mu_p + \mu_n)$

(a) For silicon

Conductivity, $\sigma = e\ n_l\,(\mu_n + \mu_p)$

$= 1.602 \times 10^{-19} \times 1.75 \times 10^{10}\ (1500 + 500)$

$= 1.602 \times 1.75 \times 10^{-9} \times 2000$

$= 1.602 \times 1.75 \times 2 \times 10^{-6}$ mho/cm.

$= 1.602 \times 3.50 \times 10^{-6}$ mho/cm.

$= 5.607 \times 10^{-6}$ mho/cm

Resistivity, $\rho = \dfrac{1}{\sigma} = \dfrac{1}{5.607 \times 10^{-6}}$ ohm-cm

$= \dfrac{10^6}{5.607}$

$= 0.178 \times 10^6$ ohm-cm

(b) For germanium,

Conductivity, $\sigma = en_I (\mu_n + \mu_p)$

$= 1.602 \times 10^{-19} \times 2.65 \times 10^{13} (3500 + 1500)$

$= 1.602 \times 10^{-6} \times 2.65 \times 5000$

$= 1.602 \times 2.65 \times 5 \times 10^{-3}$

$= 21.226 \times 10^{-3}$ mho/cm

Resistivity, $\rho = \dfrac{1}{\sigma} = \dfrac{1}{21.226 \times 10^{-3}}$ mho/cm

$= \dfrac{10^3}{21.226}$ ohm-cm

$= 0.0471 \times 10^3$ ohm-cm

$= 47.1$ ohm-cm

2. A specimen of intrinsic germanium at 300 °K having a concentration of carriers of $2.6 \times 10^{13}/\text{cm}^3$ is doped with impurity atoms of one for every million of germanium atoms. Assuming that all impurity atoms are ionised and the concentration of germanium atoms is $5 \times 10^{22}/\text{cm}^3$, determine the resistivity of the doped material assuming electron mobility to be 3600 cm^2/volt. sec.

Solution

Number of free electrons contributed by the impurity $= \dfrac{5 \times 10^{22}}{10^6} = 5 \times 10^{16}/\text{cm}^3$. Compared to this, the charge carriers present in the pure germanium are neglected.

Hence, the conductivity, $\sigma \approx N_D q \, \mu_n$ where N_D is the number of donor atoms.

Hence, $\sigma = 5 \times 10^{16} \times 1.602 \times 10^{-19} \times 3600$

$= 28.84$ mho/cm

$$\text{Resistivity, } \rho = \frac{1}{\sigma}$$

$$= \frac{1}{28.84} = 34.7 \times 10^{-3} \text{ ohm} - \text{cm}$$

3. An a.c. voltage of peak value 30 V is connected in series with a silicon diode and load resistance of 500 Ω. If the forward resistance of diode is 20 Ω, find (i) peak current through diode, (ii) peak output voltage. What will be these values if the diode is assumed to be ideal?

Solution

V_{in} = Peak input voltage = 30 V
r_f = Forward resistance = 20 Ω
R_L = Load resistance = 500 Ω
V_o = Potential barrier voltage = 0.7 V

The diode will conduct during the +ve half-cycles of a.c. input voltage only. The equivalent circuit is shown in fig. 7.25 below.

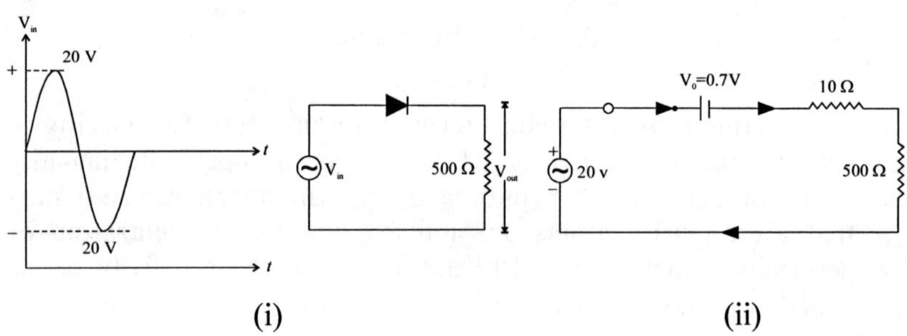

Fig. 7.25

(i) The peak current through the diode will occur at the instant when the input voltage reaches positive peak *i.e.*

$$V_{in} = V_F = 30 \text{ V}$$

$$\therefore \quad V_F = V_o + (I_f)_{peak}\,(r_f + R_L), \qquad \text{...(1)}$$

$$\text{or,} \quad (I_f)_{peak} = \frac{V_F - V_0}{r_f + R_L} = \frac{30 - 0.7}{20 + 500} = \frac{29.3}{520} A$$

$$= 0.0563\text{A} = 56.3 \text{ mA}$$

(ii) Peak output voltage $= (I_f)_{peak} \times R_L$
$= 56.3 \times 500 \times 10^{-3}$ V
$= 563 \times 50 \times 10^{-3}$ V
$= 28150 \times 10^{-3}$ V
$= 28.15$ V

Ideal diode : For an ideal diode, $V_o = 0$ and $r_f = 0$ in equn. (1)

So, $V_F = \left(I_f\right)_{peak} \times R_L$

or $(I_f)_{peak} = \dfrac{V_F}{R_L} = \dfrac{30}{500} = \dfrac{3}{50}$

$= 0.06$ A
$= 60 \times 10^{-3}$ A
$= 60$ mA.

Peak output voltage $= (I_f)_{peak} \times R_L$
$= 60 \text{ mA} \times 500\ \Omega$
$= 30$ V

4. An a.c. voltage of peak value 10 V is connected in series with a silicon diode and a load resistance of 1 KΩ. If the forward resistance of the diode is 10Ω, find (i) peak current through the diode, (ii) peak output voltage, assuming the junction voltage drop is 0.7 V.

Solution

V_{in} = peak input voltage = 10 V
R_L = Load resistance = 1 KΩ = $1 \times 10^3\ \Omega$
r_f = forward resistance of the diode = 10 Ω
V_o = junction voltage drop = 0.7 V

(i) The peak current through the diode will occur at the instant when the input voltage reaches positive peak *i.e.*,

$V_{in} = V_F = 10$V

$\therefore\ V_F = V_o + (I_f)_{peak}\ (r_f + R_L)$ ...(1)

or, $(I_F)_{peak} = \dfrac{V_F - V_0}{r_f + R_L} = \dfrac{10-0.7}{10+1000}$

$= \dfrac{9.3}{1010} = 9.2$ mA

(ii) Peak output voltage $= (I_f)_{peak} \times R_L$

$$= 9.2 \times 10^{-3} \times 10^{3}$$
$$= 9.2 \text{ V}$$

5. Calculate the barrier potential for the junction at 95°C if its value at 25°C is 0.7 volt.

Solution

$$Q_2 = 95°C, \quad Q_1 = 25°C, \quad V_{B1} = 0.7 \text{ V}$$
$$V_{B2} = ?$$

It is observed that for both Ge and Si, barrier potential decreases by about 2 mV/°C, *i.e.*,

$$\Delta V_B = -0.002 \ \Delta\theta$$

Here,
$$\Delta V_B = -0.002 \times (\theta_2 - \theta_1)$$
$$= -0.002 \ (95 - 25)$$
$$= -0.002 \times 70 \text{ V}$$
$$= -0.140 \text{ V}$$
$$\therefore V_{B2} = 0.7 - 0.14 = 0.56 \text{ volt.}$$

6. The saturation current density of p-n junction Ge diode is 200 mA/m² at 27°C. Find the voltage to be applied across the junction to have a forward current density of 10^4 amp/m² to flow.

Solution

The diode current for Ge diode at temperature T is given by

$$I = I_0\left[e^{Ve/KT} - 1\right], \quad \text{...(1)}$$

or in terms of current density, we have

$$J = J_0\left[e^{Ve/KT} - 1\right] \quad \text{...(2)}$$

where,
$$J = \frac{I}{A}, \quad \text{...(3)}$$

K = Boltzman Constant = 1.38×10^{-23} Joule/K

Given Saturation current density $J_0 = 200 \times 10^{-3}$ amp/m²

$J = 10^4$ amp/m², $T = 27 + 273 = 300$ °K

$$\therefore \quad \frac{J}{J_0} = \frac{10^4}{200 \times 10^{-3}} = 5 \times 10^4$$

$$= e^{Ve/KT} - 1$$

or, $e^{Ve/KT} \approx 5 \times 10^4$

or, $\frac{Ve}{KT} = \log_e 50000 = 2.3026 \times 4.6990$

$= 10.82$

$$V = \frac{10.82 \times KT}{e} = \frac{10.82 \times 1.38 \times 10^{-23} \times 300}{1.6 \times 10^{-19}}$$

$= 0.28$ volt.

7. The current through a p-n junction diode is 55 mA at a forward bias voltage of 3.0 V. If the temperature is 27°C, find the static and dynamic resistance of the diode.

Solution

$V = 3$ V, $I = 55$ mA

The static resistance

$$R_{dc} = \frac{V}{I} = \frac{3}{55 \times 10^{-3}} = 54.5 \text{ ohms}$$

The dynamic resistance

$$R_{ac} \approx \frac{26}{I} = \frac{26}{55} = 0.47\ \Omega$$

8. A Zener diode has a breakdown voltage of 9.1 volts with a maximum power dissipation of 364 milliwatts. What is the maximum current the diode can withstand?

Solution

$V_Z = 9.1$ volts, $P_{max} = 364$ mW

$I_Z = ?$

Here, $P_{max} = V_z \cdot I_z$

or, $(I_z)_{max} = \frac{P_{max}}{V_z} = \frac{364 \times 10^{-3}\ W}{9.1\ V}$

$= 40$ mA

9. Calculate the current flowing in the circuit given below which has two oppositely connected ideal diodes in parallel.

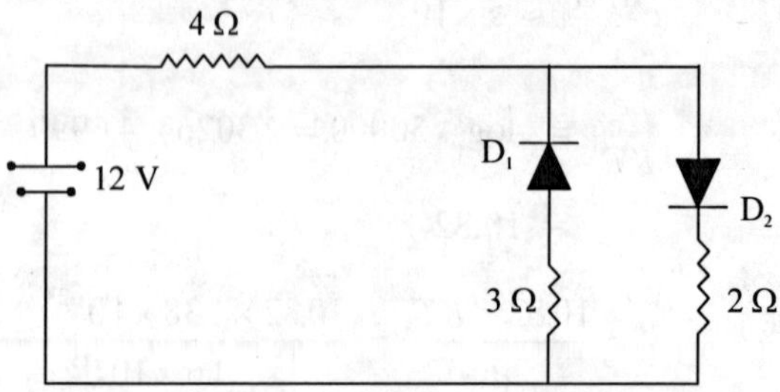

Fig. 7.26

Solution

Since D_1 is reverse biased, it acts like an open switch. Hence, no current flows through D_1 and 3 Ω resistor. As D_2 is forward biased, it acts like a closed switch. Consequently, current flows through it and 2 Ω resistor.

So, $$I = \frac{12\,V}{(4+2)\Omega} = \frac{12\,V}{6\,\Omega}$$

$$= 2 \text{ amp.}$$

10. Estimate the dynamic resistance of a junction diode at 27°C for a current of 1 mA.

Solution

$I = 1$ mA, $R_{ac} = ?$

The dynamic resistance is given by

$$R_{ac} = \frac{26}{I}$$

Where, I is in mA.

Here, $$R_{ac} = \frac{26}{1} = 26 \text{ ohms}$$

11. Calculate the static and dynamic resistance of a p-n junction germanium diode if temperature is 27°C and I_s = 1 μA for an applied forward bias of 0.2 volt.

Solution

$$I_s = 1\ \mu A = 10^{-6}\ A$$

$$V = 0.2\ V,\quad R_{dc} = ?,\ R_{ac} = ?$$

The static resistance

$$R_{dc} = \frac{V}{I_s} = \frac{0.2\ V}{10^{-6}\ A}$$

$$= 0.2 \times 10^6\ \Omega$$

$$= 2 \times 10^5\ \Omega$$

The dynamic resistance $R_{ac} \approx \dfrac{26}{10^{-6}\,A} = \dfrac{26}{10^{-3}\ mA}$

$$= 26 \times 10^3\ \Omega$$

❑ ❑ ❑

8

Rectifiers

Q. 1. With a neat circuit diagram describe the functions of a half-wave rectifier using a crystal diode. Derive expressions for efficiency and ripple factor of a half-wave rectifier.

Ans. *Half-wave rectifier :* A diode has rectifying action. It can be used as a half-wave rectifier as well as a full-wave rectifier. In half-wave rectification, the rectifier conducts current only during the +ve half cycles of the input a.c. supply. The –ve half cycles of a.c. supply are suppressed. Therefore, current always flows in one direction through the load at the interval of every half-cycle. The output d.c. becomes as follows :

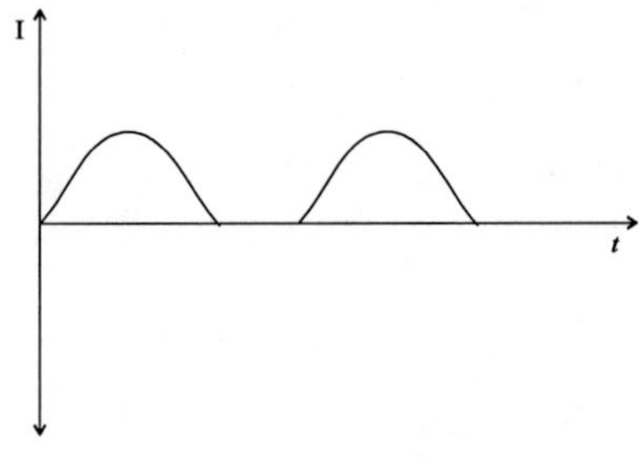

Fig. 8.1. D.C. Output

Generally, rectifier converts input a.c. to output d.c. A load resistance connected across the output gives the d.c. output voltage across it.

Description of the circuit

A half-wave rectifier circuit is shown in fig. 8.2 with a single crystal diode. The a.c. supply to be rectified is applied in series with the diode through a transformer. The output d.c. voltage is obtained across the load resistance R_L.

Circuit diagram of half-wave rectifier

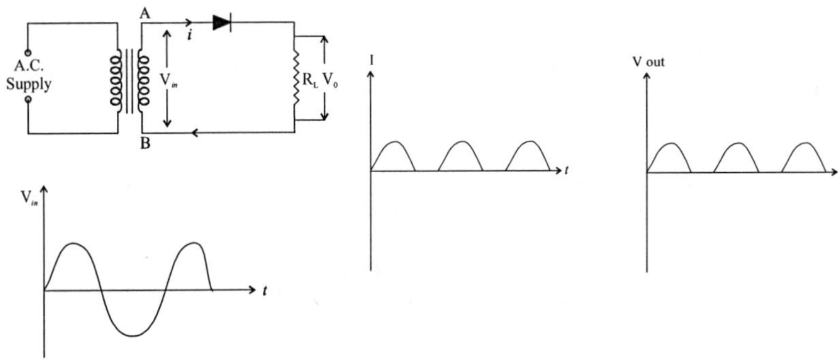

(i) (ii)

Fig. 8.2

Function

The a.c. voltage across the secondary *AB* of the transformer fluctuates between +ve and –ve half cycles. During the +ve half cycle of input a.c. voltage, end *A* becomes +ve w.r.t. end *B*. As a result of which the diode becomes forward biased and hence there is conduction of current. During –ve half cycle of input a.c., end '*A*' is –ve w.r.t. end *B*. This makes the diode reverse biased and there is no current conduction. Therefore, current flows through the diode during +ve half cycles of input a.c. voltage only. As a result of which the current through R_L becomes unidirectional, *i.e.*, the output across the load is d.c.

Efficiency of half-wave rectifier

The ratio of d.c. power output to the applied input a.c. power is known as rectifier efficiency. Mathematically we can write

$$\text{Rectifier efficiency, } \eta = \frac{\text{d.c. power output}}{\text{Input a.c. power}} = \frac{P_{d.c.}}{P_{a.c.}} \qquad ...(1)$$

Efficiency is always expressed in terms of percentage.

Let us consider the a.c. voltage in the form of a simple harmonic wave represented by

$$V = V_m \sin \theta$$

where V_m is the amplitude of a.c. or d.c. value of a.c. or steady value of a.c. supply.

Let r_f and R_L be the diode resistance and load resistance respectively. The diode conducts during +ve half cycle of a.c. supply only.

d.c. power

The output current is pulsating direct current. Therefore, in order to find d.c. power, average current has to be found out.

$$I_{av} = I_{dc} = \frac{1}{2\pi}\int_0^{\pi} i\,d\theta = \frac{1}{2\pi}\int_0^{\pi} \frac{Vm\sin\theta}{r_f + R_L}\,d\theta$$

$$= \frac{V_m}{2\pi\,(r_f + R_L)}\int_0^{\pi}\sin\theta\,d\theta$$

$$= \frac{V_m}{2\pi(r_f + R_L)}[-\cos\theta]_0^{\pi}$$

$$= \frac{V_m}{2\pi(r_f + R_L)}\times 2$$

$$= \frac{V_m}{(r_f + R_L)}\times\frac{1}{\pi}$$

$$= \frac{I_m}{\pi}\left(\because I_m = \frac{V_m}{r_f + R_L}\right)$$

$$\text{d.c. power, } P_{d.c.} = I^2{}_{d.c.}\times R_L = \left(\frac{I_m}{\pi}\right)^2\times R_L$$

a.c. power input

The a.c. power input is given by

$$P_{ac} = I_{rms}^2\,(r_f + R_L)$$

For a half-wave rectified wave, $I_{rms} = \dfrac{I_m}{2}$

$$\therefore\ P_{ac} = \left(\frac{I_m}{2}\right)^2\times\left(r_f + R_L\right)$$

$$\therefore \text{ Rectifier efficiency} = \frac{\text{d.c. output power}}{\text{a.c. input power}}$$

$$= \frac{(I_m / \pi)^2 \times R_L}{(I_m / 2)^2 (r_f + R_L)}$$

$$= \frac{0.406 \ R_L}{r_f + R_L}$$

$$= \frac{0.406}{1 + \frac{r_f}{R_L}} \qquad ...(2)$$

The efficiency will be max^m^ if r_f is negligible as compared to R_L.

$\therefore$ Max. rectifier efficiency = 40.6%

This shows that in half-wave rectification, a max^m^ of 40.6% of a.c. power is converted into d.c. power.

Ripple factor

The output of a rectifier consists of a d.c. component as well as an a.c. component, known as ripple. But the presence of a.c. in the output of the rectifier is not desirable. The smaller the value of the a.c. component in the rectifier output, the more effective is the rectifier.

The ratio of rms value of a.c. component to the d.c. component in the rectifier output is known as ripple factor. Mathematically, we can write,

$$\text{Ripple factor} = \frac{\text{r.m.s. value of a.c. component}}{\text{value of d.c. component}}$$

$$= \frac{I_{ac}}{I_{dc}} \qquad ...(3)$$

Expression for ripple factor

By definition, the effective (*i.e.*, rms) value of total load current is given by,

$$I_{rms} = \sqrt{I_{dc}^2 + I_{ac}^2}$$

$$I_{ac} = \sqrt{I_{rms}^2 - I_{dc}^2}$$

Dividing throughout by I_{dc} we get,

$$\frac{I_{ac}}{I_{dc}} = \frac{1}{I_{dc}} \sqrt{I_{rms}^2 - I_{dc}^2}$$

But I_{ac}/I_{dc} is the ripple factor

$$\text{Ripple factor } = \frac{1}{I_{dc}} \sqrt{I_{rms}^2 - I_{dc}^2}$$

$$= \sqrt{\left(\frac{I_{rms}}{I_{dc}}\right)^2 - 1}$$

For half-wave rectification

In half-wave rectification,

$$I_{rms} = \frac{I_m}{2}, \; I_{dc} = \frac{I_m}{\pi}$$

$$\text{Ripple factor } = \sqrt{\left(\frac{I_m / 2}{I_m / \pi}\right)^2 - 1} = 1.21$$

It is clear that a.c. component exceeds the d.c. component in the output of a half-wave rectifier. This results in greater pulsations in the output. Therefore, half-wave rectifier is ineffective for conversion of a.c. into a.c.

Q. 2. With a neat circuit diagram describe the functions of a full-wave bridge rectifier. Find expressions for efficiency and ripple factor of the full-wave rectifier.

Ans.

(i) Rectifier

A rectifier is an electronic device which converts a.c. to d.c.

There are two types of rectifiers, namely :

(a) Half-wave rectifier

(b) Full-wave rectifier

Again there are two circuits for a full-wave rectifier, such as

(i) Center tap full-wave rectifier

(ii) Bridge type full-wave rectifier.

(ii) Full-wave bridge rectifier

A full-wave bridge rectifier consists of 4 diodes D_1, D_2, D_3 and D_4 which are connected to form a bridge as shown in the fig. 8.3.

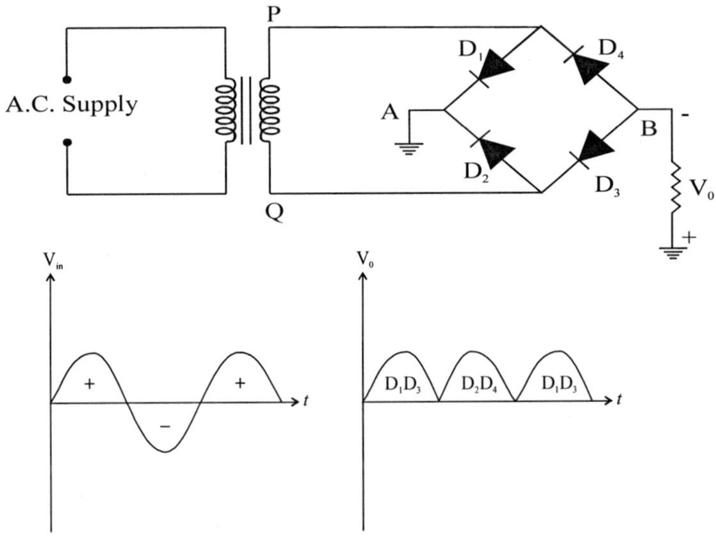

Fig. 8.3

The a.c. supply to the rectifier is applied to the diagonally opposite ends of the bridge through a transformer. Between other two junctions of the bridge the load resistance 'R_L' is connected.

Function

During the +ve half cycle of secondary voltage, the end 'P' of the secondary winding becomes +ve and the end 'Q' –ve. This makes diodes D_1 & D_3 forward biased while diodes D_2 & D_4 are reverse biased. Therefore only diodes D_1 & D_3 conduct. These two diodes will be in series through the load 'R_L'.

During the –ve half cycle of secondary voltage, the end p becomes –ve & end Q becomes +ve. This makes diodes D_2 & D_4 forward biased where D_1 & D_3 are reverse biased. Therefore only diodes D_2 & D_4 conduct. These two diodes will be in series through the load 'R_L'. The current always flows (for both +ve & –ve half cycle of a.c.) from A to B through the load in the same direction. Consequently, a.c. output is obtained across the load R_L.

Efficiency

Fig. 8.4 shows the process of full-wave rectification. Let $V = V_m \sin \theta$ be the a.c. voltage to be rectified. Let r_f and R_L be the diode resistance and load resistance respectively. Obviously, the rectifier will conduct current through the load in the same direction for both half cycles of input a.c. voltage. The instantaneous current 'i' is given by :

$$i = \frac{V}{r_f + R_L} = \frac{V_m \sin\theta}{r_f + R_L}$$

d.c. output power

The output current is pulsating direct current. Therefore, in order to find the a.c. power, average current has to be found out from the elementary knowledge of electrical engineering.

$$I_{dc} = \frac{2\ I_m}{\pi}$$

$$\therefore \text{ d.c. power output, } P_{dc} = I_{dc}^2 \times R_L = \left(\frac{2\ I_m}{\pi}\right)^2 \times R_L \qquad ...(1)$$

$$\downarrow$$

$$(VI_{dc} = RI_{dc} \,.\, I_{dc})$$

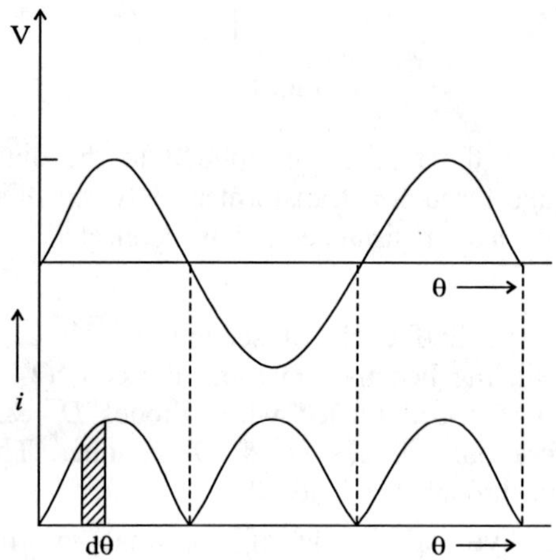

Fig. 8.4

a.c. input power

The a.c. input power is given by

$$P_{ac} = I_{rms}^2 \times \left(r_f + R_L\right)$$

For a full-wave rectified waves,

$$I_{rms} = \frac{I_m}{\sqrt{2}}$$

$$P_{ac} = \left(\frac{I_m}{\sqrt{2}}\right)\left(r_f + R_L\right) \qquad ...(2)$$

$\therefore$ Full wave rectification efficiency,

$$\eta = \frac{P_{dc}}{P_{ac}}$$

$$= \frac{(2\,I_m/\pi)^2\,R_L}{\left(\frac{I_m}{\sqrt{2}}\right)^2 (r_f + R_L)}$$

$$= \frac{8}{\pi^2} \times \frac{R_L}{r_f + R_L} = \frac{0.812\,R_L}{r_f + R_L} = \frac{0.812}{1 + \frac{r_f}{R_L}}$$

The efficiency will be max^m if r_f is negligible as compared to R_L.

∴ Max^m efficiency = 81.2%

This is double the efficiency due to half-wave rectifier. Therefore, a full-wave rectifier is twice as effective as a half-wave rectifier.

Ripple factor

The output of a rectifier consists of a d.c. component and an a.c. component (also known as ripple). The a.c. component is undesirable and accounts for the pulsations in the rectifier output. The effectiveness of a rectifier depends upon the magnitude of a.c. component in the output; the smaller this component, the more effective is the rectifier.

The ratio of rms value of a.c. component to the d.c. component in the rectifier output is known as ripple factor, *i.e.*,

$$\text{Ripple factor} = \frac{\text{r.m.s. value of a.c. component}}{\text{value of d.c. component}}$$

Therefore, ripple factor is very important in deciding the effectiveness of a rectifier. The smaller the ripple factor, the lesser the effective a.c. component and hence more effective is the rectifier.

Mathematical analysis

The output current of a rectifier contains d.c. and a.c. component. By def^n, the effective (*i.e.*, rms) value of total load current is given by :

$$I_{rms} = \sqrt{I^2{}_{dc} + I^2{}_{ac}}$$

or, $$I_{ac} = \sqrt{I^2_{rms} - I^2_{dc}}$$

Dividing throughout by I_{dc}, we get

$$\frac{I_{ac}}{I_{dc}} = \frac{1}{I_{dc}} \sqrt{I_{rms}^2 - I_{dc}^2}$$

But (I_{ac}/I_{dc}) is the ripple factor

$$\therefore \text{ Ripple factor } = \frac{1}{I_{dc}} \sqrt{I_{rms}^2 - I_{dc}^2}$$

$$= \sqrt{\left(\frac{I_{rms}}{I_{dc}}\right)^2 - 1}$$

For full-wave rectification

In full-wave rectification,

$$I_{rms} = \frac{I_m}{\sqrt{2}}, \quad I_{dc} = \frac{2\,I_m}{\pi}$$

$$\therefore \text{ Ripple factor } = \sqrt{\left(\frac{I_m/\sqrt{2}}{2\,I_m/\pi}\right)^2 - 1} = 0.48$$

i.e., $\dfrac{\text{Effective a.c. component}}{\text{d.c. component}} = 0.48$

This shows that in the output of a full-wave rectifier, the d.c. component is more than the a.c. component. Consequently, the pulsations in the output will be less than in half-wave rectifier. For this reason, full-wave rectification is invariably used for conversion of a.c. into d.c.

Q. 3. What is a filter circuit? Why it is necessary in a rectifier circuit?

Ans. *Filter Circuit* : A filter circuit is a device which removes the a.c. component of rectifier output but allows the d.c. component to reach the load. Thus, a filter circuit actually filters out the a.c. component from the d.c.

Necessity

Generally, a rectifier is required to produce pure d.c. supply for using various places in the electronic circuits. However, the output of a rectifier has pulsating character *i.e.*, it contains a.c. and d.c. components. The a.c. component is undesirable and must be kept away from the load. To do so, a filter circuit is used which removes (or filters out) the a.c. component and allows only the d.c. component to reach the load.

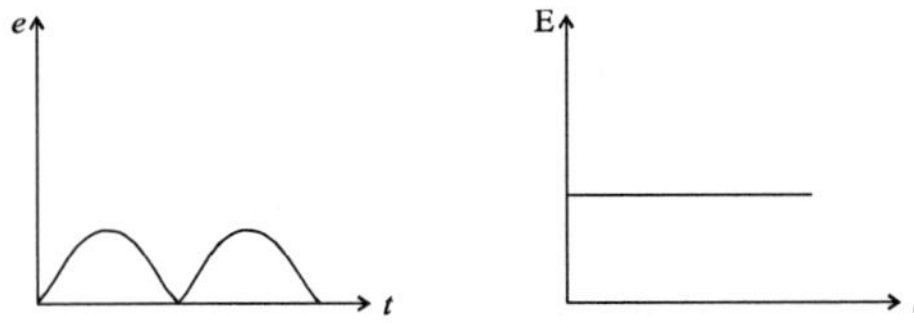

(i) Rectifier Output

(ii) Pure D.C. Output

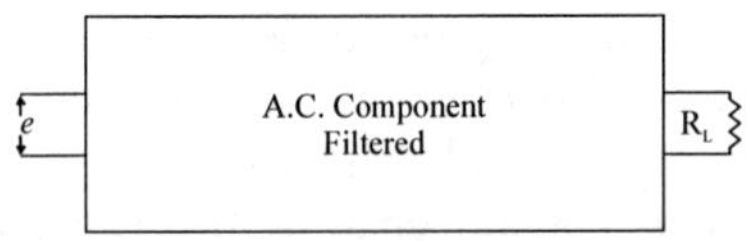

(iii) Filter Circuit

Fig. 8.5

Obviously, a filter circuit should be installed between the rectifier and the load as shown in fig. 8.5. A filter circuit is generally a combination of inductors (L) and capacitor (C). The filtering of L & C depends upon the basic electrical principles. A capacitor passes a.c. readily but does not pass d.c. at all. On the other hand, an inductor opposes a.c. but allows d.c. to pass through it. It then becomes clear that suitable network of L and C can effectively remove the a.c. component, allowing the d.c. component to reach the load.

Q. 4. Describe the actions of the following filter circuits :

(i) Shunt capacitor filter or capacitor filter (C filter)
(ii) Series inductor filter or the inductor filter (L filter)
(iii) The choke input filter or L.C. filter or L-section filter
(iv) Capacitor input filter or C-L-C filter or π-section filter.

Ans.

(i) Shunt capacitor filter or capacitor filter (C filter)

This filter circuit consists of a capacitor 'C' connected across the rectifier output in parallel with load R_L. The output of the rectifier is fed across the capacitor (fig. 8.6).

Action

The increase in the rectifier voltage charges the capacitor and also supplies current to the load (R_L). At the end of 1/4th of a cycle, the capacitor is charged to the peak value V_m of the rectifier voltage.

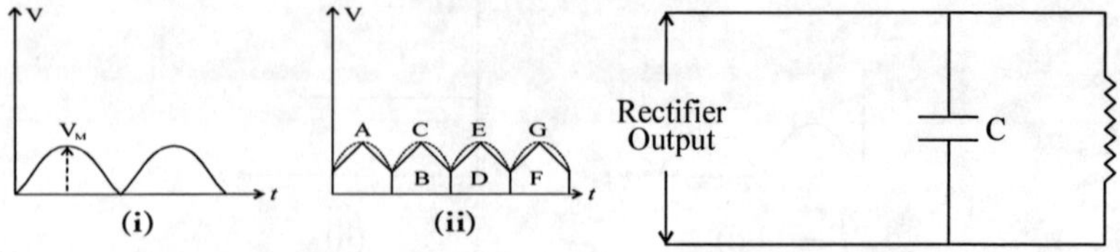

Fig. 8.6. Capacitor filter

Now, the rectifier voltage begins to fall and the capacitor discharges through the load. Consequently, the voltage across it decreases. The voltage across load will fall only slightly because immediately the next voltage peak comes and again charges the capacitor. This process is repeated again and again resulting the output voltage wave form like ABCDEFG. It should be noted here that very small amount of ripple is left in the output.

A capacitor offers infinite reactance to d.c. Because we have $f = 0$ for d.c.

$$\therefore \; X_c = \frac{1}{\omega C} = \frac{1}{2\pi f C} = \frac{1}{0} = \infty \quad (\because f = 0)$$

Hence the reactance offered by a capacitor to d.c. is infinite. As a result of which, a capacitor doesn't allow d.c. to pass through it.

Uses

The shunt capacitor filter circuit is widely used for small load currents (upto 50 mA). It is also commonly employed in transistor radio battery eliminators. Its cost is low, size is small, weight is low and it has good characteristics. For these reasons a capacitor filter circuit is popular.

(ii) Series inductor filter or the inductor filter or L-filter

An inductor filter consists of choke (inductance) in series with the load. The output of the full-wave rectifier is fed across the inductor filter (fig. 8.7).

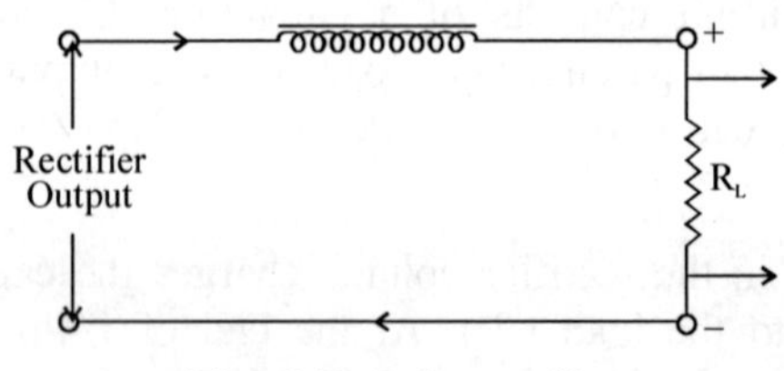

Fig. 8.7. Inductor filter

Action

An inductance opposes the flow of alternating current through it. Because its reactance,

$$X_L = 2\pi fL$$

where f is the frequency of the a.c. and L is the inductance.

For d.c., $f = 0$, so $X_L = 0$

That means, d.c. can pass easily through the inductor but a.c. can't.

The current shape with inductor filter is shown in fig. 8.8 below.

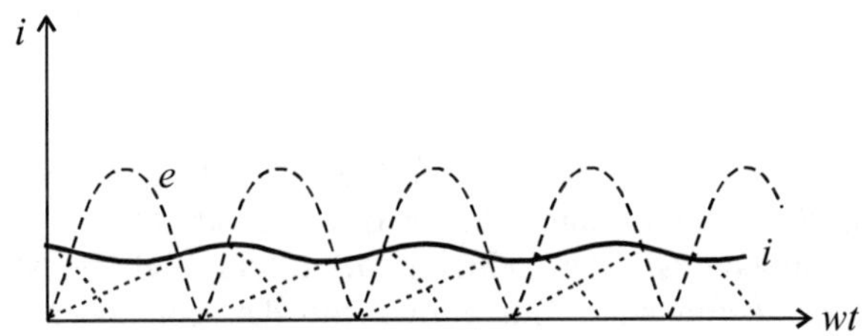

Fig. 8.8. Current shape with filter inductor

Here the ripple factor is given by

$$r = \frac{R_L}{1600\,L} \quad \text{for } f = 60 \text{ Hz.}$$

Thus the ripple factor depends on the load and inductance 'L'. If 'L' is more ripple factor is less.

Uses

This filter is useful for small resistances (heavy load).

(iii) The choke input filter or L-C filter or L-section filter

A choke input filter is the combination of inductor filter and capacitor filter. In inductor filter, the ripple factor increases with the increase in the load. But the ripple factor decreases in case of capacitor filter. The combination of these two filters into L-C filter would make ripple independent of load.

Fig. 8.9 shows a typical choke input filter circuit. It consists of a choke 'L' connected in series with the rectifier output and a filter capacitor 'C' across the load. Here, a single filter section is shown, but several identical sections are often used to reduce the ripple effectively.

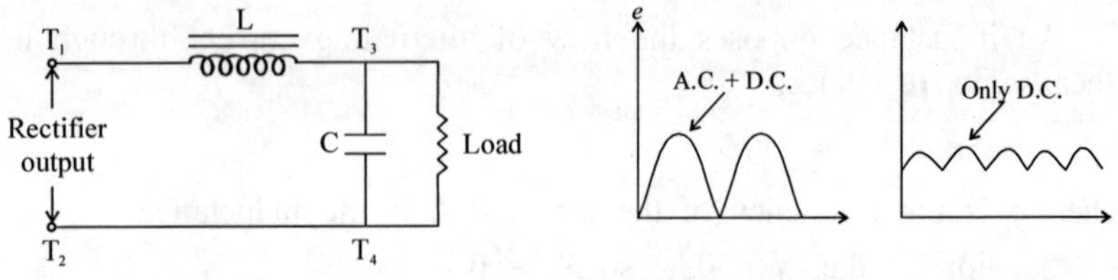

(a) (b) Input (c) Output

Fig. 8.9

Action

The output of the full-wave rectifier is applied across the terminals T_1, T_2 of the filter circuit. The rectifier output contains a.c. as well as d.c. components. The inductor '*L*' offers high resistance to the flow of a.c. component. But allows the d.c. component to pass. As a result of which the a.c. component appears across the choke (inductor) while the entire d.c. component passes through it towards the load. At the terminal T_3, the rectifier output contains the d.c. component and the remaining part of a.c. component. The capacitor '*C*' offers low reactance and hence bypasses the a.c. component but prevents the d.c. component to flow through it. Therefore, only d.c. component reaches the load. In this way a.c. component has been filtered out by the choke input filter and only d.c. component is obtained across the load.

(iv) Capacitor input filter or C-L-C filter or Pi-section filter

This filter consists of a capacitor filter followed by a L-C filter. The addition of the input capacitor filter with full-wave diode rectifier results in a high load voltage which is higher than that obtained from a L-C filter. The circuit of *Pi* filter is shown in fig. 8.10.

Action

The pulsating output from the full wave rectifier is applied across the input terminals T_1 and T_2 of the filter. The three circuit components C_1, L and C_2 exhibit filtering action, *i.e.*, separation of a.c. from d.c.

The capacitor C_1 offers low reactance to a.c. component of rectifier output while it offers infinite reactance to the d.c. component. As a result of which capacitor C_1 bypasses a.c. and blocks d.c. Thus d.c. component flows through the choke *L*.

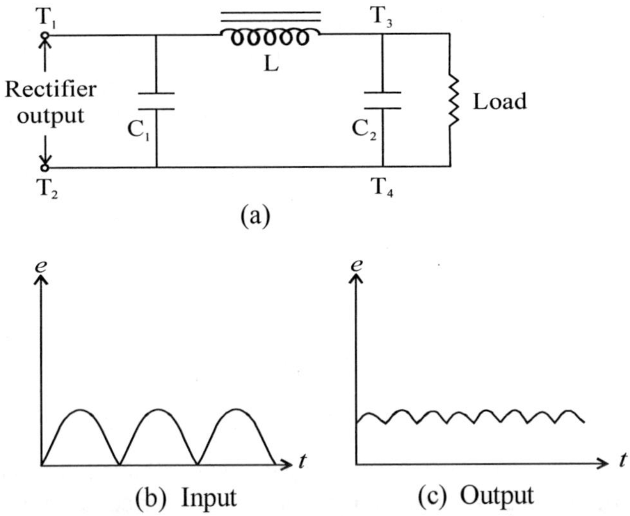

Fig. 8.10

The inductor L offers high reactance to the a.c. component but almost zero reactance to the d.c. component. Therefore it allows the d.c. component to flow through it while blocks a.c.

The filter capacitor C_2 bypasses the a.c. component, if any present from that obtained from choke 'L'. Therefore only pure d.c. will appear across the load.

EXERCISES

Numerical Problems with Solutions

1. The applied input a.c. power to a half-wave rectifier is 120 watts. The d.c. output power obtained is 48 watts. Find the rectification efficiency. What happens to the rest of the 72 watts?

Solution

(i) P_{in} = 120 Watts, P_{out} = 48 Watts, η = ?

Rectification efficiency 'η' is given by

$$\eta = \frac{\text{d.c. output power}}{\text{a.c. input power}}$$

$$= \frac{48}{120} = \frac{4}{10} = 0.4 \quad 40\%$$

(ii) 40% efficiency of rectification does not mean that 60% of power is lost in the rectifier circuit. In fact, a crystal diode consumes little power due to its small internal resistance. The 120 watts a.c. power is contained as 60 watts in positive half-cycles and 60 watts in negative half-cycles. The 60 watts in the negative half-cycles are not supplied at all. Only 60 watts in the positive half-cycles are converted into 48 watts.

$$\therefore \text{ power efficiency } = \frac{48}{60} \times 100$$

$$= \frac{4}{5} \times 100$$

$$= 80\%$$

Although 120 watts of a.c. power was supplied, the half-wave rectifier accepted only 60 watts and converted it into 48 watts d.c. power. Hence, it is right to say that efficiency of rectification is 40% and not 80% which is power efficiency.

2. A half-wave rectifier uses a transformer of turn ratio 4 : 1. If the primary voltage is 240 V (rms), calculate (i) d.c. output voltage (ii) peak inverse voltage. Assume the diode to be ideal.

Solution

$$\text{Turn ratio,} \quad K = \frac{N_p}{N_s} = \frac{4}{1}$$

$$\text{RMS primary voltage } = V_{in} = 240 \text{ V}$$

$\therefore$ Maximum primary voltage,

$$(V_{in})_{\max} = \sqrt{2} \times (V_{in})_{rms}$$

$$= \sqrt{2} \times 240 = 240 \times 1.414$$

$$= 339.36 \text{ V}$$

Maximum secondary voltage,

$$(V_{out})_{\max} = (V_{in})_{\max} \times \frac{N_s}{N_p}$$

$$= 339.36 \times 1/4$$

$$= 84.84 \text{ V}$$

(i) $$I_{dc} = \frac{I_m}{\pi}$$

$$\therefore V_{dc} = \frac{I_m}{\pi} \times R_L = \frac{(V_{out})_{max}}{\pi}$$

$$= \frac{84.84}{\pi} = \frac{84.84}{3.142} = 27 \text{ V}$$

(ii) During the negative half-cycle of a.c. supply, the diode is reverse biased and hence conducts no current. Hence, the maximum secondary voltage appears across the diode.

∴ Peak inverse voltage = 84.84 V.

3. A half-wave rectifier uses a transformer of turn ratio 2 : 1. If the primary voltage is 240 V (rms) and the load resistance is 500 Ω, calculate (i) d.c. output voltage (ii) peak inverse voltage.

Solution

Turn ratio, $K = \frac{N_p}{N_s} = \frac{2}{1}$

RMS primary voltage = V_{in} = 240 V

R_L = 500 Ω

∴ Maximum primary voltage,

$$(V_{in})_{max} = \sqrt{2} \times V_{in}$$

$$= \sqrt{2} \times 240$$

$$= 1.414 \times 240 = 339.36 \text{ V}$$

Maximum secondary voltage,

$$(V_{out})_{max} = (V_{in})_{max} \times \frac{N_s}{N_p}$$

$$= 339.36 \times \frac{1}{2}$$

or, V_m = 169.68 V

$$I_m = \frac{V_m}{R_L} = \frac{169.68}{500} = 0.339 \text{ A}$$

$$I_{dc} = \frac{I_m}{\pi} = \frac{0.339}{3.142} = 0.1078 \text{ A}$$

(i) d.c. output voltage $= I_{dc} \times R_L$

$= 0.1078 \times 500$

$= 53.9$ V

(ii) The maximum secondary voltage appears across the diode as the diode conducts no current during the negative half-cycle of a.c. supply because the diode is reverse biased in this half-cycle.

$\therefore$ Peak Inverse voltage = 169.68 V

4. A crystal diode having internal resistance $r_f = 30\ \Omega$ is used for half-wave rectification. If the applied voltage is $V = 60 \sin \omega t$ and load resistance is $R_L = 800\ \Omega$. Calculate (i) I_m, I_{dc}, I_{rms} (ii) a.c. power input and d.c. power output (iii) d.c. output voltage (iv) efficiency of rectification.

Solution

$$r_f = 30\ \Omega,\quad R_L = 800\ \Omega$$

$$V = 60 \sin \omega t$$

$\therefore$ Maximum voltage, $V_m = 60\ V$

(i)
$$I_m = \frac{V_m}{r_f + R_L} = \frac{60}{30+800} = \frac{60}{830}$$

$$= 72\ mA$$

$$I_{dc} = \frac{I_m}{\pi} = \frac{72\ mA}{3.142} = 22.9\ mA$$

$$I_{rms} \frac{I_m}{\sqrt{2}} = \frac{72\ mA}{1.414} = 50.9\ mA$$

(ii) a.c. power input $= (I_{rms})^2 \times (r_f + R_L)$

$= (50.9 \times 10^{-3})^2\ (30 + 800)$

$= (50.9)^2 \times 10^{-6} \times 830$

$= 2.15$ watts

d.c. power output $= I^2_{dc} \times R_L$

$= (22.9 \times 10^{-3})^2 \times 800$

$= (22.9)^2 \times 10^{-6} \times 800$

$= 524.41 \times 8 \times 10^{-4}$

$= 0.4195$ watts.

(iii) d.c. output voltage $= I_{de} \times R_L$
$= 22.9 \times 10^{-3} \times 800$
$= 22.9 \times 8 \times 10^{-1}$
$= 18.32$ volts

(iv) Efficiency of rectification $= \dfrac{0.4195}{2.15} \times 100$
$= 19.5\%$

5. A half-wave rectifier uses a vacuum diode whose plate resistance is 100 Ωs. What should be the value of load resistance so that the rectification efficiency is 40%.

Solution

$r_p = 100$ ohms, $R_L = ?$
$\eta = 40\% = 0.40$

For a half-wave rectifier, efficiency

$$\eta = \frac{0.406}{1+\left(\frac{r_P}{R_L}\right)}$$

$$0.4 = \left(\frac{0.406}{1+\frac{r_P}{R_L}}\right) = \frac{0.406}{1+\frac{100}{R_L}}$$

or, $$0.4\left(1+\frac{100}{R_L}\right) = 0.406$$

or, $$1+\frac{100}{R_L} = \frac{0.406}{0.4} = 1.015$$

or, $$\frac{100}{R_L} = 0.015$$

or, $$R_L = \frac{100}{0.015} = \frac{100}{15\times 10^{-3}}$$

$$= \frac{100\times 10^{3}}{15} = 6666.66 \text{ ohms}$$

6. A half-wave rectifier is used to supply 50 V d.c. to a resistive load of 1000 Ω. The diode has a resistance of 20 Ω. Compute a.c. voltage needed.

Solution

Output d.c. voltage = V_{dc} = 50 V
diode resistance = r_f = 20 Ω
load resistance = R_L = 1000 Ω

Let V_m be the maximum value of a.c., voltage required.

$$V_{dc} = I_{dc} \times R_L = \frac{I_m}{\pi} \times R_L$$

$$= \frac{V_m}{\pi(r_f + R_L)} \times R_L$$

or, $$50 = \frac{V_m}{\pi(20+1000)} \times 1000$$

or, $$V_m = \frac{50\pi \times 1020}{1000}$$

$$= \frac{51000\pi}{1000} = 51 \times 3.142$$

$$= 160.24 \text{ V}$$

Hence, a.c. voltage of maximum value 160.24 V is required.

7. A full-wave rectifier uses two junction diodes of forward resistance 20 ohms. If the load resistance is 980 ohms and maximum secondary voltage appearing across each diode is 20 volts, find (i) the rectification efficiency (ii) the maximum load current (iii) the mean load current (iv) the rms load current.

Solution

r_f = 20 ohms, R_L = 980 ohms, E_o = 20 V
η = ? I_o = ? I_{dc} = ?, I_{rms} = ?

(i) For a full-wave rectifier,

$$\eta = \frac{81.2}{1+\frac{r_f}{R_L}} = \frac{81.2}{1+\frac{20}{980}}$$

$$= \frac{81.2 \times 980}{1000} = \frac{81.2 \times 49}{50} \; 79.576\%$$

(ii) Maximum load current,

$$I_0 = \frac{E_0}{r_f + R_L} = \frac{20}{20+980} = \frac{20}{1000}$$

$$= 0.02 \text{ amp}$$

(iii) Mean load current,

$$I_{dc} = \frac{2\,I_0}{\pi} = \frac{2 \times 0.02}{3.142}$$

$$= 0.0127 \text{ amp}$$

(iv) RMS load current,

$$I_{rms} = \frac{I_0}{V_2} = \frac{0.02}{1.414} = 0.014 \text{ amp}$$

8. A full-wave rectifier uses two diodes, the internal resistance of each diode may be assumed constant at 30 Ω. The transformer rms secondary voltage from centre tap to each end of secondary is 50 V and load resistance is 970 Ω. Compute (i) the mean load current (ii) the rms value of load current, and (iii) the rectification efficiency.

Solution

$$r_f = 30\ \Omega,\quad R_L = 970\ \Omega,\quad V_{rms} = 50\ V$$

$$I_{dc} = ?,\quad I_{rms} = ?,\quad \eta = ?$$

Maximum a.c. voltage, $V_m = V_{rms} \times \sqrt{2}$

$$= 50\ V \times 1.414 = 70.7\ V$$

Maximum load current,

$$I_m = \frac{V_m}{r_f + R_L}$$

$$= \frac{70.7\ V}{(30 + 970)\Omega} = \frac{70.7}{1000}\ A$$

$$= 70.7 \text{ mA}$$

(i) Mean load current,

$$I_{dc} = \frac{2\,I_m}{\pi} = \frac{2 \times 70.7}{\pi}$$

$$= 45 \text{ mA}$$

(ii) RMS value of load current,

$$I_{rms} = \frac{I_m}{\sqrt{2}}$$

$$= \frac{70.7}{\sqrt{2}} = 50 \text{ mA}$$

(iii) Rectification efficiency,

$$\eta = \frac{0.812}{1+\frac{r_f}{R_L}} = \frac{0.812}{1+\frac{30}{970}}$$

$$= \frac{0.812 \times 970}{1000}$$

$$= \frac{0.812 \times 970 \times 100}{1000}\%$$

$= 0.812 \times 97\% = 78.76\%$

9. In the centre-tap circuit shown in Fig 8.11 below, the diodes are assumed to be ideal *i.e.* having zero internal resistance. Find (i) d.c output voltage (ii) peak inverse voltage (iii) rectification efficiency.

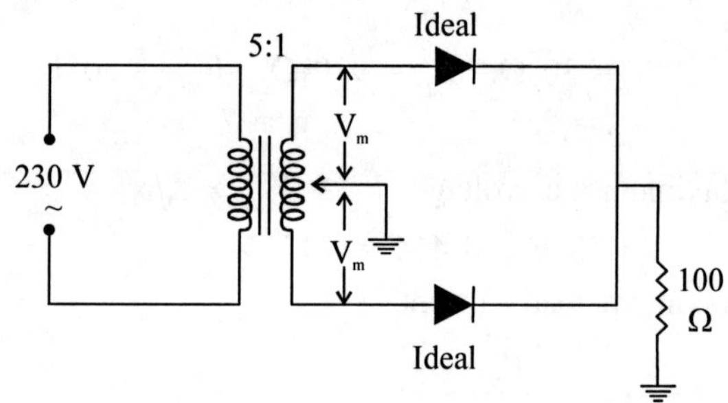

Fig. 8.11

Solution

$R_L = 100\ \Omega$, $r_f = 0$ for ideal diode

Primary to secondary turns, $\dfrac{N_p}{N_s} = \dfrac{5}{1}$

RMS primary voltage $= 230$ V

RMS secondary voltage $= 230 \times \dfrac{1}{5} = 46$ V

Maximum voltage across secondary $= 46\sqrt{2}$ V

$= 65$ V

Maximum voltage across half secondary winding

$$= V_m = \frac{65}{2} = 32.5 \text{ V}$$

(i) Average current, $I_{dc} = \dfrac{2\,V_m}{\pi R_L} = \dfrac{65}{\pi \times 100}$

$= 0.207$ A

$\therefore$ d.c. output voltage, $V_{dc} = I_{dc} \times R_L$

$= 0.207 \times 100$

$= 20.7$ V

(ii) The peak inverse voltage is equal to the maximum secondary voltage, *i.e.*

$PIV = 65$ V

(iii) Rectification efficiency $= \dfrac{0.812}{1+\frac{r_f}{R_L}}$

$= 0.812 \qquad (\because\ r_f = 0)$

$= 81.2\%$

10. In the rectifier circuit shown in Fig. 8.12 below, the maximum voltage across half of secondary winding is 50 V. Find (i) the average load voltage (ii) peak inverse voltage (iii) output frequency. Assume the diodes to be ideal.

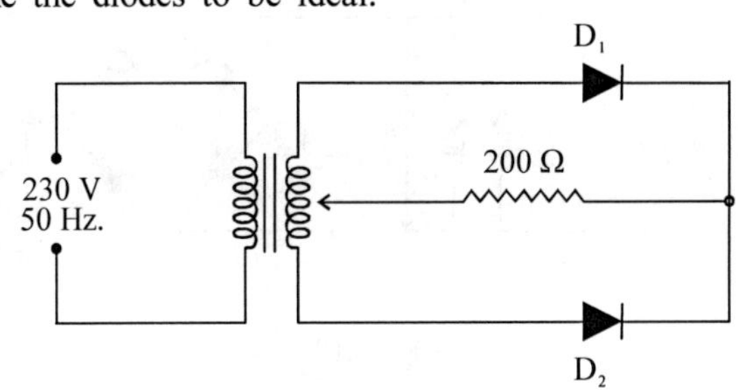

Fig. 8.12

Solution

$V_m = 50$ V, $r_f = 0$ for ideal diode

$R_L = 200\ \Omega$, $f_{in} = 50$ H_z, $V_{in} = 230$ V

(i) Average load current, $I_{dc} = \dfrac{2V_m}{\pi R_L} = \dfrac{100}{\pi \times 200}$

$= \dfrac{1}{2\pi} = \dfrac{1}{3.142 \times 2} = \dfrac{0.318}{2}$ A

$= 0.159$ A

Average load voltage, $V_{dc} = I_{dc} \times R_L$

$$= \frac{0.318}{2} \times 200$$

$$= \frac{0.318}{2} \times 2$$

$$= 31.8 \text{ V}$$

(ii) The peak inverse voltage is equal to the maximum secondary voltage, *i.e.*, $PIV = 2\ V_m = 2 \times 50 = 100$ V.

(iii) In full-wave rectification, there are two output pulses for each complete cycle of the input a.c. voltage. Therefore, the output frequency is twice that of the a.c. supply frequency, *i.e.*,

$$f_{out} = 2 f_{in} = 2 \times 50 = 100 \text{ Hz}$$

11. In the Full-wave bridge rectifier circuit shown in Fig. 8.13 below, the diodes are taken to be ideal. Compute (i) d.c. output voltage (ii) peak inverse voltage (iii) output frequency.

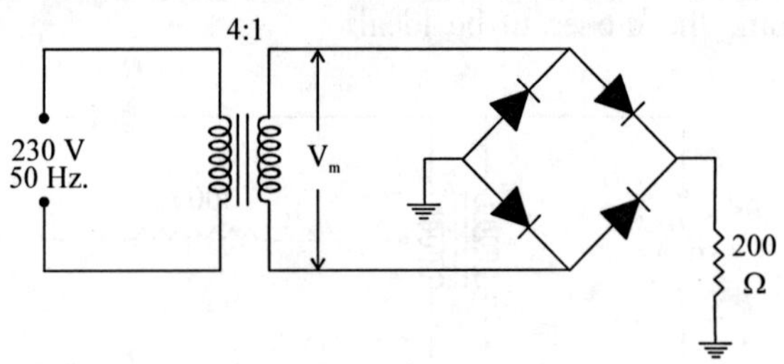

Fig. 8.13

Solution

$$f_{in} = 50 \text{ Hz}, \quad R_L = 200\ \Omega$$

Primary to secondary turns ratio, $\frac{N_p}{N_s} = \frac{4}{1}$

RMS primary Voltage $= 230$ V

$\therefore$ RMS secondary voltage $= 230 \times \frac{N_s}{N_p}$

$$= 230 \times \frac{1}{4}$$

$$= 57.5 \text{ V}$$

Maximum voltage across secondary,

$$V_m = V_{rms} \times \sqrt{2}$$
$$= 57.5 \times 1.414$$
$$= 81.3 \text{ V}$$

(i) Mean current, $I_{dc} = \dfrac{2V_m}{\pi R_L} = \dfrac{2 \times 81.3}{\pi \times 200}$

$$= 0.26 \text{ A}$$

∴ d.c. output voltage,

$$V_{dc} = I_{dc} \times R_L$$
$$= 0.26 \times 200 \text{ V}$$
$$= 52 \text{ V}$$

(ii) The peak inverse voltage is equal to the maximum secondary voltage, *i.e.*

$$PIV = 81.3 \text{ V}$$

(iii) In full-wave rectification, there are two output pulses for each complete cycle of the input a.c. voltage. So, the output frequency is twice that of the a.c. supply frequency, *i.e.*,

$$f_{out} = 2 \times f_{in} = 2 \times 50 = 100 \text{ Hz}$$

12. In the full-wave bridge type rectifier circuit shown in Fig. 8.14 below, the maximum secondary voltage is 136 V. Find (i) the d.c. load voltage (ii) peak inverse voltage (iii) output frequency.

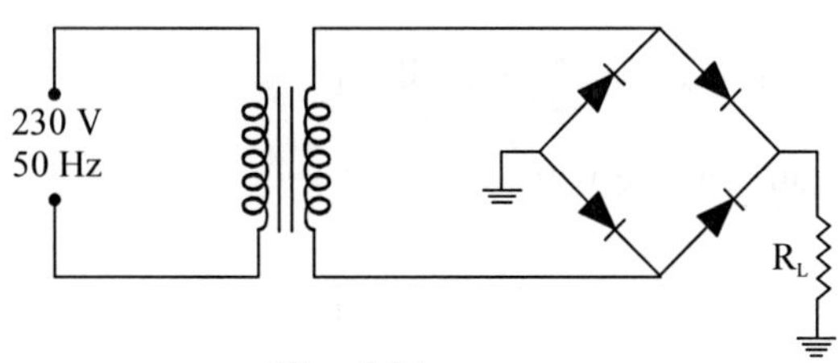

Fig. 8.14

Solution

$$f_{in} = 50 \text{ Hz}, \; V_m = 136 \text{ V}$$

(i) $V_{dc} = I_{dc} \times R_L = \dfrac{2V_m}{\pi R_L} \times R_L$

$$= \frac{2V_m}{\pi} = \frac{2 \times 136}{3.142}$$

$= 86.569$ V

≈ 86.6 V

(ii) $PIV = 136$ V

(iii) $f_{out} = 2f_{in} = 2 \times 50 = 100$ Hz

13. The four diodes used in a bridge rectifier circuit have forward resistances which may be taken to be constant at 1 Ω and infinite reverse resistance. The a.c. supply is 230 V rms and the load resistance is 580 Ω. Compute (i) the mean load current and (ii) power dissipated in each diode.

Solution

$r_f = 1\ \Omega,\ V_{rms} = 230\ \text{V},\ R_L = 580\ \Omega$

(i) Maximum a.c. voltage, $V_m = V_{rms} \times \sqrt{2}$

$= 230 \times 1.414$

$= 325$ V

At any instant in the bridge rectifier, two diodes in series are conducting. Hence, total circuit resistance $= 2\ r_f + R_L$

Maximum load current,

$$I_m = \frac{V_m}{2r_f + R_L} = \frac{325}{2 \times 1 + 580}$$

$$= \frac{325}{582} = 0.558 \text{ A}$$

Mean load current,

$$I_{dc} = \frac{2I_m}{\pi} = \frac{2 \times 0.558}{3.142}$$

$= 0.355$ A

(ii) Since each diode conducts only half a cycle, diode rms current is

$$I_{rms} = \frac{I_m}{2} = \frac{0.558}{2} = 0.279 \text{ A}$$

Power dissipated in each diode,

$$P = I_{rms}^2 \times r_f$$

$$= (0.279)^2 \times 1$$
$$= 0.778 \text{ watts.}$$

14. A power supply A delivers 20 V d.c. with a ripple of 1.0 V rms while the power supply B delivers 50 V d.c. with a ripple of 2.0 mV. Which power supply serves as a better power supply?

Solution

For power supply A, we have

$$\text{Ripple factor } = \frac{V_{ac(rms)}}{V_{dc}} = \frac{1}{20} \times 100 = 5\%$$

For power supply B, we have

$$\text{Ripple factor } = \frac{V_{ac(rms)}}{V_{dc}} = \frac{2 \times 10^{-3}}{50} = \frac{10^{-3}}{25} \times 100$$
$$= 0.004\%$$

As the power supply 'B' has lower ripple factor, it is a better power supply.

15. A voltage of 200 Sin 100 πt is applied to a half-wave vacuum tube rectifier without filter. The diode resistance is 1 KΩ and the load resistance is 9,000 Ωs. Compute (i) peak, mean and rms values of current (ii) d.c. power output (iii) a.c. power input (iv) rectifier efficiency; (v) ripple factor.

Solution

$V_m = 200$ V, $r_p = 1$ KΩ $= 1000$ Ω

$R_L = 9{,}000$ Ωs

(i) I_m = peak value of current $= \dfrac{V_m}{r_p + R_L}$

$$= \frac{200}{1000 + 9000} = \frac{200}{10{,}000}$$

$$= \frac{2}{100} = \frac{1}{50} = .02 \text{ amp} = 20 \text{ mA}$$

$$I_{av} = I_{dc} = \frac{I_m}{\pi} = \frac{20\ mA}{3.142} = 6.36 \text{ mA}$$

$$I_{rms} = \frac{I_m}{2} = \frac{20}{2} = 10 \text{ mA}$$

(ii) d.c. power output,

$$P_{dc} = I_{dc}^2 \times R_L = (6.36 \times 10^{-3})^2 \times 9000 \text{ watts}$$
$$= (6.36)^2 \times 10^{-6} \times 9 \times 10^3 \text{ watts}$$
$$= 364 \times 10^{-3} \text{ watts} = 0.364 \text{ watts.}$$

(iii) a.c. power input,

$$P_{ac} = I_{rms}^2 \left(r_p + R_L\right)$$
$$= (10)^2 \times 10^{-6} \ (1000 + 9000)$$
$$= 10^2 \times 10^{-6} \times 10{,}000$$
$$= 10^2 \times 10^{-6} \times 10^4 = 1 \text{ watt}$$

(iv) Rectifier efficiency,

$$\eta = \frac{P_{dc}}{P_{ac}} \times 100\%$$

$$= \frac{0.364}{1} \times 100 = 36.4\%$$

(v) The ripple factor

$$r = \sqrt{\frac{I_{rms}^2}{I_{dc}^2} - 1}$$

$$= \sqrt{\frac{(10)^2}{(6.36)^2} - 1}$$

$$= \sqrt{\left(\frac{10}{6.36}\right)^2 - 1}$$

$$= \sqrt{(1.57)^2 - 1} = \sqrt{1.4649}$$
$$= 1.21$$

16. A power supply is delivering 100 watts to a load of 10 KΩ. Find the a.c. voltage present across the load if the ripple factor is 0.1%.

Solution

$$P = 100 \text{ watts}, \quad R_L = 10 \text{ K}\Omega = 10 \times 10^3 \ \Omega\text{s}$$

$$r = 0.1\% = \frac{0.1}{100}, \quad V_{ac} = ?$$

Power is given by

$$P = \frac{V_{dc}^2}{R_L}$$

or, $$V_{dc} = \sqrt{P \times R_L}$$

$$= \sqrt{100 \times 10^4} = \sqrt{10^6}$$

$$= 10^3 \text{ volts}$$

The ripple factor is given by

$$r = \frac{V_{ac}}{V_{dc}} = \frac{0.1}{100}$$

or, $$V_{ac} = \frac{0.1}{100} \times V_{dc}$$

$$= \frac{1}{1000} \times 1000 = 1 \text{ volt}$$

17. The circuit of a Zener diode is shown in Fig. 8.15 below. Calculate (i) the input voltage (ii) the voltage drop across series resistance and (iii) the current through Zener diode.

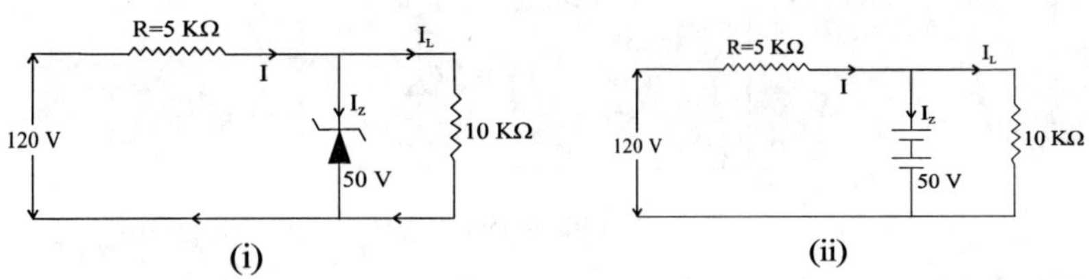

Fig. 8.15

Solution

(i) Output voltage = V_z = 50 V

(ii) Voltage drop across R

$$= V_{in} - V_z$$

$$= 120 \text{ V} - 50 \text{ V}$$

$$= 70 \text{ V}$$

(iii) Load current, $$I_L = \frac{V_Z}{R_L}$$

$$= \frac{50\,V}{10\,K\Omega}$$

$$= \frac{50 \times 10^{-3}}{10}\ \text{A}$$

$$= 5\ \text{mA}$$

Current through R, $I = \dfrac{70\,V}{5\,K\Omega}$

$$= 14\ \text{mA}$$

Applying Kirchhoff's first law,

$$I = I_L + I_z$$

or, $I_z = I - I_L$

$$= 14 - 5 = 9\ \text{mA}$$

$\therefore$ Zener Current = 9 mA

18. What value of series resistance is required when three 10 watt, 10 volt, 1000 mA Zener diodes are connected in series to obtain a 30 volt regulated output from a 45 volt d.c. power source?

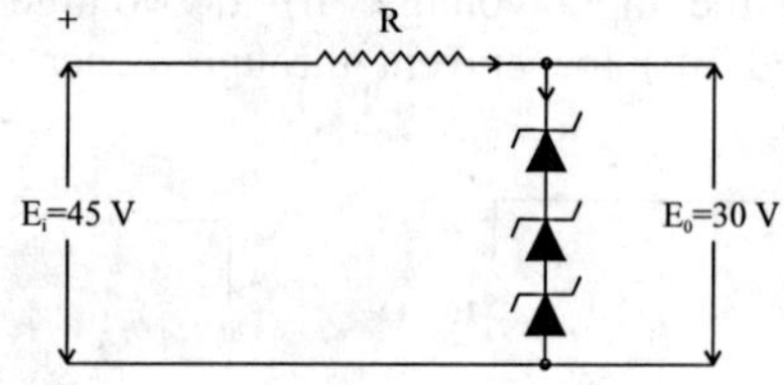

Fig. 8.16

Solution

When there is no load, the Zeners carry the maximum current.

$$V_z = 10\ \text{V},\ I_z = 1000\ \text{mA}$$

$$E_{in} = 45\text{V},\ E_{out} = 10 + 10 + 10 = 30\text{V}$$

Let *R* ohms be the required series resistance. Voltage across *R*

$$= E_{in} - E_{out} = 45 - 30 = 15\ \text{V}$$

$$\text{R} = \frac{15V}{1000mA} = 15\ \text{Ohms}$$

❑ ❑ ❑

9
Transistors

Q. 1. What is a transistor?

Ans. A transistor consists of two p-n junctions cemented together having three leads for connection. The cementing of two p-n junctions are done in two ways and accordingly two types of transistors are obtained, namely,

(i) n.p.n transistor

(ii) p.n.p transistor

The symbols of these two transistors are shown below in fig. 9.1.

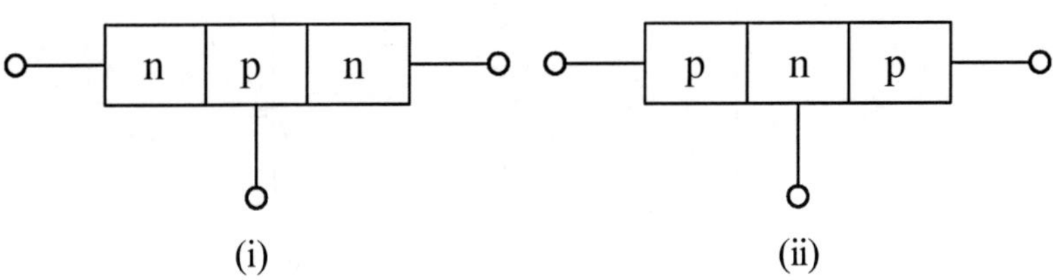

Fig. 9.1

2. Write short notes on :

(a) pnp transistor

(b) npn transistor

Ans.

(a) pnp transistor

A pnp transistor is formed by two p-sections separated by a thin section of n-type semiconductor a shown in fig. 9.2.

The three terminals of the pnp transistor are emitter, collector and base. The emitter is always forward biased w.r.t. base. The emitter

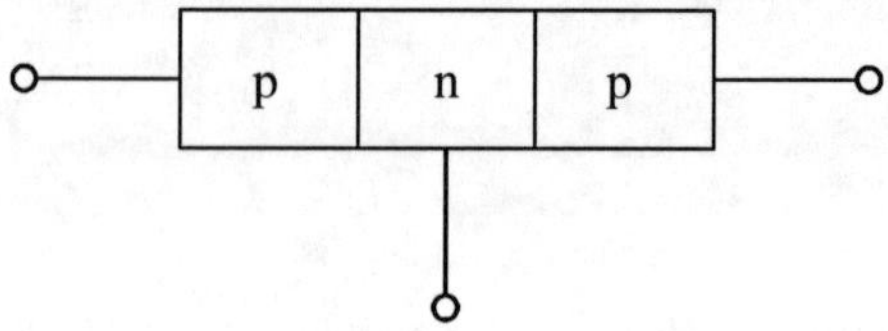

Fig. 9.2

here supplies holes (+ve charge carriers) to its junction with the base. The collector is always reverse biased. It collects the charges and hence removes them from its junction with the base. The collector of pnp transistor collects the holes that flow in the output circuit. The middle section is the base. The base-emitter junction is forward biased offering low resistance in the emitter circuit. The base-collector junction is reverse biased providing high resistance in the collector circuit.

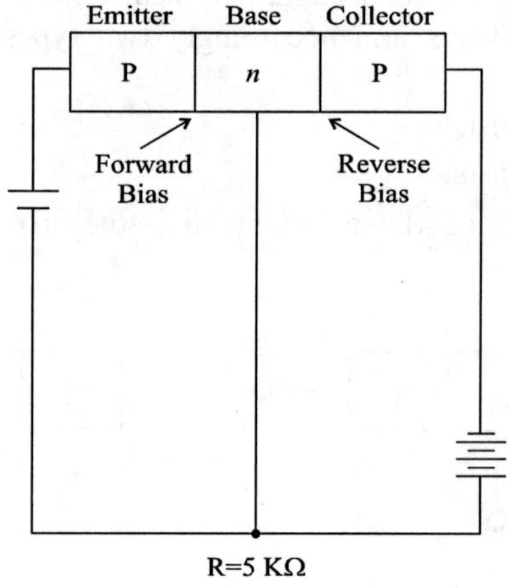

Fig. 9.3

Function

Fig. 9.4 shows the basic connection of pnp transistor. The forward bias in the emitter base junction causes the holes in the p-type emitter to flow towards the base. This gives emitter current I_E. The holes cross the n-type base and they tend to combine with the electrons. As the base is lightly doped and very thin, only a few holes (5%) combine with the electrons. The remainder (more than 95%) flow to the collector and form collector current I_c. In this way, almost the entire emitter current flows in the collector circuit. It should be noted that the current conduction in pnp transistor is due to holes. However, in the external connecting wires, the current is still by electrons.

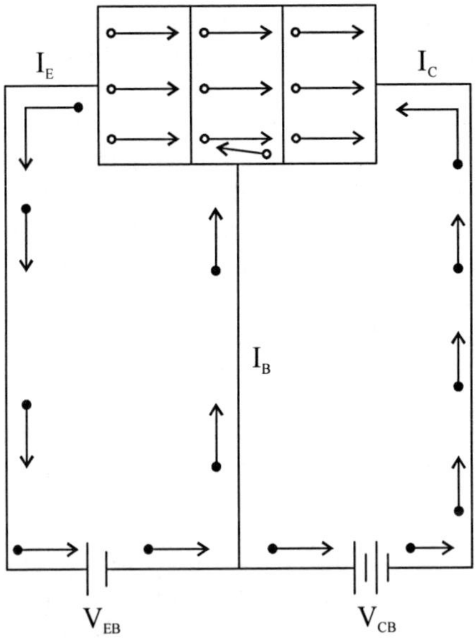

Fig. 9.4. Basic connection of pnp transistor

Symbol of pnp transistor

The pnp transistor has the following circuit symbol. The arrow in the emitter indicates the direction of the conventional current flow.

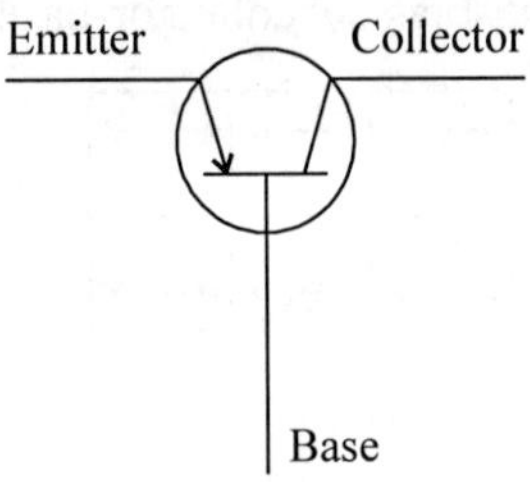

Fig. 9.5

(b) npn transistor

A npn transistor is composed of two n-type semiconductors separated by a thin section of p-type. This is shown in fig. 9.6.

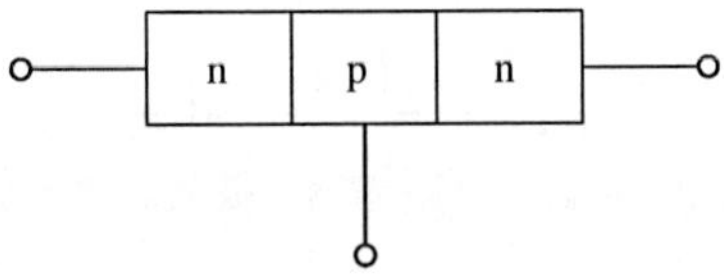

Fig. 9.6

The npn transistor has 3 sections, called emitter, base and collector. The middle section is the base which forms two junctions between the emitter and the collector. The emitter is always forward biased w.r.t. base. The emitter of npn transistor has forward bias and supplies free electrons to its junction with the base. The collector is always reverse biased. It is in the output section. The collector of npn transistor has reverse bias and collects electrons. The base is always in between the emitter and the collector. The base-emitter junction is forward biased offering low resistance for the flow of electrons. The base-collector junction is reverse biased and provides high resistance in the collector circuit.

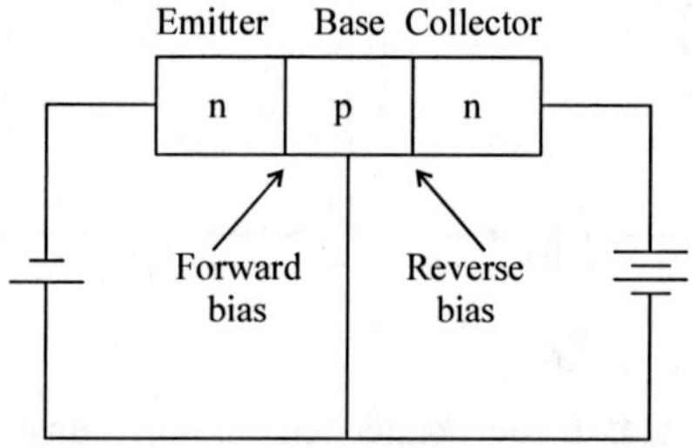

Fig. 9.7

Function

Fig. 9.7 shows the npn transistor with forward bias to emitter-base junction and reverse bias to collector-base junction. The forward

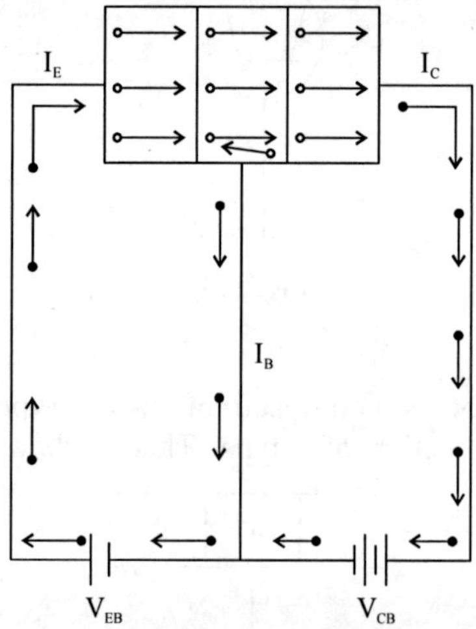

Fig. 9.8. Basic connection of npn transistor

bias causes the electrons in the n-type emitter to flow towards the base. This forms the emitter current I_E. As these electrons flow through the p-type base, they tend to combine with holes. As the base is very thin and lightly doped, only a few electrons (<5%) combine with holes to give base current I_B. The remainder (>95%) electrons go to the collector to form collector current I_c. In this way, almost the entire emitter current flows in the collector circuit. It is clear that the emitter current is the sum of collector current and base current *i.e.*, $I_E = I_B + I_C$.

Symbol of npn transistor

The symbol of npn transistor used in electronic circuits is shown in fig. 9.9. The arrow mark in the emitter indicates the direction of conventional current flow with forward bias. The conventional current in case of npn transistor flows out of the emitter as indicated by the out going arrow.

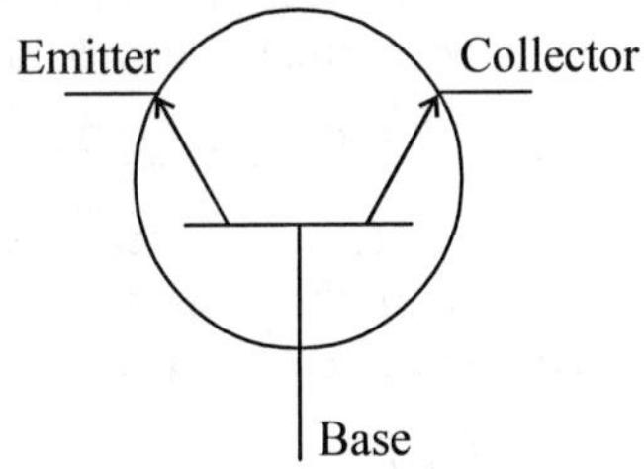

Fig. 9.9

Q. 3. Name three transistor connections.

Ans. Three transistor connections are :

(a) common base connection
(b) common emitter connection
(c) common collector connection

Q. 4. Define current amplification factor α. Establish the relation

$$I_c = \frac{\alpha}{1-\alpha} I_B + \frac{I_{CBO}}{1-\alpha}.$$

Ans.

Current amplification factor (α)

The current amplification factor 'α' is defined as the ratio of change in collector current to the change in emitter current at constant collector base voltage. Mathematically, it can be expressed as

$$\alpha = \frac{\Delta I_C}{\Delta I_E} \text{ at constant } V_{CB} \qquad ...(1)$$

The current amplification factor 'α' is always less than unity. This is because the resistance of the output circuit is much higher than the resistance of the input circuit. So ΔI_C is always less than ΔI_E. Hence the ratio $\frac{\Delta I_C}{\Delta I_E}$ is less than unity. The value of 'α' can be increased (but not more than unity) by decreasing the base current. This is achieved by making the base thin and lightly doped. The values of 'α' remain within 0.9 and 0.99.

Expression for collector current

Entire emitter current can't reach the collector. Due to electron-hole combinations occurring in base area, base current is developed. This base current is a small part of emitter current. Moreover, as the collector-base junction is reverse biased, therefore, some leakage current flows due to minority carriers. Thus it follows that total collector current is the sum of the part of emitter current reaching the collector and the leakage current. So we can write

total collector current, $I_C = \alpha I_E + I_{leakage}$ ...(2)

It is clear that if $I_E = 0$ (*i.e.*, emitter circuit is open), a small leakage current still flows in the collector circuit. This $I_{leakage}$ is abbreviated as I_{CBO}, meaning collector-base current with emitter open.

$$\therefore \quad I_C = \alpha I_E + I_{CBO} \qquad ...(3)$$

Now, $I_E = I_C + I_B$

$$\therefore \quad I_C = \alpha (I_C + I_B) + I_{CBO}$$

$$\Rightarrow I_C - \alpha I_C = \alpha I_B + I_{CBO}$$

$$\Rightarrow I_C (1 - \alpha) = \alpha I_B + I_{CBO}$$

$$\Rightarrow I_C = \frac{\alpha}{1-\alpha} I_B + \frac{I_{CBO}}{1-\alpha} \qquad ...(4)$$

Relation (3) or (4) can be used to find I_c. It is further clear from these relations that the collector current of a transistor can be controlled by either the emitter or base current.

Q. 5. How can you draw transistor characteristics in CB connection. Explain with the help of circuit diagram.

Ans. The complete behaviour of a transistor in an electrical circuit can be described with the help of the various current-voltage curves and relations. The curves so obtained are known as transistor characteristics. In case of common base connection, the most important characteristics are input characteristics and output

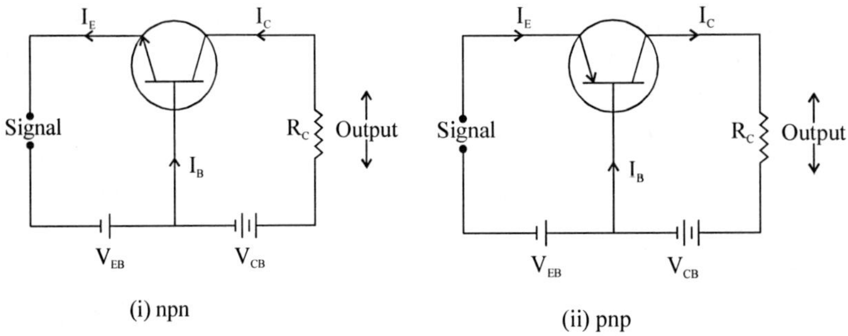

Fig. 9.10

characteristics. These characteristics can be drawn with the help of the circuit connections for npn and pnp transistors.

Input characteristics

It is the curve between emitter current I_E and emitter-base voltage V_{EB} at constant collector-base voltage V_{CB}. The emitter current is generally taken along Y-axis and emitter-base voltage along X-axis. Fig. 9.11 shows the input characteristics of a typical transistor in *CB* arrangement.

The following points may be noted from these characteristics :

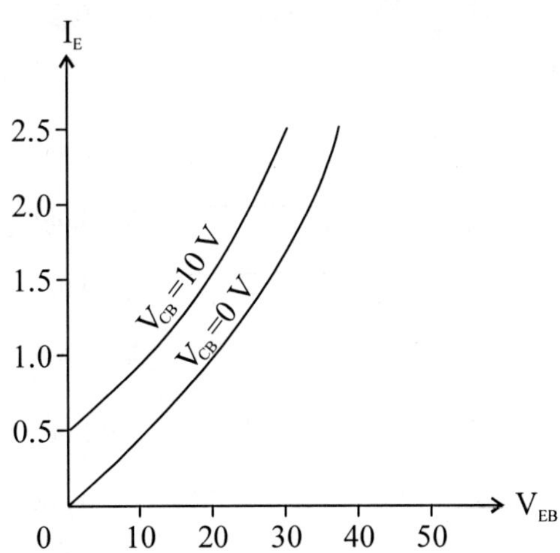

Fig. 9.11. Emitter-base voltage versus emitter current

(i) The emitter current I_E increases rapidly with small increase in emitter-base voltage V_{EB}. It means that input resistance is very small.

(ii) The emitter current is almost independent of collector-base voltage V_{CB}. This leads to the conclusion that emitter current (and hence collector current) is almost independent of collector.

Input resistance

It is the ratio of change in emitter-base voltage (ΔV_{EB}) to the resulting change in emitter current (ΔI_E) at constant collector-base voltage (V_{CB}) *i.e.*,

$$\text{Input resistance, } r_i = \frac{\Delta V_{EB}}{\Delta I_E} \text{ at constant } V_{CB}.$$

In fact, input resistance is the opposition offered to the signal current. As a very small V_{EB} is sufficient to produce a large flow of emitter current I_E, therefore, input resistance is quite small, of the order of a few ohms.

Output characteristics

It is the curve between collector current I_C and collector-base voltage V_{CB} at constant emitter current I_E. Generally, collector current is taken along Y-axis and collector-base voltage along X-axis. Fig. 9.12 shows the output characteristics of a typical transistor in CB arrangement.

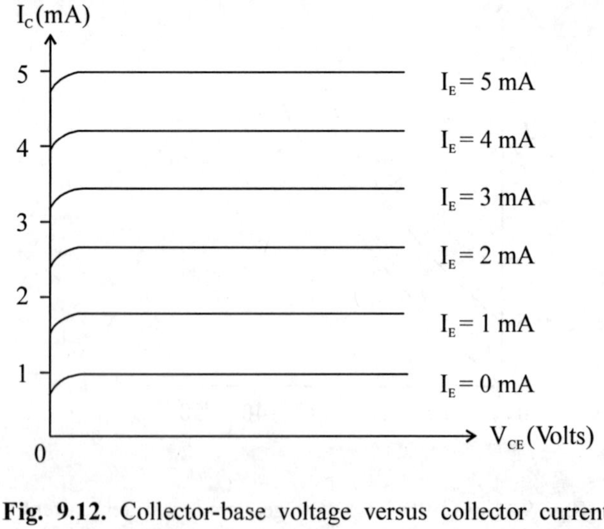

Fig. 9.12. Collector-base voltage versus collector current

The following points may be noted from the characteristics :

(i) The collector current I_C varies with V_{CB} only at very low voltage (< 1V). The transistor is never operated in this region.

(ii) When the value of V_{CB} is raised above 1-2V, the collector current becomes constant as indicated by straight horizontal curves. It means that now I_C is independent of V_{CB} and depends upon I_E only. This is consistent with the theory that the emitter current flows almost entirely to the collector terminal. The transistor is always operated in this region.

(iii) A very large change in collector-base voltage produces only a tiny change in collector current. This means that output resistance is very high.

Output resistance

It is the ratio in collector base voltage (ΔV_{CB}) to the resulting change in collector current (ΔI_C) at constant emitter current *i.e.*,

$$\text{output resistance, } r_0 = \frac{\Delta V_{CB}}{\Delta I_C} \text{ at constant } I_E$$

The output resistance of CB circuit is very high, of the order of several tens of kilo-ohms. This is not surprising because the collector current changes very slightly with the change in V_{CB}.

Q. 6. Define current amplification factor β for CE transistor connection. Establish a relation between α and β.

Ans. *Current amplification factor (β) :* The ratio of change in collector current (ΔI_C) to the change in base current (ΔI_B) is known as base current amplification factor *i.e.*,

$$\beta = \frac{\Delta I_C}{\Delta I_B}$$

In almost any transistor, less than 5% of emitter current flows as the base current. Therefore, the value of β is generally greater than 20. Usually, its value ranges from 20 to 500. This type of connection is frequently used as it gives appreciable current gain as well as voltage gain.

Relation between β and α

A simple relation exists between α and β. This can be derived as follows :

$$\beta = \frac{\Delta I_C}{\Delta I_B} \qquad \text{...(1)}$$

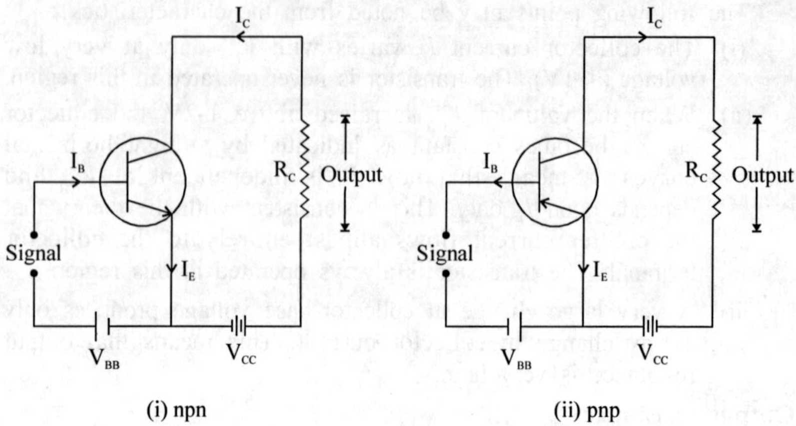

Fig. 9.13

$$\alpha = \frac{\Delta I_C}{\Delta I_E} \qquad ...(2)$$

Now, $I_E = I_B + I_C$

$$\Rightarrow \Delta I_E = \Delta I_B + \Delta I_C$$

$$\Rightarrow \Delta I_B = \Delta I_E - \Delta I_C$$

Substituting the value of ΔI_B in equn. (1) we get

$$\beta = \frac{\Delta I_C}{\Delta I_E - \Delta I_C} \qquad ...(3)$$

Dividing the numerator and denominator of RHS of equn. (3) by ΔI_E, we get

$$\beta = \frac{\Delta I_C / \Delta I_E}{\frac{\Delta I_E}{\Delta I_E} - \frac{\Delta I_C}{\Delta I_E}} = \frac{\alpha}{1-\alpha} \quad \left(\because \alpha = \frac{\Delta I_C}{\Delta I_E}\right)$$

$$\Rightarrow \beta = \frac{\alpha}{1-\alpha}$$

It is clear that as α approaches unity, β approaches infinity. In other words, the current gain in common emitter connection is very high. It is due to this reason that this circuit arrangement is used in about 90 to 95 per cent of all transistor applications.

Q. 7. Find expressions for collector current in CE transistor connection.

Ans. *Expressions for collector current :* In common emitter circuit, I_B is the input current and I_C is the output current.

We know, $I_E = I_B + I_C$ ...(1)

and $I_C = \alpha\, I_E + I_{CBO}$ ...(2)

From equn. (2) we get,

$$I_C = \alpha\, I_E + I_{CBO} = \alpha\,(I_B + I_C) + I_{CBO}$$

or, $I_C\,(1 - \alpha) = \alpha\, I_B + I_{CBO}$

or, $$I_C = \frac{\alpha}{1-\alpha}\, I_B + \frac{1}{1-\alpha}\, I_{CBO} \quad ...(3)$$

From equn. (3), it is apparent that if $I_B = 0$ (*i.e.*, base circuit is open), the collector current will be the current to the emitter. This is abbreviated as I_{CEO} meaning collector-emitter current with base open.

$$I_{CEO} = \frac{1}{1-\alpha} I_{CBO}$$

Substituting the value of $\frac{1}{1-\alpha} I_{CBO} = I_{CEO}$ in equn. (3), we get,

$$I_C = \frac{\alpha}{1-\alpha} I_{CEO} + I_{CEO}$$

or, $I_C = \beta I_B + I_{CEO}$ $\left(\because\ \beta = \frac{\alpha}{1-\alpha}\right)$

Q. 8. With a neat circuit diagram describe the procedure of drawing the characteristic curves of CE transistor connection.

Ans. The input and output characteristics of common emitter transistor connection can be drawn with the help of following circuit arrangement (fig. 9.14).

Input characteristics

It is the curve between base current I_B and base-emitter voltage V_{BE} at constant collector-emitter voltage V_{CE}.

The input characteristics of a CE connection can be obtained with the help of the circuit shown in fig. 9.14. The base current I_B is

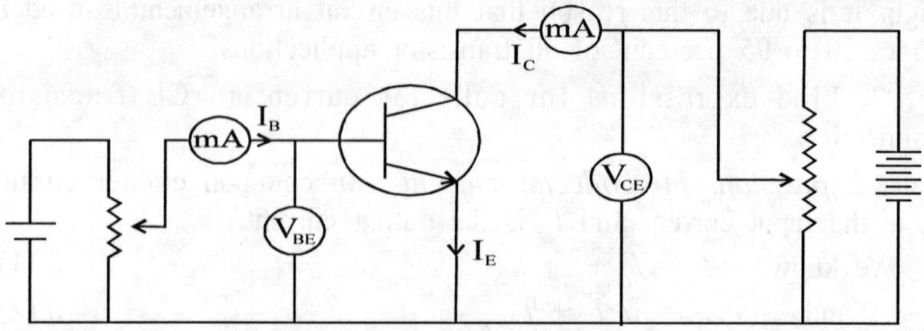

Fig. 9.14. CE npn transistor connection

recorded for different values of V_{BE} with constant V_{CE}. A graph is plotted between I_B and V_{BE} for different constant values of V_{CE} (1V, 10V etc.). The input characteristics so obtained are shown in fig. 9.15.

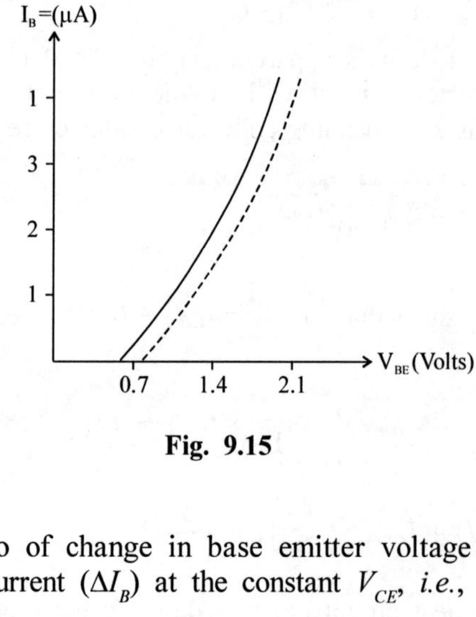

Fig. 9.15

Input resistance

It is the ratio of change in base emitter voltage (ΔV_{BE}) to the change in base current (ΔI_B) at the constant V_{CE}, *i.e.*,

$$\text{Input resistance, } r_i = \frac{\Delta V_{BE}}{\Delta I_B} \text{ at constant } V_{CE}$$

The value of input resistance for a CE circuit is of the order of a few hundred ohms.

Output characteristics

It is the curve between collector current I_C and collector-emitter voltage V_{CE} at constant base current I_B.

With the help of circuit arrangement of fig. 9.14, output characteristics of a CE transistor connection are drawn. Keeping the base current I_B constant, the collector-current I_C is recorded for different values of V_{CE}. A graph is plotted between I_C & V_{CE} with I_B constant. The procedure is repeated for different constant values of I_B. Output characteristic curves are shown in fig. 9.16.

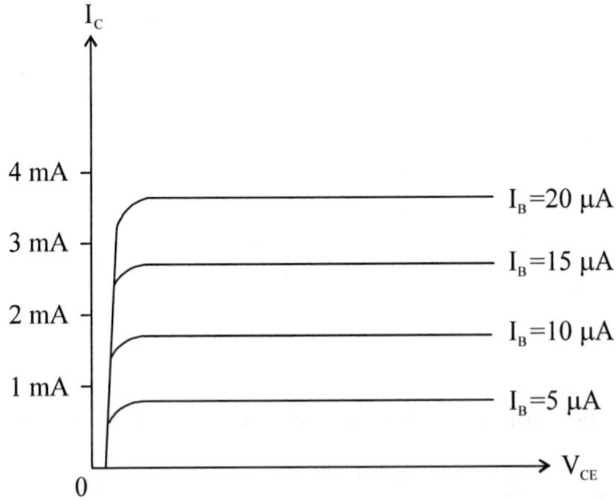

Fig. 9.16

Output resistance

It is the ratio of change in collector-emitter voltage (ΔV_{CE}) to the change in collector current (ΔI_c) at constant I_B, *i.e.*,

$$\text{Output resistance, } r_0 = \frac{\Delta V_{CE}}{\Delta I_C} \text{ at constant } I_B$$

The output resistance of a CE circuit is less than that of CB circuit. Its value is of the order of 50 KΩ.

Q. 9. Define current amplification factor 'γ' for common collector transistor connection. Relate γ with α.

Ans. *Current amplification factor (γ) :* In common collector transistor connection collector is common to input and output circuit. Both npn and pnp transistor connection with collector common to base and emitter are shown in fig. 9.17. The input current is I_B and the output current is I_E. The current amplification factor is defined as the ratio of change in emitter current (ΔI_E) to the change in base current (ΔI_B).

$$\gamma = \frac{\Delta I_E}{\Delta I_B}$$

This circuit provides about the same current gain as the common emitter circuit as $\Delta I_E = \Delta I_C$. However, its voltage gain is always less than 1.

Relation between γ and α

$$\gamma = \frac{\Delta I_E}{\Delta I_B} \qquad \text{...(1)}$$

$$\alpha = \frac{\Delta I_C}{\Delta I_E} \qquad \text{...(2)}$$

Now, $I_E = I_B + I_C$

or, $\Delta I_E = \Delta I_B + \Delta I_C$

or, $\Delta I_B = \Delta I_E - \Delta I_C$

Substituting the value of ΔI_B in equn. (1) we get

$$\gamma = \frac{\Delta I_E}{\Delta I_E - \Delta I_C}$$

Dividing the numerator and denominator on RHS by ΔI_E, we get,

$$\gamma = \frac{\Delta I_E / \Delta I_E}{\frac{\Delta I_E}{\Delta I_E} - \frac{\Delta I_C}{\Delta I_E}} = \frac{1}{1-\alpha} \qquad \left(\because \alpha = \frac{\Delta I_C}{\Delta I_E}\right)$$

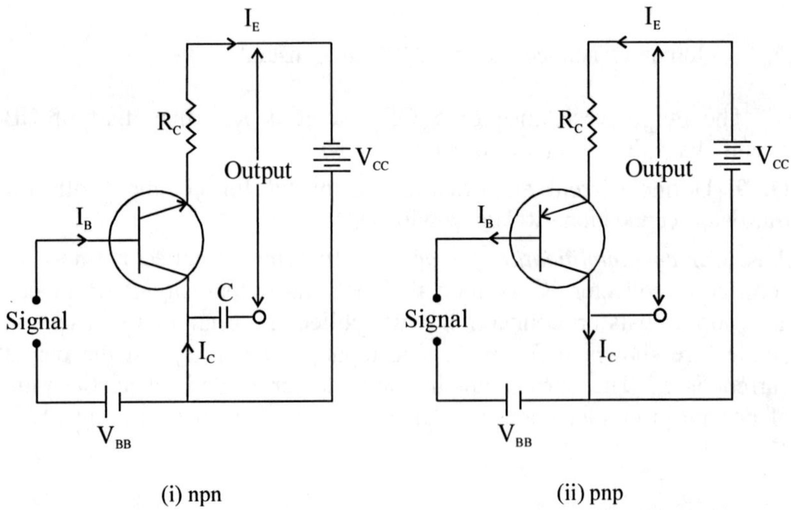

Fig. 9.17

$$\gamma = \frac{1}{1-\alpha} \qquad ...(3)$$

But we have, $\beta = \dfrac{\alpha}{1-\alpha}$

$$\Rightarrow \quad (1 - \alpha)\,\beta = \alpha$$

$$\Rightarrow \quad \beta - \alpha\,\beta - \alpha = 0$$

$$\Rightarrow \quad \beta = \alpha\,(\beta + 1) \quad \Rightarrow \alpha = \frac{\beta}{\beta+1} \qquad ...(4)$$

Substituting equn. (4) in equn. (3) we get,

$$\gamma = \frac{1}{1-\alpha} = \frac{1}{1-\frac{\beta}{1+\beta}}$$

$$= \frac{1+\beta}{1+\beta-\beta} = 1+\beta$$

$$\Rightarrow \quad \gamma = 1 + \beta \qquad ...(5)$$

Q. 10. Find expressions for emitter current in common collector transistor connection.

Ans. We know, $I_C = \alpha\, I_E + I_{CBO}$

Also $\quad I_E = I_B + I_C$

$$= I_B + (\alpha I_E + I_{CBO})$$

$$\therefore \quad I_E\,(1 - \alpha) = I_B + I_{CBO}$$

or, $$I_E = \frac{I_B}{1-\alpha} + \frac{I_{CBO}}{1-\alpha}$$

or, $$I_E = (\beta + 1)\, I_B + (\beta + 1)\, I_{CBO}$$

Q. 11. Mention the uses of common collector transistor connection.

Ans. The common collector circuit has very high input resistance (about 750 KΩ) and very low output resistance (about 25 Ω). As a result of which the voltage gain provided by this circuit is always less than unity. Therefore this circuit arrangement is seldom used for amplification. However, this circuit is primarily used for impedance matching, *i.e.*, for driving a low impedance load from a high impedance source.

Q. 12. Compare the characteristics of CB, CE & CC transistor connections.

Ans. The comparison of various characteristics of the three connections is given below in the tabular form.

S.No.	Characteristics	Common base	Common emitter	Common collector
1.	Input resistance	Low (about 100 Ω)	Low (about 750 Ω)	Very high (about 750 KΩ)
2.	Output resistance	Very high (about 450 KΩ)	High (about 45 KΩ)	Low (about 50 Ω)
3.	Voltage gain	About 150	About 500	Less than 1
4.	Applications	For high frequency applications	For audio frequency applications	For impedance matching

Q. 13. Out of the three transistor connections, which one is the most efficient? Why?

Ans. Out of the three transistor connections, the common emitter circuit is the most efficient. It is used in about 90 to 95 per cent of all transistor applications. The main reasons for this are :

(i) High current gain

In common emitter connection, I_C is the output current and I_B is the input current. The collector current here is expressed as

$$I_C = \beta I_B + I_{CEO}$$

As the value of β is very large, therefore, the output current I_C is much more than the input current I_B. This increases the current gain effectively. The current gain of CE arrangement ranges from 20 to 500.

(ii) High voltage and power gain

As we have seen above, CE arrangement has high current gain. This, is turn, increases the voltage and power gain of CE circuit. In comparison to CB and CC circuits, the common emitter connection has the highest voltage and power gain. For this reason the CE transistor connection is wildly used for amplifying purposes.

(iii) Moderate output to input impedance ratio

In a common emitter circuit, the ratio of output impedance to input impedance is small (about 50). Therefore, this circuit arrangement is suitable for coupling between various transistor stages.

On the other hand, the ratio of output impedance to input impedance is very large in CB and CC transistor connections. Hence coupling doesn't become so effective in these cases. This results in gross mismatching.

Q. 14. Discuss on input and output impedance of three transistor connections.

Ans. *Common base transistor connection :* The input impedance in this case is low, about 50 Ω. The effective input impedance remains practically R_i and is not affected by the resistance connected between the emitter and the base.

The output impedance R_o in this configuration is of the order of 2 mHz. Since the load resistance is parallel to the output impedance, the effective output impedance R_e is calculated as,

$$\frac{1}{R_e} = \frac{1}{R_0} + \frac{1}{R_L}$$

$$= \frac{R_0 + R_L}{R_0 R_L}$$

$$\Rightarrow R_e = \frac{R_0 R_L}{R_0 + R_L} \qquad ...(1)$$

R_e is very nearly equal to the load impedance.

Common-emitter transistor connection

The input impedance in CE connection is relatively higher. It is of the order of 700 – 2000 Ω. The output impedance with no load is about 5×10^5 Ω. This is lower than that in common base configuration. Since the load resistance is usually of the order of 10 KΩ, the effective output impedance is equal to the load resistance.

Common-collector transistor connection

The input impedance of common collector transistor connection is dependent on the value of the load resistance in the emitter circuit. The input impedance of this configuration is calculated with the help of the circuit arrangement shown in fig. 9.18.

$$\text{Input impedance } R_i = \frac{\text{Input voltage}}{\text{Input current}}$$

$$= \frac{V_i}{I_B}$$

$$V_i = V_{EB} + V_o$$

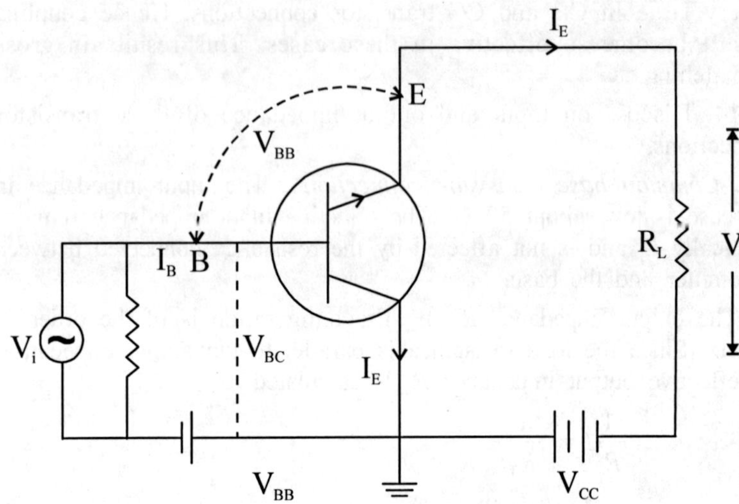

Fig. 9.18

Where V_o is always less than V_i

So, $$R_i = \frac{V_{EB} + V_0}{I_B}$$

$$= \frac{V_{EB}}{I_B} + \frac{V_0}{I_B} \qquad \text{...(2)}$$

Now, $\frac{V_{EB}}{I_B}$ is the internal resistance between base and emitter, denoted by r_{BE}. As this junction is forward biased r_{BE} has a very low value. On the other hand, $\frac{V_0}{I_B}$ depends upon the circuit as well as the transistor.

Putting $\dfrac{I_E}{I_B} = \beta + 1$

We have, $\dfrac{V_0}{I_B} = \dfrac{V_0}{\frac{I_E}{\beta+1}} = \dfrac{(\beta+1)\, V_0}{I_E} = (\beta+1)\, R_L$

Thus, $R_i = r_{BE} + (\beta + 1)\, R_L$ ...(3)

As r_{BE} is very small, it can be neglected in comparison with the other term. So

$$R_i = (\beta + 1)\, R_L \qquad \text{...(4)}$$

This shows that a high value of load resistance is required in order to obtain a high input impedance.

The output impedance of this configuration is low. In comparison to other configurations, the output impedance of common collector transistor circuit is the lowest.

Q. 15. What do you mean by d.c. and a.c. load lines in transistor analysis?

Ans. In drawing output characteristic in common emitter transistor connection, the collector current is plotted against different collector emitter voltages. From the output characteristics, the collector current can be obtained at any desired collector-emitter voltage. The other method which is more convenient is called the load line method. This method can be used to solve the above problems in the transistor analysis.

D.C. load line

Let us consider a common emitter npn transistor circuit shown in fig. 9.19(a) below where no signal is applied. Therefore, d.c. conditions prevail in the circuit. The output characteristics of the circuit are shown in fig. 9.19(b).

The value of collector-emitter voltage V_{CE} at any time is given by,

$$V_{CE} = V_{CC} - I_C R_C \qquad \text{...(1)}$$

As V_{CC} and R_C are fixed values, therefore, it is a first degree equation and can be represented by a straight line on the output characteristics. This is known as d.c. load line and determines the locus of $V_{CE} - I_C$ points for any given value of R_C. To add load line, we need two end points of the straight line. These two points can be located as under :

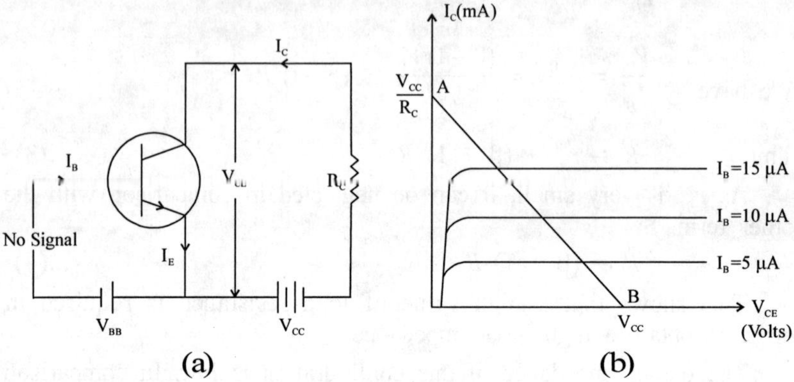

Fig. 9.19

(i) When the collector current $I_C = 0$, then collector-emitter voltage is maximum and is equal to V_{CC} *i.e.*,

$$\text{Max. } V_{CE} = V_{CC} - I_C R_C$$
$$= V_{CC} \qquad (\because I_C = 0)$$

This gives the first point B (OB = V_{CC}) on the collector-emitter voltage axis as shown in fig. 9.19(b).

(ii) When collector-emitter voltage $V_{CE} = 0$, the collector current is maximum and is equal to V_{CC}/R_C *i.e.*,

$$V_{CE} = V_{CC} - I_C R_C$$

or, $$0 = V_{CC} - I_C R_C$$

or, $$I_C = \frac{V_{CC}}{R_C}$$

$$\text{Max. } I_C = \frac{V_{CC}}{R_C}$$

This gives the second point A (OA = V_{CO}/R_C) on the collector current axis as shown in fig. 9.19(b). By joining these two points, d.c. load line AB is constructed.

A.C. load line

It is a line drawn on the output characteristic of a transistor circuit which gives the value of i_C and V_{CC} corresponding to an applied a.c. signal for a fixed value of V_{CC} and load.

The d.c. load line is drawn taking into account the resistance connected to the collector. This we have regarded so far as the load resistance. But in a practical amplifier, the output of the amplifier derived between the collector and ground is applied into some other circuit or equipment through a coupling device which is usually a capacitor. The effective load is therefore parallel combination of the impedance (It may be reactive or resistive) of the equipment or the circuit and the collector resistance. Therefore, when a.c. signal is applied, load R_L is replaced by the effective or load resistance R_{eff} or R_{AC}.

Thus to draw the load line under a.c. conditions (when signal is applied) R_{eff} is used to calculate the maximum collector current. The two fixed point being located, the load line is drawn. It is obvious that the slope of a.c. load line will differ from that of the d.c. load line. But the operating point is common to both of them. It is the a.c. load line which is used in the design and analysis of a practical amplifier.

The line AB constructed in fig. 9.19(b) above is called a load line because the resistance R_C connected to the device is called load or load resistance for the circuit.

Q. 16. What do you mean by operating point or Q point?

Ans. *Operating point or Q-point.* The zero signal values of I_C and V_{CE} are known as the operating points.

It is called operating point because the variations of I_C and V_{CE} take place about this point when signal is applied. It is also called quiescent (silent) point or Q-point because it is the point on $I_C - V_{CE}$ characteristic when the transistor is silent *i.e.*, in the absence of signal.

Let us suppose that in the absence of signal the base current is 5μA. Then I_C and V_{CE} conditions in the circuit must be represented by some point on I_B = 5μA characteristic. But I_C and V_{CE} conditions in the circuit should also be represented by some point on the d.c. load line AB. The point Q where the load line and the characteristic intersect is the only point which satisfies both these conditions. Therefore, the point Q describes the actual state of affairs in the circuit in the zero signal conditions and is called the operating point. Referring to fig. 9.20 for I_B = 5μA, the zero signal values are

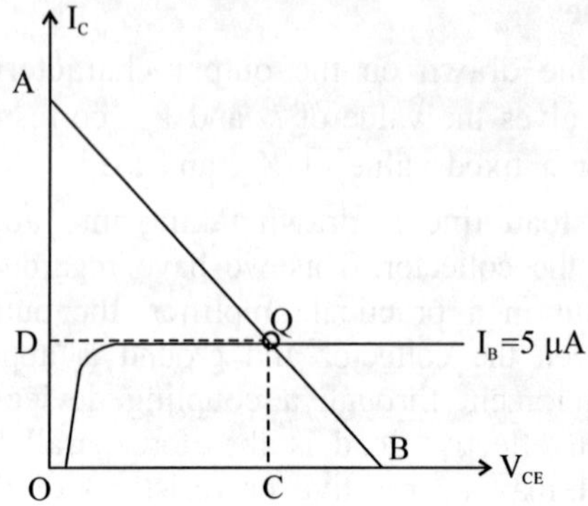

Fig. 9.20

$$V_{CE} = \text{OC volts}$$
$$I_C = \text{OD mA}$$

It follows, therefore, that the zero signal values of I_C and V_{CE} (*i.e.*, operating point) are determined by the point where a.c. load line intersects the proper base current curve.

Q. 17. What do you mean by transistor biasing?

Ans. *Transistor biasing* : The proper flow of zero signal collector current and the maintenance of proper collector-emitter voltage during the passage of signal as known as transistor biasing.

Explanation

For faithful amplification, a transistor amplifier must satisfy three basic conditions namely :

(i) Proper zero signal collector current
(ii) Proper base-emitter voltage at any instant
(iii) Proper collector-emitter voltage at any instant

When these conditions are fulfilled, it is known as transistor biasing. The basic purpose of transistor biasing is to keep the base-emitter function properly forward biased and collector-base junction properly reverse biased during the application of signal. The circuit which provides transistor biasing is known as biasing circuit. It should be noted that transistor biasing is very much necessary for the proper operation of transistor in a circuit.

Q. 18. What do you mean by stabilisation of operating point? Why it is necessary?

Ans. The transistor parameters such as β, V_{BE} etc. vary with the change of temperature and with the replacement of transistor. The inherent variations of transistor parameters may change the operating point, resulting in unfaithful amplification. It is therefore very important for stabilising the operating point which should be independent of variations in transistor parameters.

Stabilisation

The process of making operating point independent of temperature changes or variations in transistor parameters is known as stabilisation.

Need for stabilisation

Stabilisation of the operating point is needed because of the following reasons :

(i) Temperature dependence of I_C.

(ii) Individual variations.

(iii) Thermal runway.

(i) Temperature dependence of I_C

The collector current I_C is given by,

$$I_C = \beta I_B + I_{CEO} = \beta I_B + (\beta + 1)\ I_{CBO}$$

The collector leakage current I_{CBO} is greatly influenced (especially in germanium transistor) by temperature changes. A rise of 10°C doubles the collector leakage current which may be as high as 0.2 mA for low powered germanium transistors. As biasing conditions in such transistors are generally so set that zero signal I_C = 1 mA, therefore, the change in I_C due to temperature variations can't be tolerated. This necessitates to stabilise the operating point *i.e.*, to hold I_C constant in spite of temperature variations.

(ii) Individual variations

The value of β and V_{BE} are not exactly the same for any two transistors even of the same type. Further, V_{BE} itself decreases when temp. increases. When a transistor is replaced by another of the same type, these variations change the operating point. This necessitates to stabilise the operating point *i.e.*, to hold I_C constant irrespective of individual variations in transistor parameters.

(iii) Thermal runway

The collector current for a CE configuration is given by,

$$I_C = \beta I_B + (\beta + 1)\ I_{CBO} \qquad ...(1)$$

The collector leakage current I_{CBO} is strongly dependent on temperature. The flow of collector current produces heat within the transistor. This raises the transistor temp. and if no stabilisation is done, the collector leakage current I_{CBO} also increases. It is clear from equn. (1) that if I_{CBO} increases, the collector current I_C increases by $(\beta + 1)\ I_{CBO}$. The increased I_C will raise the temp. of the transistor, which in turn will cause I_{CBO} to increase. This effect is cumulative and in a matter of seconds, the collector current may become very large, causing the transistor to burn out.

The self-destruction of an unstabilised transistor is known as thermal runway.

In order to avoid thermal runway and consequent destruction of transistor, it is very essential that operating point is stabilised *i.e.*, I_C is kept constant. In practice, this is done by causing I_B to decrease automatically with temp. increase by circuit modification. Then decrease in βI_B will compensate for the increase in $(\beta + 1)\ I_{CBO}$. Keeping I_C nearly constant. In fact, this is what is always aimed at while building and designing a biasing circuit.

Q. 19. What are the essentials of a transistor biasing circuit?

Ans. The essentials of a transistor biasing circuit in order to give faithful amplification are :

(i) It should ensure proper zero signal collector current.

(ii) It should ensure that V_{CE} doesn't fall below 0.5 volt for silicon transistors at any instant.

(iii) I should ensure the stabilisation of operating point.

Q. 20. Define stability factor. Find an expression for stability factor in CE configuration.

Ans. *Stability factor* : The rate of change of collector current I_C w.r.t. the collector leakage current I_{CO} at constant β and I_B is called stability factor.

$$\text{Stability factor, } S = \frac{dI_C}{dI_{CO}} \text{ at constant } I_B \text{ and } \beta$$

Expression for stability factor in CE connection

The general expression of stability factor for a CE configuration can be obtained as under,

$$I_C = \beta\ I_B + (\beta + 1)\ I_{CO}$$

Differentiating above expression w.r.t. I_C, we get

$$I = \beta \frac{dI_B}{dI_C} + (\beta + 1)\ \frac{dI_{CO}}{dI_C}$$

or, $$I = \beta\frac{dI_B}{dI_C} + \frac{(\beta+1)}{S} \qquad \left[\because \frac{dI_{CO}}{dI_C} = \frac{1}{S}\right]$$

or, $$S = \frac{\beta+1}{1-\beta\frac{(dI_B)}{(dI_C)}}$$

Q. 21. Discuss different methods of transistor biasing.

Ans. There are three different methods of transistor biasing. These are :

(i) Base resistor method
(ii) Feedback resistor method
(iii) Potential divider method

(i) Base resistor method

In this method, a high resistance R_B (several hundred KΩ) is connected between the base and +ve end of supply for npn transistor (fig. 9.21) and between base and –ve end of supply for npn transistor. Here the required zero signal base current is provided by V_{CC} and it flows through R_B. It is because now base is +ve w.r.t. emitter *i.e.*, base emitter junction is forward biased. The required value of zero signal base current I_B (and hence $I_C = \beta I_B$) can be made to flow by selecting the proper value of base resistor R_B.

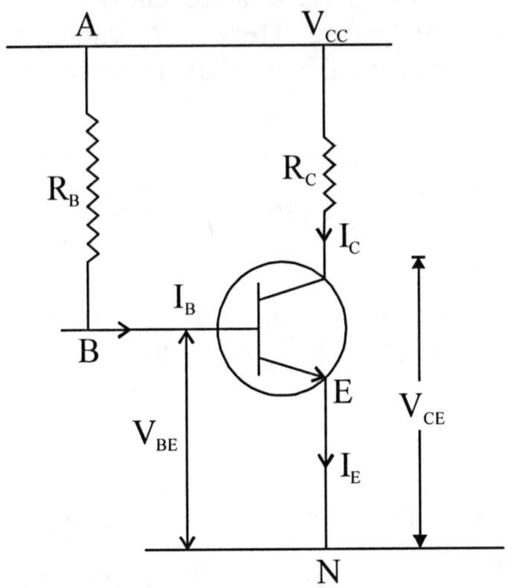

Fig. 9.21. npn CE transistor connection

Circuit analysis

It is required to find the value of R_B so that required collector current flows in the zero signal conditions. Let I_C be the required zero signal collector current.

$$I_B = \frac{I_C}{\beta}$$

Considering the closed circuit ABENA and applying Kirchhoff's voltage law, we get

$$V_{CC} = I_B R_B + V_{BE}$$

or, $$I_B R_B = V_{CC} - V_{BE}$$

$$\therefore \quad R_B = \frac{V_{CC} - V_{BE}}{I_B} \qquad \text{...(1)}$$

As V_{CC} and I_B are known and V_{BE} can be seen from the transistor manual, therefore, value of R_B can be readily found from equn. (1).

Since V_{BE} is generally quite small as compared to V_{CC}, the former can be neglected with little error. It then follows from equn. (1) that :

$$R_B = \frac{V_{CC}}{I_B}$$

It may be noted that V_{CC} is a fixed known quantity and I_B is chosen at some suitable value. Hence, R_B can always be found directly, and for this reason, this method is sometimes called fixed bias method.

$$\text{Stability factor, } S = \frac{\beta + 1}{1 - \beta \dfrac{(dI_B)}{(dI_C)}}$$

In fixed-bias method of biasing, I_B is independent of I_C so that $dI_B/dI_C = 0$. Putting the value of $dI_B/dI_C = 0$ in the above expression, we have,

stability factor, $S = \beta + 1$

Thus the stability factor in a fixed bias is $(\beta + 1)$. This means that I_C changes $(\beta + 1)$ times as much as any change in I_{CO}.

Advantages

(i) This biasing circuit is very simple as only one resistance R_B is required.

(ii) Biasing conditions can easily be set and the calculations are simple.

(iii) There is no loading of the source by the biasing circuit since no resistor is employed across base-emitter junction.

Disadvantages

(i) This method provides poor stabilisation. It is because there is no means to stop a self-increase in collector current due to temp. rise and individual variations. For example, if β increases due to transistor replacement, then I_C also increases by the same factor as I_B is constant.

(ii) The stability factor is very high. Therefore, there are strong chances of thermal runway.

Due to these disadvantages, this method of biasing is rarely employed.

(ii) Feedback resistor method

Transistor biasing with feedback resistor is done by connecting one end of R_B to the base and the other end to the collector as shown in the fig. 9.22. Here, the required zero signal base-current is determined by the collector-base voltage V_{CB}. As V_{CB} forward biases,

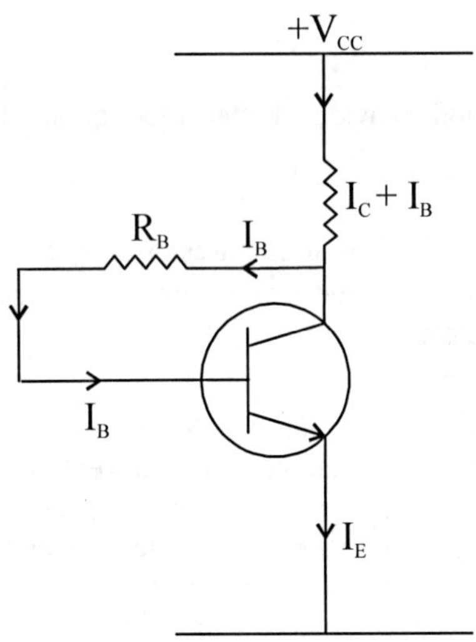

Fig. 9.22

the base-emitter junction and hence base current I_B flows through R_B. Consequently collector current flows in the circuit with zero signal.

Circuit analysis

The required value of R_B needed to give the zero signal current I_C can be determined as follows. Referring to fig. 9.22,

$$V_{CC} = I_C R_C + I_B R_B + V_{BE}$$

$$\text{or,} \quad R_B = \frac{V_{CC} - V_{BE} - I_C R_C}{I_B}$$

$$= \frac{V_{CC} - V_{BE} - \beta I_B R_C}{I_B} \qquad (\because I_C = \beta I_B)$$

Alternatively

$$V_{CE} = V_{BE} + V_{CB}$$

$$\text{or,} \quad V_{CB} = V_{CE} - V_{BE}$$

$$\therefore \quad R_B = \frac{V_{CB}}{I_B} = \frac{V_{CE} - V_{BE}}{I_B} \qquad \text{where } I_B = \frac{I_C}{\beta}$$

It can be shown mathematically that stability factor S for this method of biasing is less than (β + 1) *i.e.*,

Stability factor, $S < (\beta + 1)$

Therefore this method provides better thermal stability than the fixed bias.

Advantages

(i) It is a simple method as it requires only one resistance R_B.

(ii) This circuit provides some stabilisation of the operating point as discussed below :

$$V_{CE} = V_{BE} + V_{CB}$$

Suppose the temperature increases. This will increase collector leakage current and hence the total collector current. But as soon as collector current increases, V_{CE} decreases due to greater drop across R_C. The result is that V_{CB} decreases *i.e.*, lesser voltage is available across R_B. Hence the base current I_B decreases. The smaller I_B tends to decrease the collector current to original value.

Disadvantages

(i) The circuit does not provide good stabilisation because stability factor is fairly high though it is lesser than that of fixed bias.

Therefore, the operating point does change, although to lesser extent, due to temperature variations and other effects.

(ii) This circuit provides a negative feedback which reduces the gain of the amplifier as explained hereafter. During the +ve half-cycle of the signal, the collector current increases. The increased collector current would result in greater voltage drop across R_C. This will reduce the base current and hence collector current.

(iii) Potential divider method

This method of transistor biasing provides stabilisation to a transistor and hence it is widely used. Here, two resistances R_1 and R_2 are connected across the supply voltage V_{CC} and provide necessary biasing. The emitter resistance R_E maintains stabilisation. Since R_1 and R_2 form the potential divider, the name "potential divider" is used. The voltage drop across R_2 supplies forward biasing to the base emitter junction. Consequently, the base current is caused and hence collector current flows with zero signal. The circuit arrangement with potential divider R_1 and R_2 is shown in fig. 9.23.

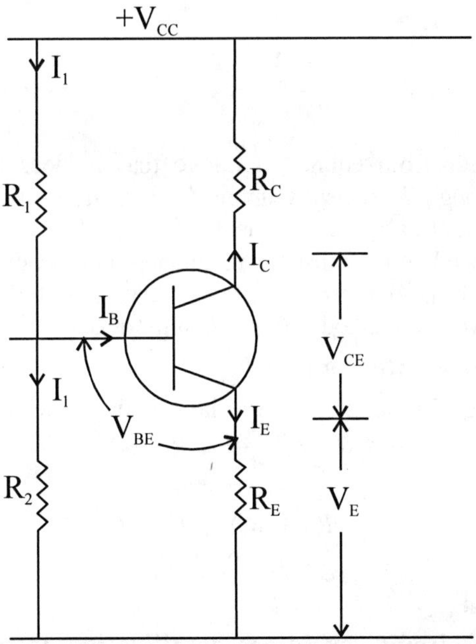

Fig. 9.23

Circuit analysis

Suppose that the current flowing through resistance R_1 is I_1. As base current I_B is very small, therefore, it can be assumed with reasonable accuracy that current flowing through R_2 is also I_1.

(i) Collector current I_C

$$I_1 = \frac{V_{CC}}{R_1 + R_2}$$

$\therefore$ voltage across resistance R_2

$$V_2 = \left(\frac{V_{CC}}{R_1 + R_2}\right) R_2$$

Applying Kirchhoff's voltage law to the base circuit of fig. 9.23.

$$V_2 = V_{BE} + V_E$$

or, $$V_2 = V_{BE} + I_E R_E$$

or, $$I_E = \frac{V_2 - V_{BE}}{R_E}$$

Since $I_E = I_C$

$$I_C = \frac{V_2 - V_{BE}}{R_E} \qquad ...(1)$$

It is clear from equn. (1) above that I_C does not at all depend upon β. Though I_C depends upon V_{BE} but in practice $V_2 >> V_{BE}$, so that I_C is practically independent of V_{BE}. Thus I_C in this circuit is almost independent of transistor parameters and hence good stabilisation is ensured. It is due to this reason that potential divider bias has become universal method for providing transistor biasing.

(ii) Collector-emitter voltage V_{CE}

Applying Kirchhoff's voltage law to the collector side,

$$V_{CC} = I_C R_C + V_{CE} + I_E R_E$$

$$= I_C R_C + V_{CE} + I_C R_E \qquad (\because I_E = I_C)$$

$$= I_C (R_C + R_E) + V_{CE}$$

$$\therefore V_{CE} = V_{CC} - I_C (R_C + R_E)$$

Stabilisation

In this circuit, excellent stabilisation is provided by R_E. Consideration of equn. (1) reveals this fact,

$$V_2 = V_{BE} + I_C R_E$$

Suppose the collector current I_C increases due to rise in temperature. This will cause the voltage drop across emitter resistance R_E to increase. As voltage drop across R_2 (*i.e.*, V_2) is independent of I_C, therefore, V_{BE} decreases. This in turn causes I_B to decrease. The reduced value of I_B tends to restore I_C to the original value.

Stability factor

It can be shown mathematically that stability factor of the circuit is given by,

$$\text{stability factor,} \quad S = \frac{(\beta+1)\ (R_T + R_E)}{R_T + R_E + \beta R_E}$$

$$= (\beta+1) \times \frac{1+\frac{R_T}{R_E}}{\beta+1+\frac{R_T}{R_E}}$$

$$\text{where} \quad R_T = \frac{R_1 R_2}{R_1 + R_2}$$

If the ratio R_T/R_E is very small, then R_T/R_E can be neglected as compared to 1 and the stability factor becomes :

$$\text{Stability factor} = (\beta+1) \times \frac{1}{\beta+1} = 1$$

This is the smallest possible value of S and leads to the maximum possible thermal stability.

Low value of R_T can be obtained by making R_2 very small. When R_2 has low value, the current drawn from V_{CC} will have large value. This puts restrictions on the choice of R_T. Increasing the value of R_E requires greater V_{CC} in order to maintain the same value of zero signal collector current. As a result of which the ratio R_T/R_E can't be made very small from design considerations. The value of R_T/R_E can't be neglected as compared to one (1). In actual practice the stability factor of the circuit may be around 10.

EXERCISES

Numerical Problems with answers

1. In a common base connection, the emitter current is 1.5 mA. Find the base and collector currents if the base current amplification factor is 0.94.

Solution

$$\alpha = 0.94,\ I_E = 1.5 \text{ mA}$$
$$I_B = ?,\ I_C = ?$$

Here, $\alpha = \dfrac{I_C}{I_E}$ or $I_C = \alpha I_E$

$$\therefore I_C = 0.94 \times 1.5 \text{ mA}$$
$$= 1.410 \text{ mA} = 1.41 \text{ mA}$$

Base Current $I_B = I_E - I_C$

$$= 1.5 - 1.41$$
$$= 0.09 \text{ mA}$$

2. In a common base connection, $I_E = 1$ mA, $I_C = 0.95$ mA. Calculate the value of I_B.

Solution

$$I_E = 1 \text{ mA},\quad I_C = 0.95 \text{ mA}$$
$$I_E = I_B + I_C \quad \text{or} \quad I_B = I_E - I_C$$

or $I_B = 1 \text{ mA} - 0.95 \text{ mA}$

$$= 0.05 \text{ mA}$$

3. In a common base connection, the collector current is 0.95 mA and base current is 0.05 mA. Calculate the value of base current amplification factor α.

Solution

$$I_C = 0.95 \text{ mA},\ I_B = 0.05 \text{ mA}$$
$$\alpha = ?$$

Here, $I_E = I_B + I_C$

$$= 0.05 + 0.95 = 1 \text{ mA}$$

$$\alpha = \frac{I_C}{I_E} = \frac{0.95}{1} = 0.95$$

4. In a common base connection, the collector current is 0.9 mA and the emitter current is 1 mA. Find the current amplification factor α.

Solution

$$I_C = 0.9 \text{ mA},\ I_E = 1 \text{ mA}$$
$$\alpha = ?$$

Here, $\alpha = \dfrac{I_C}{I_E} = \dfrac{0.9}{1} = 0.9$

5. In a transistor if $I_C = 4.9$ mA and $I_E = 5$ mA. What is the value of α ?

Solution

$$I_C = 4.9 \text{ mA}, \; T_E = 5 \text{ mA}$$
$$\alpha = ?$$

Here, $$\alpha = \frac{I_C}{I_E} = \frac{4.9}{5} = 0.98$$

6. In a transistor circuit $I_E = 1$ mA, $I_C = 0.94$ mA. What is the value of $I_B = ?$

Solution

$$I_E = 1 \text{ mA}, \; I_C = 0.94 \text{ mA}$$
$$I_B = ?$$

We have, $$I_E = I_B + I_C$$

or, $$I_B = I_E - I_C$$
$$= 1 - 0.94$$
$$= 0.06 \text{ mA}$$

7. In a common base connection, the emitter current is 1 mA. If the emitter circuit is open, the collector current is 50 μA. Find the total collector current when $\alpha = 0.92$.

Solution

$$I_E = 1 \text{ mA}, \; \alpha = 0.92, \; I_{CBO} = 50 \; \mu\text{A}$$

Here, $$\text{I}_\text{C} = \alpha \, I_E + I_{CBO}$$
$$= 0.92 \times 1 + 50 \times 10^{-3}$$
$$= 0.92 + 50 \times 10^{-3}$$
$$= 0.92 + 0.05$$
$$= 0.97 \text{ mA}$$

8. Find the value of β if $\alpha = 0.99$.

Solution

$$\alpha = 0.99, \quad \beta = ?$$

We have $$\beta = \frac{\alpha}{1-\alpha}$$

$$= \frac{0.99}{1-0.99} = \frac{0.99}{0.01}$$

$$= 99$$

9. Calculate the value of β if : (i) $\alpha = 0.9$ and (ii) $\alpha = 0.98$.

Solution

$$\text{(i)}\quad \beta = \frac{\alpha}{1-\alpha} = \frac{0.9}{1-0.9} = \frac{0.9}{0.1} = 9$$

$$\text{(ii)}\quad \beta = \frac{\alpha}{1-\alpha} = \frac{0.98}{1-0.98} = \frac{0.98}{0.02} = 49$$

10. In a transistor $\beta = 50$ and $I_B = 10\ \mu A$. Find I_E.

Solution

$$\beta = 50,\ I_B = 10 \times 10^{-3}\text{ mA} = 0.01\text{ mA},\ I_E = ?$$

We have $\quad \beta = \dfrac{I_C}{I_B}$

or, $\quad I_C = \beta\, I_B = 50 \times 10 \times 10^{-3}$ mA
$= 500 \times 10^{-3}$ mA
$= 0.5$ mA

Using the relation, $I_E = I_B + I_C$
$= .01 + 0.50$
$= 0.51$ mA

11. In a transistor, $\beta = 45$, the voltage across 5 KΩ resistance which is connected in the collector circuit is 5 volts. Find the base current.

Solution

$$R_C = 5\text{ K}\Omega = 5 \times 10^3\ \Omega$$

Voltage drop across $R_C = 5$ volts

$$\beta = 45,\ I_B = ?$$

We have $\quad I_C = \dfrac{5}{5\times10^3}$ A $= 10^3$ A $= 1$ mA

But, $\quad \beta = \dfrac{I_C}{I_B}$ or $I_B = \dfrac{I_C}{\beta} = \dfrac{1}{45} = 0.022$ mA

12. In a transistor circuit, collector load is 2 KΩ whereas quiescent current (Zero signal collector current) is 1 mA.

(i) What is the operating point if $V_{CC} = 10\ V$?

(ii) What will be the operating point if $R_C = 4$ KΩ ?

Solution

$V_{CC} = 10V, \quad I_C = 1 \text{ mA}$

(i) When collector load $R_C = 2 \text{ K}\Omega$

$$V_{CE} = V_{CC} - I_C R_C$$
$$= 10 - 1 \times 10^{-3} \times 2 \times 10^3$$
$$= 10 - 2 = 8V$$

$\therefore$ Operating point is 8V, 1 mA.

(ii) When collector load $R_C = 4 \text{ K}\Omega$

$$V_{CE} = V_{CC} - I_C R_C$$
$$= 10 - 1 \times 10^{-3} \times 4 \times 10^3$$
$$= 10 - 4 = 6V$$

$\therefore$ Operating point is 6V, 1 mA.

13. For a transistor, $\gamma = 41$. Calculate α and β.

Solution

$$\gamma = \beta + 1 \quad \text{or} \quad \beta = \gamma - 1 = 41 - 1 = 40$$

Again, $\gamma = \dfrac{1}{1-\alpha}$

or, $1-\alpha = \dfrac{1}{\gamma}$

or, $\alpha = 1-\dfrac{1}{\gamma}$

$$= \frac{\gamma-1}{\gamma} = \frac{41-1}{41}$$

$$= \frac{40}{41} = 0.975$$

14. Find the value of 'β' for a transistor for which β = 0.987.

Solution

We know, $\beta = \dfrac{\alpha}{1-\alpha}$

$$= \frac{0.987}{1-0.987}$$

$$= \frac{0.987}{0.013}$$

$$= 75.92$$

15. In a common base circuit at V_{CB} = 3V, a change in emitter current from 13.55 to 20.15 mA produces a change in collector current from 13.4 to 19.9 mA. Compute the current gain.

Solution

$$V_{CB} = 3V, \quad \Delta I_E = 20.15 - 13.55$$
$$= 6.60 \text{ mA}$$
$$\Delta I_C = 19.9 - 13.4 = 6.5 \text{ mA}$$
$$\alpha = ?$$

We have
$$\alpha = \frac{\Delta I_C}{\Delta I_E}$$
$$= \frac{6.5}{6.6} = 0.9848$$

16. In the common base transistor amplifier, the input resistance is 200 ohm and load resistance 100 Kilo-ohm. Compute the voltage gain, if α = 0.95.

Solution

$$\alpha = 0.95, \quad R_L = 100 \times 10^3 \text{ ohm}$$
$$R_i = 200 \text{ ohm}, \quad \text{voltage gain} = ?$$

We have, voltage gain $= \alpha \times \dfrac{R_L}{R_i}$

$$= 0.95 \times \frac{100 \times 10^3}{200}$$
$$= \frac{95 \times 1000}{200} = 475$$

17. In a common base connection, input resistance is 250 ohms and the load resistance is 500 kilo-ohms. Calculate voltage gain and power gain if α = 0.96.

Solution

$$\alpha = 0.96, \quad Ri = 250 \text{ ohms}$$
$$R_L = 500 \times 10^3 \text{ ohms},$$
Voltage gain = ?, Power gain = ?

We have

$$\text{Voltage gain} = \alpha \times \frac{R_L}{R_i}$$
$$= 0.96 \times \frac{500 \times 10^3}{250}$$

$$= 0.96 \times 2 \times 10^3$$
$$= 96 \times 20$$
$$= 1920$$

$$\text{Power gain} = \alpha^2 \times \frac{R_L}{R_i}$$

$$= (0.96)^2 \times \frac{500 \times 10^3}{250}$$

$$= 1843.2$$

18. A transistor is operated in common emitter configuration at V_{CC} = 2V, such that a change in base current from 100 μA to 200 μA produces a change in the collector current from 9 mA to 16.5 mA. Calculate the value of β.

Solution

$$V_{CC} = 2\text{V}, \quad \Delta I_B = 200 - 100 = 100\ \mu\text{A}$$
$$\Delta I_C = 16.5 - 9 = 7.5\ \text{mA}$$

We know, $\beta = \dfrac{\Delta I_C}{\Delta I_B}$

$$= \frac{7.5 \times 10^{-3}\,\text{A}}{100 \times 10^{-6}\,\text{A}}$$

$$= 7.5 \times 10^{-3} \times 10^4$$
$$= 75$$

19. In a common emitter circuit, the input resistance is 6 KΩs and the output resistance is 5.1 KΩs. If β = 65, find the voltage gain and power gain.

Solution

$$\beta = 65, \quad R_i = 0.6 \times 10^3 \text{ ohms}$$
$$R_0 = 5.1 \times 10^3 \text{ ohms}, \quad \text{voltage gain} = ?$$
$$\text{power gain} = ?$$

We have voltage gain $= \beta \times \dfrac{R_0}{R_i}$

$$= 65 \times \frac{5.1 \times 10^3}{0.6 \times 10^3}$$

$$= \frac{65 \times 1.7}{0.2} = 552.5$$

$$\text{Power gain } = \beta^2 \times \frac{R_0}{R_i} = (65)^2 \times \frac{5.1 \times 10^3}{0.6 \times 10^3}$$

$$= 35912.5$$

20. A change in 100 mV in the base-emitter voltage causes a change of 50 μA in the base current. Find the input resistance of the transistor.

Solution

$$\Delta V_{BE} = 100 \text{ mV} = 100 \times 10^{-3} \text{ V}$$
$$\Delta I_B = 50 \ \mu\text{A} = 50 \times 10^{-6} \text{ A}$$

$$\therefore \text{ Input resistance, } R_i = \frac{\Delta V_{BE}}{\Delta I_B}$$

$$= \frac{100 \times 10^{3} V}{50 \times 10^{-6} A}$$

$$= 2 \times 10^3 \ \Omega$$

$$= 2 \text{ K}\Omega$$

21. If the collector current changes from 2 mA to 3 mA in a transistor when collector-emitter voltage is increased from 4V to 20V, what is the output resistance?

Solution

$$\Delta I_C = 3 - 2 = 1 \text{ mA} = 1 \times 10^{-3} \text{A}$$
$$\Delta V_{CE} = 20 - 4 = 16 \text{V}$$

$$\therefore \text{ Output resistance, } R_0 = \frac{\Delta V_{CE}}{\Delta I_C}$$

$$= \frac{16}{1 \times 10^{-3}} = 16 \times 10^3 \ \Omega$$

$$= 16 \text{ K}\Omega$$

22. For a common emitter transistor circuit, the collector load $R_C = 3$ KΩ and the input resistance $R_i = 1.5$ KΩ. If the current gain is 50, what is the voltage gain and power gain of the circuit?

Solution

$R_C = 3$ KΩ $= 3 \times 10^3$ Ω

$R_i = 1.5$ KΩ $= 1.5 \times 10^3$ Ω

$\beta = 50$, voltage gain = ?

power gain = ?

We have voltage gain $= \beta \times \frac{R_C}{R_i}$

$$= 50 \times \frac{3 \times 10^3}{1.5 \times 10^3}$$

$$= 100$$

Power gain $= \beta^2 \times \frac{R_C}{R_i}$

$$= \frac{(50)^2 \times 3 \times 10^3}{1.5 \times 10^3}$$

$$= 5{,}000$$

23. The maximum power dissipation of a transistor is 75 mW. If V_{CE} = 15 V, what is the maximum collector current that can be allowed without destruction of the transistor?

Solution

P_D = 75 mW = 75 × 10^{-3} W
V_{CE} = 15 V, I_C = ?

We have $P_D = V_{CE} \times I_C$

or, $I_C = \frac{P_D}{V_{CE}} = \frac{75 \times 10^{-3}\ \text{W}}{15\text{V}}$

$= 5 \times 10^{-3}$ A
$= 5$ mA

24. In a common emitter configuration, the base current is 50 μA and β is 100. Calculate emitter current.

Solution

β = 10, I_B = 50 μA = 50 × 10^{-6}A
I_E = ?

We have $\beta = \frac{I_C}{I_B}$ or $I_C = \beta I_B$

So, $I_C = 100 \times 50 \times 10^{-6}$ A = 5 mA
Now, $I_E = I_B + I_C = 50 \times 10^{-6} + 5 \times 10^{-3}$
$= (0.05 + 5)$ mA = 5.05 mA

25. A transistor is connected in a common emitter connection with a resistance of 1 KΩ joined between the collector and

emitter (Fig. 9.24). Collector supply voltage is 6V and potential drop across 1 KΩ resistance is 0.4 V. If $\alpha = 0.90$, calculate V_{CE}, I_B and I_E.

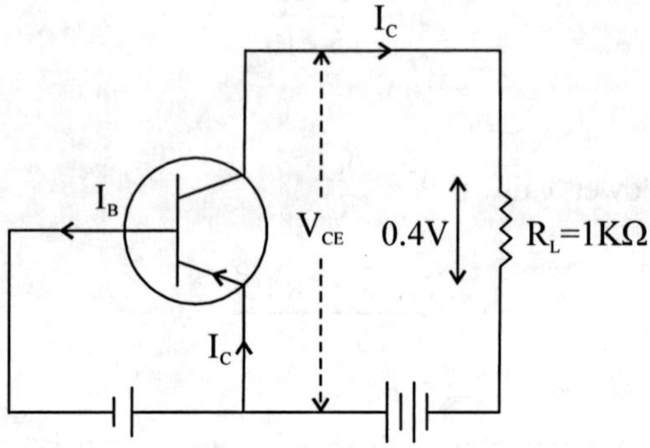

Fig. 9.24

Solution

$R_L = 1\ K\Omega = 1 \times 10^3\ \Omega,\ \ \alpha = 0.9$

$V_{CC} = 6$ V, potential drop across $R_L = 0.4$V

$V_{CE} = ?,\ \ I_B = ?,\ \ I_E = ?$

$V_{CE} = V_{CC} - 0.4\text{V} = 6\text{V} - 0.4\text{V} = 5.6\text{V}$

$$I_C = \frac{\textit{potential drop across } R_L}{R_L}$$

$$= \frac{0.4}{10^3} = 0.4 \times 10^{-3}\ \text{A} = 0.4\ \text{mA}$$

Now, $\beta = \dfrac{\alpha}{1-\alpha} = \dfrac{0.9}{1-0.9} = \dfrac{0.9}{0.1} = 9$

$$I_B = \frac{I_C}{\beta} = \frac{0.4}{9} = 0.044\ \ \text{mA}$$

$$I_E = I_C + I_B = 0.40 + 0.044 = 0.444\ \text{mA}$$

❑ ❑ ❑

10
Amplifiers

Q. 1. What do you mean by amplification? What is faithful amplification?

Ans. *Amplification :* The increase of weak signal by a triode or a transistor is called amplification.

Faithful amplification

The process of raising the strength of a weak signal without any change in shape is known as faithful amplification. Transistor biasing helps in achieving faithful amplification.

Q. 2. How does a transistor act as an amplifier?

Ans. *Transistor as an amplifier :* A transistor acts as an amplifier by increasing the strength of a weak signal. The basic circuit of a transistor amplifier is shown in fig. 10.1.

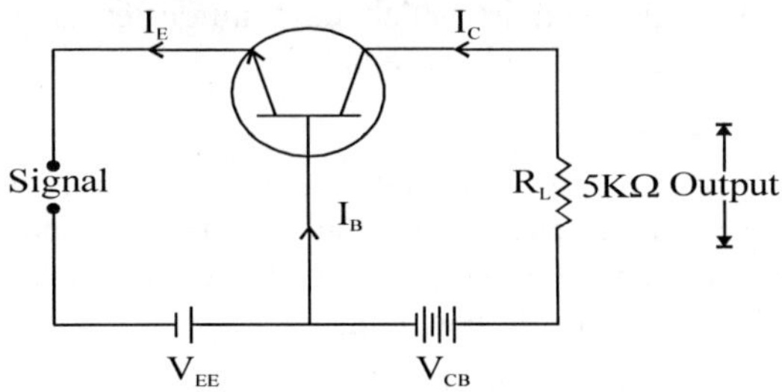

Fig. 10.1

Action

The weak signal is applied between emitter-base junction and the output is taken across the load R_L connected in the collector circuit. In order to get faithful amplification, the input circuit is forward biased. To do so, a d.c. voltage V_{EE} is applied in the input circuit in addition to the signal. This d.c. voltage is known as bias voltage and its magnitude is such that the input circuit always remains forward biased irrespective of the polarity of the signal.

Since the input circuit has low resistance, a small change in signal voltage causes an appreciable change in emitter current. This causes almost the same change in collector current due to transistor action. The collector current flowing through the high load resistance R_L produces a large voltage across it. In this way, a wean signal voltage applied in the input circuit appears in the amplified form in the collector circuit. Thus a transistor acts as an amplifier.

Q. 3. What are different types of amplifiers?

Ans. *Classification of amplifiers :* Amplifiers are classified in different ways as maintained below :

(1) According to use,
(2) According to circuits,
(3) In terms of frequency range of operation,
(4) According to transistor operation,
(5) According to the range of frequencies amplified compared with central frequency,
(6) According to circuit configuration.

(1) Classification depending on the type of use

(a) Voltage amplifier

The purpose is to get max^m possible voltage gain. The transistor used should have high β value. The load for such amplifiers may be either tuned or untuned in which case amplifier is a tuned voltage amplifier or an untuned voltage amplifier.

(b) Power amplifier

The purpose is to get maximum possible power gain. The transistor used should have also high β and high power rating *i.e.*, the transistor used for this purpose should be capable of handling large current at higher collector potential.

(2) Classification according to circuit

(a) Single stage amplifiers

Single stage amplifier uses only one amplifier and is suitable when small amplification (voltage or power) is required.

(b) Cascade amplifiers

It consists of more than one amplifier joined in such a manner that the output of one amplifier is fed into the input of the next amplifier. This process achieves sufficiently high gain.

The cascade amplifiers are further classified according to method of coupling or linking the two adjacent stages as follows :

(i) Resistance

Capacitance (RC) coupled amplifiers using resistor as the load and a capacitor as the coupling element.

(ii) Inductance

Capacitance (LC) coupled amplifiers using an inductor as the load and a capacitor as the coupling element.

(iii) Transformer coupled amplifier using the transformer as the load and also for coupling.

(iv) Direct

Couple (DC) amplifiers using direct coupling between the stages.

(3) Classification according to frequency range of operation

For convenience, the signal frequencies to be amplified are designated as follows :

Audio frequency (AF) – 20 Hz to 20 KHz.

Radio frequency (RF) – frequency > 100 KHz.

Video frequency (VF) – 50 Hz – 6 mHz, the upper limit may be still higher.

Amplifiers useful for amplifying signals whose frequency remain within the appropriate range are named as AF amplifier, RF amplifier, VF amplifier.

However, amplifiers suitable only for very low frequency range (VLF), 3KHz – 30KHz or low frequency range (LF), 30 KHz – 300 KHz are also known as VLF or LF amplifiers respectively.

A type of fixed frequency RF amplifier used in Radio, T.V. and Radar receivers is known as Intermediate Frequency (IF) amplifier.

(4) Classification according to transistor operation

This classification is based on the duration of transistor conduction period in relation to one cycle of the input signal, such as class A, class B, class AB and class C.

(5) Classification according to the bandwidth

(a) Narrowband amplifiers

In this type of amplifiers, the band of frequencies to be amplified is small in comparison with the central frequency. Ordinarily, RF tuned amplifiers are narrow band amplifiers.

(b) Wide band amplifiers

The band of frequencies amplified is large compared with the central frequency in this type of amplifiers. R-C coupled amplifiers, LC coupled amplifiers, Transformer coupled amplifiers, pulse amplifiers, video amplifiers etc. come under this category.

(6) Classification according to circuit configuration

Transistor amplifiers are classified as :

(a) common emitter (b) common base and (c) common collector type.

Q. 4. Discuss the factors upon which the performance of a transistor amplifier depends?

Ans. The performance of a transistor amplifier depends upon input resistance, output resistance, effective collector load, current gain, voltage gain and power gain. We shall now discuss these factors w.r.t. a common emitter connection circuit which is universally adopted.

(i) Input resistance

It is the ratio of small change in base emitter voltage (ΔV_{BE}) to the resulting change in base current (ΔI_B) at constant collector-emitter voltage *i.e.*,

$$\text{Input resistance, } R_i = \frac{\Delta V_{BE}}{\Delta I_B}$$

Fig. 10.2 shows the general form of an amplifier. The input voltage V_{BE} causes an input current I_B.

∴ Input resistance,

$$R_i = \frac{\Delta V_{BE}}{\Delta I_B} = \frac{V_{BE}}{I_B}$$

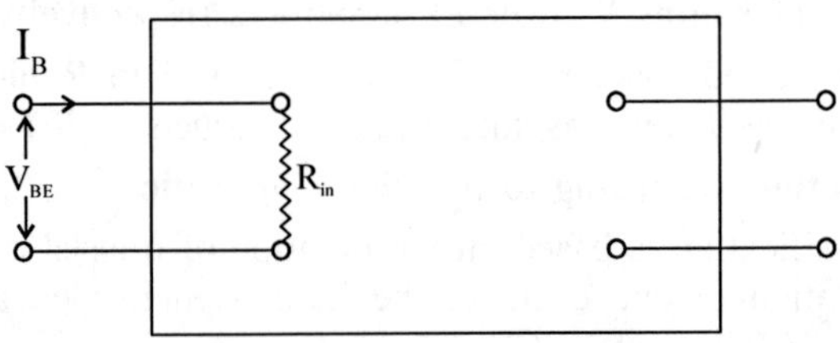

Fig. 10.2. Amplifier

(ii) Output resistance

It is the ratio of change in collector-emitter voltage (ΔV_{CE}) to the resulting change in collector current (ΔI_C) at constant base current *i.e.*,

$$\text{Output resistance, } R_0 = \frac{\Delta V_{CE}}{\Delta I_C}$$

The output characteristics reveal that collector current changes very slightly with the change in collector emitter voltage. Therefore, output resistance of a transistor amplifier is very high of the order of several hundred kilo-ohms. The physical explanation of high output resistance is that collector-base junction is reverse biased.

(iii) Effective collector load

It is the total load as seen by the a.c. collector current.

In case of single stage amplifiers, the effective collector load is a parallel combination of R_c and R_o as shown in fig. 10.3(i).

Effective collector load, $R_{AC} = R_c \mid\mid R_o$

$$= \frac{R_c \times R_o}{R_c + R_o}$$

As output resistance R_o is several times R_c, therefore, R_c can be neglected as compared to R_o

$$R_{AC} = \frac{R_e \times R_o}{R_o} = R_c$$

It follows, therefore, that for a single stage amplifier, effective load is equal to collector load R_C.

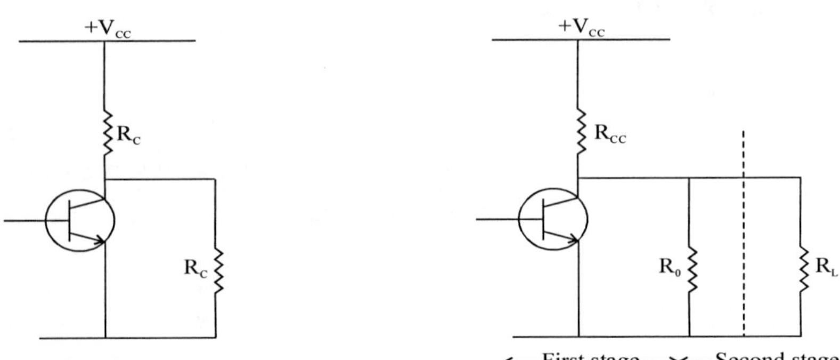

Single stage
(i)

Two stage
(ii)

Fig. 10.3

However, in a multistage amplifier (*i.e.*, having more than one amplification stage), the input resistance R_i of the next stage also comes into picture as shown in fig. 10.3. Therefore, effective collector load becomes.

Parallel combination of R_C, R_O and R_i *i.e.*,

$$\text{Effective collector load, } R_{AC} = R_c \,||\, R_o \,||\, R_i$$

$$= R_c \,||\, R_i$$

$$= \frac{R_c R_i}{R_c + R_i}$$

As input resistance R_i is quite small (25Ω to 500Ω), therefore, effective load is reduced.

(iv) Current gain

It is the ratio of change in collector current (ΔI_c) to the change in base current (ΔI_B) *i.e.*,

$$\text{Current gain, } \beta = \frac{\Delta I_C}{\Delta I_B}$$

The value of β ranges from 20 to 500. The current gain indicates that input current becomes β times in the collector circuit.

(v) Voltage gain

It is the ratio of change in output voltage (ΔV_{CE}) to the change in input voltage (ΔV_{BE}) *i.e.*,

$$\text{Voltage gain, } A_V = \frac{\Delta V_{CE}}{\Delta V_{BE}}$$

$$= \frac{\text{change in output current} \times \text{effective load}}{\text{change input current} \times \text{input resistance}}$$

$$= \frac{\Delta I_C \times R_{AC}}{\Delta I_B \times R_i} = \frac{\Delta I_C}{\Delta I_B} \times \frac{R_{AC}}{R_i} = \beta \times \frac{R_{AC}}{R_i}$$

For single stage, $R_{AC} = R_C$. However, for multistage, $R_{AC} = \dfrac{R_C \times R_i}{R_C + R_i}$

where R_i is the input resistance.

(vi) Power gain

It is the ratio of output signal power to the input signal power *i.e.*,

$$\text{Power gain,} \quad A_p = \frac{(\Delta I_C^2) \times R_{AC}}{(\Delta I_B^2) \times R_i} = \left(\frac{\Delta I_C}{\Delta I_B}\right) \times \frac{\Delta I_C \times R_{AC}}{\Delta I_B \times R_i}$$

$$= \text{Current gain} \times \text{voltage gain.}$$

Q. 5. Discuss the features of class A, class B, class AB and class C amplifiers.

Ans. The transistor power amplifiers handle large signals. Most of these amplifiers are driven so hard by the input large signal that collector current is either cut off or is in the saturation region during a large portion of the input cycle. Therefore, such amplifiers are classified depending upon the location of the operating point and the signal amplitude. The operating point of a transistor amplifier is determined by the output characteristic curve corresponding to a particular input current and the a.c. load line. On this basis, the power amplifiers are classified as :

(a) Class A power amplifier
(b) Class B power amplifier
(c) Class AB power amplifier
(d) Class C power amplifier

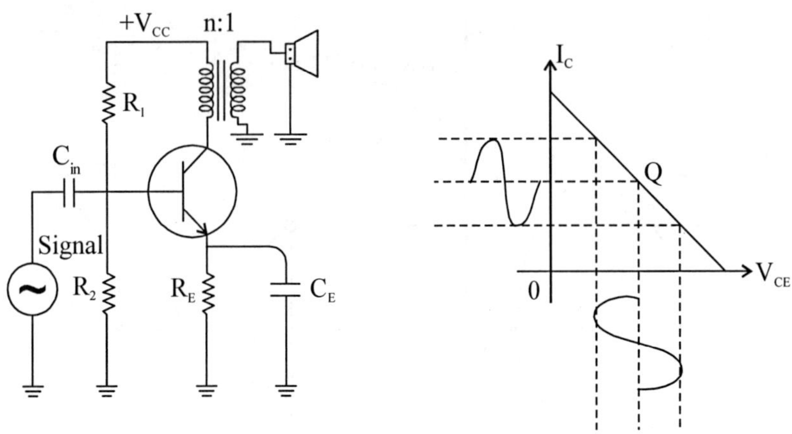

(i) (ii)

Fig. 10.4

(a) Class A power amplifier

If the collector current flows at all times during the full cycle of the signal, the power amplifier is known as class A power amplifier.

Obviously for this to happen, the power amplifier must be biased in such a way that no part of the signal is cut off fig. 10.4(i) shows circuit of class A power amplifier. Note that collector has a transformer as the load which is most common for all classes of power amplifiers. The use of transformer permits impedance matching, resulting in the transference of maximum power to the load e.g. loud speaker.

Fig. 10.4(ii) shows the class A operation in terms of a.c. load line. The operating point Q is so selected that collector current flows at all times throughout the full cycle of the applied signal. As the output wave shape is exactly similar to the input wave shape, therefore, such amplifiers have least distortion. However, they have the disadvantage of low power output and low collector efficiency (about 35%).

(b) Class B power amplifier

If the collector current flows only during the +ve half-cycle of the input signal, it is called a class B power amplifier.

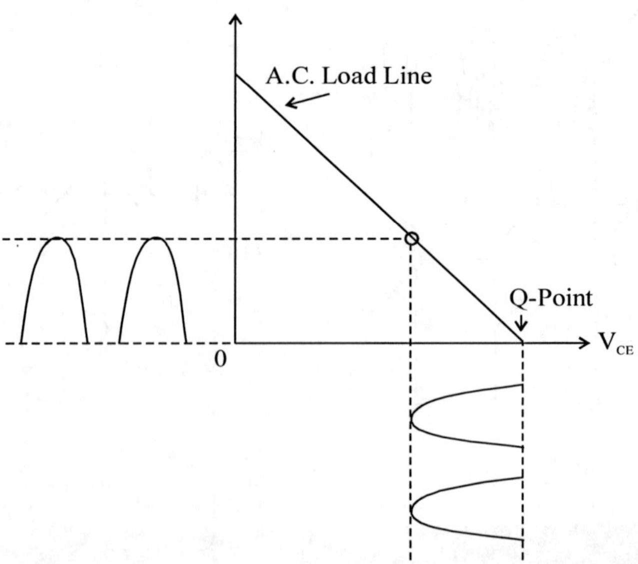

Fig. 10.5

In class B operation, the transistor bias is so adjusted that zero signal collector current is zero *i.e.*, no biasing circuit is needed at all. During the +ve half-cycle of the signal, the input circuit is forward biased and hence collector current flows. However during the negative half-cycle of the signal, the input circuit is reverse biased and no collector current flows. Fig. 10.5 shows the class B operation in terms of a.c. load line. Obviously the operating point Q shall be located at collector cut off voltage. It is easy to see that output from a class B amplifier is amplified half-wave rectification.

In a class B amplifier, the –ve half cycle of the signal is cut off and hence a severe distortion occurs. However, class B amplifiers provide higher power output and collector efficiency (50-60%). Such amplifiers are mostly used for power amplification in push-pull arrangement. In such an arrangement, 2 transistors are used in class B operation. One transistor amplifies the +ve half-cycle of the signal while the other amplifies the –ve half-cycle.

(c) Class AB power amplifier

In this type, the bias voltage and hence the operating point Q is chosen such that the operation falls intermediate between class-A and class-B. So the forward base-emitter bias will be less in this case compared to class-A amplifier discussed earlier. During the +ve half-cycle of the signal voltage the base current and hence collector current flows as usual. But as the signal voltage goes –ve, the +ve input bias decreases through still the bias continues to be forward until the magnitude of the –ve signal voltage exceeds the d.c. bias (+ve in this case). After that the input section becomes reverse biased; base current and hence collector current reduces to zero. After reaching the –ve peak, when the signal voltage starts becoming less –ve, again collector current starts flowing when the –ve signal voltage again equals the d.c. bias voltage.

The amplifiers where the input bias voltage and hence the operating point is so chosen that the collector current flows for more than half-cycle but less than full-cycle of the input signal ($180° < \theta < 360°$) are called class AB type of amplifiers.

(d) Class C power amplifier

If the collector current flows for less than half-cycle of the input signal, it is called class C power amplifier.

In class C amplifier, the base is given some –ve bias so that collector current does not flow just when the +ve half-cycle of the signal starts. Such amplifiers are never used for power amplification.

However, they are used as tuned amplifiers *i.e.*, to amplify a narrow band of frequencies near the resonant frequency.

Q. 6. Find an expression for collector efficiency in power amplifiers.

Ans. Power amplifiers can be compared on the basis of their collector efficiency. The greater the collector efficiency, the better is the power amplifier.

$$\text{Collector efficiency,} \quad \eta = \frac{\text{a.c. power output}}{\text{d.c. power input}}$$

$$= \frac{P_{ac}}{P_{dc}} \qquad \text{...(1)}$$

where, $P_{dc} = V_{cc} I_C$

$P_{ac} = V_{ce} I_C$

where V_{ce} is the rms value of signal output voltage and I_c is the rms value of output signal current.

$$\text{r.m.s. value} = \frac{1}{2}\left[\frac{\text{peak-to-peak value}}{\sqrt{2}}\right]$$

$$= 0.5 \times 0.707 \times \text{peak-to-peak value.}$$

In terms of peak-to-peak values (which are often convenient values in load-line work), the a.c. power output can be expressed as :

$$P_{ac} = [(0.5 \times 0.707)\ V_{ce}\ (p - p)]\ [0.5 \times 0.707)\ I_c\ (p - p)]$$

$$= \frac{V_{ce}(p-p) \times I_C\ (p-p)}{8}$$

$$\text{Collector } \eta\ \frac{V_{ce}(p-p) \times I_c\ (p-p)}{8\, V_{cc} I_c} \qquad \text{...(2)}$$

Q. 7. Find the max^{m} collector efficiency of class A power amplifier.

Ans. Let's consider the transformer coupled class A power amplifier. Here the load is transformer coupled with the collector. This arrangement permits impedance matching as well as keeps the d.c. power loss small due to small resistance of the transformer primary winding. Fig. 10.6 shows a transformer coupled class A power amplifier.

Maximum collector efficiency can be determined with the help of the load line shown in fig. 10.7.

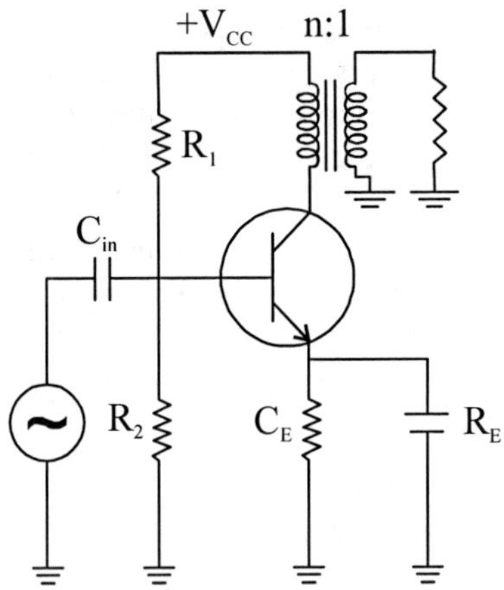

Fig. 10.6

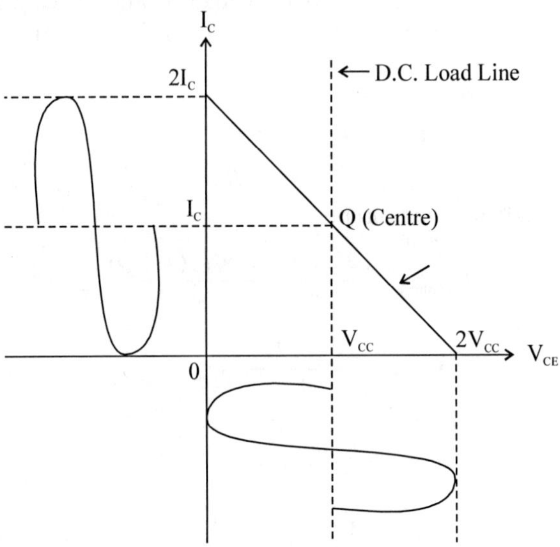

Fig. 10.7

Under zero signal conditions, the effective resistance in the collector circuit is that of the primary winding of the transformer. The primary resistance has a very small value and is assumed zero. Therefore, d.c. load line is a vertical line rising from V_{cc} as shown in fig. 10.7. When signal is applied the collector current will vary about the operating point Q.

In order to get maximum a.c. power output (and hence maximum collector η), the peak value of collector current due to signal alone should be equal to the zero signal collector current I_c. In terms of a.c. load line, the operating point Q should be located at the centre of a.c. load line.

During the peak of the +ve half cycle of the signal, the total collector current is $2I_c$ and $V_{ce} = 0$. During the negative peak of the signal, the collector current is zero and $V_{ce} = 2\ V_{cc}$.

∴ Peak-to-peak collector-emitter voltage,

$$V_{cc}\ (p - p) = 2\ V_{cc}$$

Peak-to-peak collector current

$$I_c\ (p - p) = 2\ I_c$$

$$= \frac{V_{cc}\ (p-p)}{R_L} = \frac{2\ V_{cc}}{R'_L}$$

where R'_L is the reflected value of load R_L and appears in the primary of the transformer. If η (= N_p / N_s) is the turn ratio of the transformer,

then, $R'_L = \eta^2 R_L$

d.c. power input, $P_{dc} = V_{cc} I_c$

$$= I^2_c\ R'_L \qquad (\because\ V_{cc} = I_c\ R'_L)$$

Max[m] a.c. output power,

$$P_{ac(\max)} = \frac{V_{cc}(p-p) \times I_c\,(p-p)}{8}$$

$$= \frac{2V_{cc} \times 2I_c}{8}$$

$$= \frac{1}{2}\ V_{cc} I_c \qquad \text{...(1)}$$

$$= \frac{1}{2} I_c^2\ R'_L \qquad (\because V_{cc} = I_c R'_L)$$

$$\text{Max}^{\text{m}} \text{ collector } \eta = \frac{P_{ac\,(\max)}}{P_{dc}} \times 100$$

$$= \frac{\tfrac{1}{2}\, I_c^2\ R'_L}{I_c^2\ R'_L} \times 100$$

$$= 50\%$$

Q. 8. With a neat circuit diagram discuss the actions of common base transistor amplifier. Find the input and output impedances, voltage and power gains and phase relations in CB configuration.

Ans. *Transistor as an amplifier in CB Configuration :* The transistor amplifier circuit with the base common to input and output circuit is shown in fig. 10.8 below. Various wave forms of input and output voltage are drawn in fig. 10.9(a), (b), (c), (d).

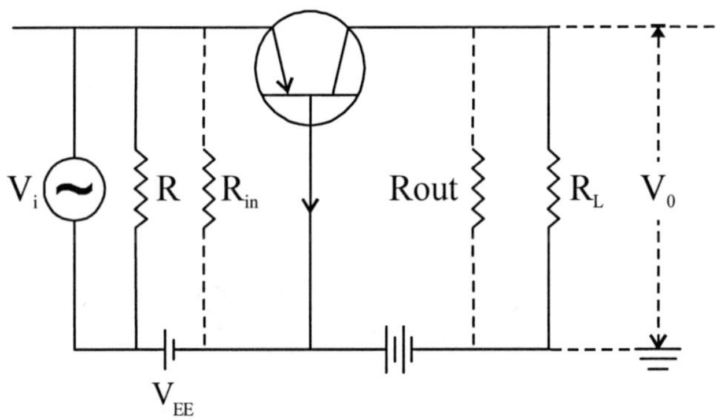

Fig. 10.8

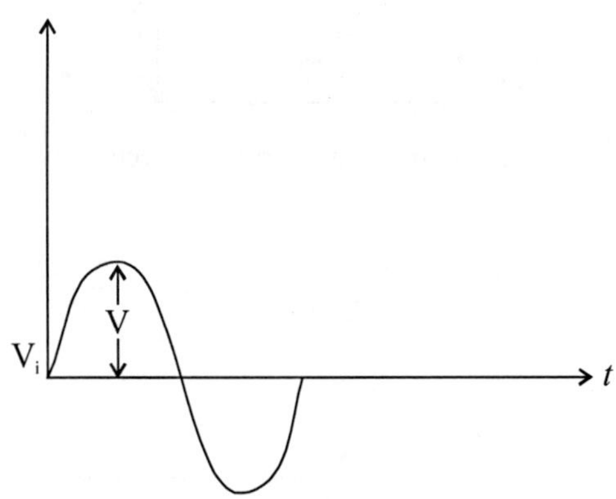

Fig. 10.9(a). Input voltage wave form

The input signal, V_i, is applied between the emitter and the base. The output, V_o, is obtained across a load R_L connected between the collector and the base. Since we are concerned with a.c. signal for amplification the terms input or output voltage or current refer to the a.c. part only in amplifiers.

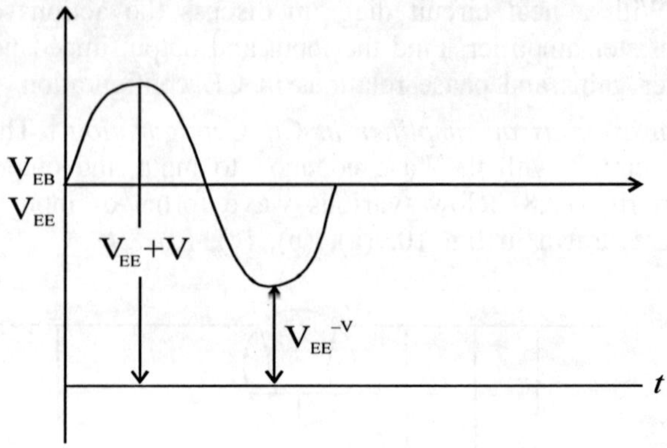

Fig. 10.9(b). Change in the emitter-base voltage due to the signal

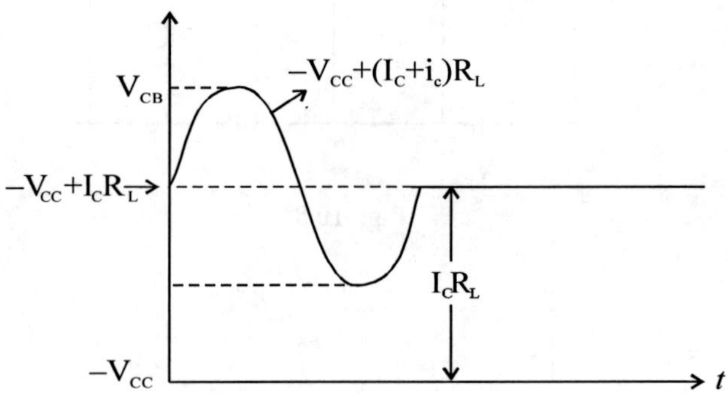

Fig. 10.9(c). Variation of collector voltage

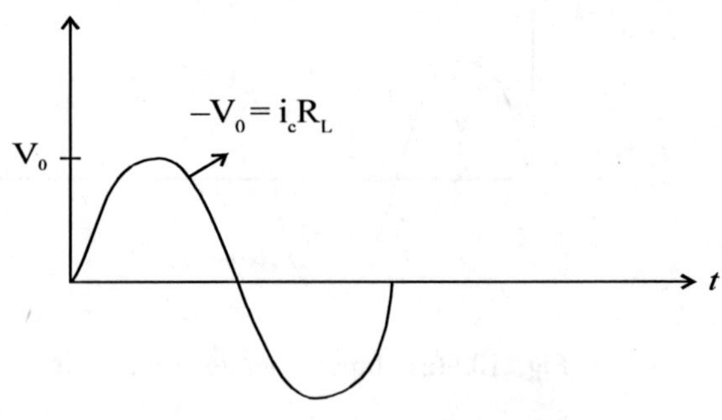

Fig. 10.9(d). Output voltage wave form

Operation

The operation of this amplifier circuit can be understood by referring to the characteristic curve, fig. 10.10 in the common base

the order of 10KΩ the effective output impedance is equal to the load resistance.

In this arrangement the base current is input current. Hence the current gain in this case is denoted by β or h_{fe} in hybrid notation.

Voltage gain

For the analysis of CB amplifier we had chosen a transistor with $\alpha = 0.99$.

Hence the corresponding $\beta = \frac{\alpha}{1-\alpha} = 99$. Assuming the input impedance of 2000Ω and a load resistance of 10000Ω.

$$\text{Voltage gain, A} = \frac{\text{Output voltage, } V_{\text{out}}}{\text{Input voltage, } V_{\text{in}}}$$

$$= \frac{i_c R_L}{i_b R_{in}} = \beta \frac{R_L}{R_{in}}$$

$$= 99 \times \frac{10000}{2000} = 495$$

The voltage gain is therefore, higher than that for the common base amplifier. Power gain, however, is considerably higher.

$$\text{Power gain} = \frac{i_c^2 R_L}{I_b^2 R_{in}} = \beta^2 \frac{R_L}{R_{in}} = (99)^2 \times \frac{10000}{2000}$$

$$= 24005$$

Because of the higher voltage and power gain the CE amplifier is most widely used.

Phase

In this configuration, however, the phase of the output signal is opposite to the phase of the input signal. The base emitter section is forward biased with base side connected to the –ve pole of the bias battery. Consider the +ve half of the signal cycle is acting at any instant. In this case, the base potential becomes less –ve; this reduces the emitter current by i_e and consequently the collector current by i_c. So the current through the load reduces and hence the voltage drop across R_L becomes less by an amount $i_e R_L$ than the normal voltage drop ($I_C R_L$) when no signal is applied. The P.D. across R_L is in opposition to V_{CC}. Since V_{CC} is –ve here, the P.D. across R_L is assigned +ve sign. So the instantaneous collector voltage V_c changes from $(-V_{CC} + I_C R_L)$ to $[-V_{CC} + (I_C - i_e)R_L]$, *i.e.*, V_c becomes more –ve or less +ve. The difference $V_c = I_c R_L$, the output voltage is therefore –ve.

On the other hand, during the –ve half-cycle of the signal, the base becomes more –ve. This results in increased emitter current and collector current. Thus the current through the load R_L increases resulting in greater drop of voltage than the no signal P.D. value. Hence the collector voltage increases or it becomes less –ve or more +ve. As such, reversal of phase of the output relative to the input takes place in this type of amplifier. Fig. 10.13 (a) represents the input waveform and the corresponding output waveform is given as fig. 10.13 (b). It is worth-while to note here that similar reversal of phase takes place in amplifiers using valves.

Q. 10. With a neat circuit diagram discuss the actions of common collector transistor amplifier. Find the input and output impedances, voltage and power gains and phase relations in CC configuration.

Ans. *Transistor as an amplifier in common collector configuration:* This circuit is shown in fig. 10.15. The input signal is applied between the base and collector. The input impedance in this case is quite high because of the reverse bias that exists between the base and the collector. In the previous two arrangements the input impedance is low.

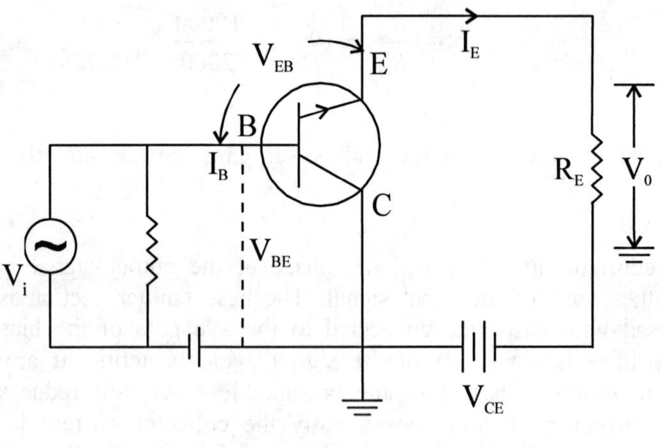

Fig. 10.15. CC amplifier circuit

Of course a resistance is connected across the input and if this resistor value is low, it will cause the input impedance to be low. However if it has a high value, we can then obtain input impedance as high as IMΩ in this circuit. On the other hand, the output impedance between the emitter and collector is low due to forward biasing, often falling below 100Ω.

Impedance

It is of interest to show here that the input impedance of a common collector amplifier is dependent on the value of the load resistance in the emitter circuit. The input impedance of the amplifier R_{in} is equal to the input voltage divided by the input current. The input voltage,

$$V_{in} = V_{eb} + V_o$$

These voltage can be added directly because they are always in phase for a resistive load.

$$\text{Thus,} \quad R_{in} = \frac{V_{eb} + V_o}{i_b} = \frac{V_{eb}}{i_b} + \frac{V_o}{i_b} \qquad ...(1)$$

Now, V_{eb}/i_b is the internal resistance between base and emitter, r_{be}; and since this junction is forward biased it has a very low value. On the other hand V_o/i_b depends upon the circuit as well as the transistor.

$$\text{Putting} \quad \frac{i_e}{i_b} = (\beta + 1)$$

$$\frac{V_o}{i_b} = \frac{(\beta + 1)}{i_e} = (\beta + 1) R_E \qquad \left(\because \frac{V_o}{i_e} = R_E \right)$$

$$\text{So,} \quad R_{in} = r_{be} + (\beta + 1)\, R_E \approx (\beta + 1)\, R_E$$

This shows that it is desirable to use as high a value of R_E as possible in order to obtain a high input impedance.

Because of high input impedance and low output impedance, this class of amplifier is specially used as a matching device between a high-output impedance stage with a low-input impedance stage.

Gain

The current gain of a common collector amplifier is slightly higher than β. This can be shown as follows :

The input current is the base current i_b. The output current flowing through the load R_E connected to the emitter, fig. 10.15 is

i_c. We know, $i_c = i_e + i_b$

Hence current gain $\gamma = \frac{i_c}{i_b} = \frac{i_e + i_b}{i_b} = \frac{i_e}{i_b} + 1$

$$= \beta + 1 \qquad ...(2)$$

Since β is high the current gain of common collector amplifier is nearly same as that of a common emitter amplifier.

The voltage gain in this circuit is always less than 1, although generally it is not much less than 1, although generally it is not much less than 1. This happens for the following reason. We have seen that the output voltage at the emitter is in phase with the input voltage at the base. Thus when the base goes +ve the emitter also goes +ve *i.e.*, they act in opposition. Thus the effective variation in the base emitter voltage will be small. Referring to fig. 10.15,

$$V_i = V_{eb} + V_o \qquad ...(3)$$

V_o is therefore always less than V_i; voltage gain approaches unity, when R_E is very high.

Power gain achieved in the common collector amplifier is large because of large current gain but it is less than the value in the other two arrangements.

Phase

Note the phase-relation between the output and input signal.

The base-emitter voltage, $V_{BE} = V_{BC} + V_{CC}$, since $V_{CC} > V_{BC}$, V_{BE} is +ve. Further P.D. across the emitter resistor R_E acts in opposition to V_{CC}. Thus effective base-emitter voltage is also reduced by this amount. So the effective base-emitter bias voltage is $V_{BE} - I_E R_E$ when $V_{BE} > I_E R_E$. Thus the base-emitter section is effectively forward biased and emitter current increases. Consider the case when the signal voltage is going +ve. For the NPN transistor base-collector voltage becomes less –ve and then V_{BE} will increase resulting in increase of the emitter current. As a result the output voltage measured across R_E will increase. Similarly when the signal voltage goes –ve, V_{BC} becomes more –ve and thus V_{BE} will decrease. So less current will flow through R_E and the output voltage will decrease. Similar situation also occurs for PNP transistor. Hence, the emitter potential follows the signal potential. Hence it is also called emitter follower amplifier. (Cathode follower amplifier using valves is analogous to the emitter follower in functioning). Thus there is no phase change.

Q. 11. With a neat circuit diagram describe the construction, operation of an R-C coupled transistor amplifier. Find the gain, frequency response curve, band width of the amplifier. Mention the advantages, disadvantages and uses of the amplfier.

Ans. The R-C coupled amplifier is usually employed for voltage amplification. Resistance and capacitance coupling is the most popular type of coupling because it is cheap and provide excellent audio fidelity over a wide range of frequency. The circuit arrangements of an R-C coupled amplifier is shown in fig. 10.16.

Construction

Fig. 10.16 shows two stages of an R-C coupled amplifier.

A coupling capacitor C_c is used to connect the output of first stage to the base (*i.e.*, input) of the second stage and so on. As the coupling from one stage to next is achieved by a coupling capacitor followed by a connection to a shunt resistor, therefore, such amplifiers are called resistance — capacitance coupled amplifiers.

The resistances R_1, R_2 and R_E form the biasing and stabilisation network. The emitter bypass capacitor offers low resistance path to the signal. Without it, the voltage gain of each stage would be lost. The coupling capacitor C_C transmits a.c. signal but blocks d.c. This prevents d.c. interference between various stages and the shifting of operating point.

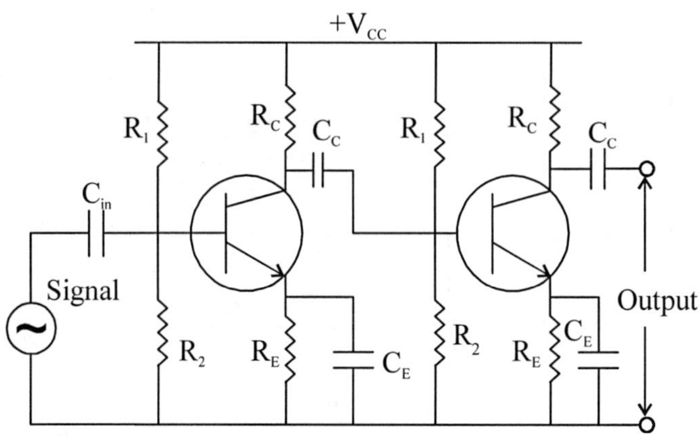

Fig. 10.16

Operation

When a.c. signal is applied to the base of the first transistor, it appears in the amplified form across its collector load R_C. The amplified signal developed across R_C is given to base of next stage through coupling capacitor C_c. The second stage does further amplification of the signal. In this way, the cascaded (one after another) stages amplify the signal and the overall gain is considerably increased.

Gain

It may be mentioned here that total gain is less than the product of the gains of individual stages. It is because when a second stage is made to follow the first stage, the effective load resistance of first stage is reduced due to the shunting effect of the input resistance of second stage. This reduces the gain of the stage which is loaded by the next stage. For instance, in a 3-stage amplifier, the gain of first and second stages will be reduced due to loading effect of next stage. However, the gain of the third stage which has no loading effect of subsequent stage, remains unchanged. The overall gain shall be equal to the product of the gains of three stages.

Frequency response

Fig. 10.17 shows the frequency response of a typical R-C coupled amplifier. It is clear that voltage gain drops off at low (< 50Hz) and high (> 20 KHz) frequencies whereas it is uniform over mid-frequency range (50 Hz to 20 KHz). This behaviour of the amplifier is briefly explained below :

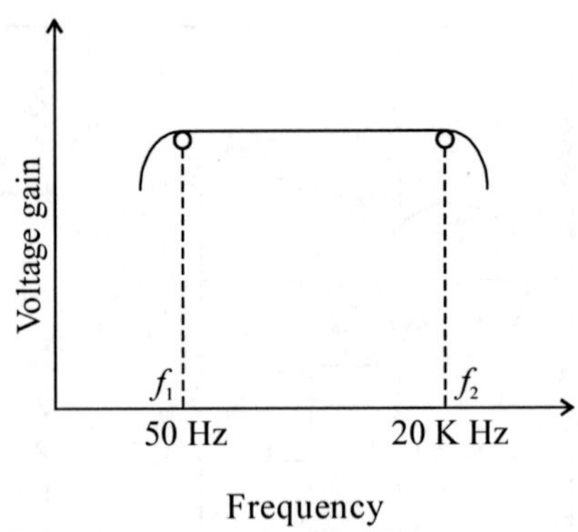

Fig. 10.17

(i) At low frequencies (<50 Hz), the reactance of coupling capacitor C_c is quite high and hence very small part of signal will pass from one stage to the next stage. Moreover, C_E cannot shunt the emitter resistance R_E effectively, because of its large reactance at low frequencies. These two factors cause a falling off voltage gain at low frequencies.

(ii) At high frequencies (> 20 KHz), the reactance C_c is very small and it behaves as a short circuit. This increases the loading effect of next stage and serves to reduce the voltage gain. Moreover at high frequency, capacitive reactance of base-emitter junction is low which increases the base-current. This reduces the current amplification factor β. Due to these two reasons, the voltage gain drops off at high frequency.

(iii) At mid frequencies (50 Hz to 20 KHz), the voltage gain of the amplifier is constant. The effect of coupling capacitor in this frequency range is such so as to maintain a uniform voltage gain. Thus, as the frequency increases in this range, reactance of C_C decreases which tends to increase the gain. However, at the same time, lower reactance means higher loading of first stage and hence lower gain. These two factors almost cancel each other resulting in a uniform gain at mid frequency.

Bandwidth and gain-bandwidth product

The bandwidth of the amplifier, $\beta = f_2 - f_1$

Since $f_1 << f_2$, the bandwidth is practically dependent on f_2. For a transistor, the gain-bandwidth product has a fixed value, f_T. Thus if the gain of the amplifier increases, its bandwidth decreases and *vice versa.*

Advantages

(i) It has excellent frequency response. The gain is constant over the audio frequency range which is the region of most importance for speech, music etc.

(ii) It has lower cost since it employs resistors and capacitors which are cheap.

(iii) The circuit is very compact as the modern resistors and capacitors are small and extremely light.

Disadvantages

(i) The R-C coupled amplifiers have low voltage and power gain. It is because the low resistance presented by the input of each stage to the preceding stage decreases the effective load resistance (R_{AC}) and hence the gain.

(ii) They have the tendency to become noisy with age, particularly in moist climates.

(iii) Impedance matching is poor. It is because the output impedance of R-C coupled amplifier is several hundred ohms whereas that of a speaker is only a few ohms. Hence, little power will be transferred to the speaker.

Uses

The R-C coupled amplifiers have excellent audio fidelity over a wide range of frequency. Therefore, they are widely used as voltage amplifiers *e.g.*, in the initial stages of public address system. If other type of coupling (*e.g.*, transformer coupling) is employed in the initial stages, this results in frequency distortion which may be amplified in next stages. However, because of poor impedance matching R-C coupling is rarely used in the final stage.

Q. 12. With a neat circuit diagram describe the construction and operation of class B push-pull amplifier. Find efficiency and output impedance rating by transformer coupling. What are the merits and demerits of the amplifier?

Ans. *Construction :* Push-pull amplifier is a power amplifier and is frequently employed in the output stages of electronic circuits. Fig. 10.18 shows the circuit of a class-B push-pull amplifier.

Two transistors Tr_1 and Tr_2 placed back to back are employed. Both transistors are operated in class B operation *i.e.*, collector current is nearly Zero in the absence of the signal. The centre-tapped secondary of driver transformer T_1 supplies equal and opposite voltages to the base circuits of two transistors.

The output transformer T_2 has the centre-tapped primary winding. The supply voltage V_{CC} is connected between the bases and this centre tap. The loudspeaker is connected across the secondary of this transformer.

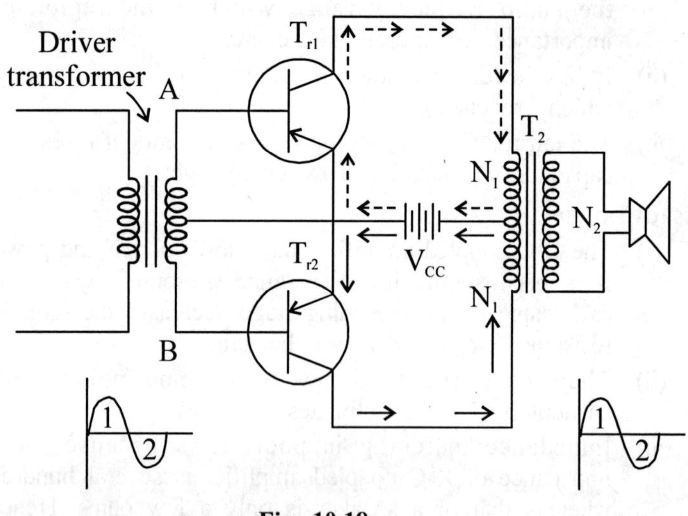

Fig. 10.18

Circuit operation

The input signal appears across the secondary AB of driver transformer. Suppose during the first half-cycle (marked 1) of the signal, end A becomes +ve and end B –ve. This will make the base-emitter junction of Tr_1 reverse biased and that of Tr_2 forward biased. The circuit will conduct current due to Tr_2 only, and is shown by solid arrows. Therefore, this half-cycle of the signal is amplified by Tr_2 and appears in the lower half of the primary of output transformer. In the next half-cycle of the signal, Tr_1 is forward biased whereas Tr_2 is reverse biased. Therefore, Tr_1 conducts and is shown by dotted arrows. Consequently, this half-cycle of the signal is amplified by Tr_1 and appears in the upper half of the output transformer primary. The centre-tapped primary of the output transformer combines two collector currents to form a sine wave output in the secondary.

Efficiency

We know, for one transistor

$$V_{c(\max)} = V_{cc}$$

and $$i_{c(\max)} = I_c$$

Since in class-B operation each transistor operates for half-cycle, these values represent the peak values for both the transistors of the push-pull amplifier.

Hence, $$P_o = V_{(rms)} = \frac{V_{C(\max)}}{\sqrt{2}} \cdot \frac{i_{c(\max)}}{\sqrt{2}}$$

$$= \frac{i_{c\,(\max)}\,R'_L}{\sqrt{2}} \cdot \frac{i_{c\,(\max)}}{\sqrt{2}}$$

$$= \frac{i^2_{c\,(\max)}\,R'_L}{2} = \frac{i_c^2\,R'_L}{2}$$

The mean value or d.c. component of current over half-cycle is $I_{dc} = I_c/\pi$

Power taken from the battery by one transistor $= \dfrac{V_{cc}I_c}{\pi}$

Hence P_{in} (for both transistor) $= \dfrac{2V_{cc}I_c}{\pi}$

$$= \frac{2I_c\,R'_c \times I_c}{\pi} = \frac{2I_c^2\,R'_L}{\pi}$$

$$\text{Hence efficiency } (\eta) = \frac{P_{out}}{P_{in}} \times 100$$

$$= \frac{\dfrac{I_c^2 R'_L}{2}}{\dfrac{2I_c^2 R'_L}{\pi}} \times 100$$

$$= \frac{\pi}{4} \times 100 = 78.5\%$$

Output impedance

It may be noted here that push-pull arrangement also permits a maximum transfer of power to the load through impedance matching. If R_L is the resistance appearing across secondary of output transformer, then resistance R'_L of primary shall become :

$$R'_L = \left(\frac{2N_1}{N_2}\right)^2 R_L$$

where $N_1 =$ Number of turns between either end of primary winding and centre tap

$N_2 =$ Number of secondary turns.

Advantages

(i) The efficiency of the circuit is quite high ($\simeq$ 75%) due to class B operation.

(ii) A high a.c. output power is obtained.

Disadvantages

(i) Two transistors have to be used.

(ii) It requires two equal and opposite voltages at the input. Therefore, push-pull circuit requires the use of driver stage to furnish these signals.

(iii) If the parameters of the two transistors are not the same, there will be unequal amplification of two halves of the signal.

(iv) The circuit gives more distortion.

(v) Transformers used are bulky and expensive.

Q. 13. What do you mean by distortion in amplifiers ? Discuss various types of distortion in amplifiers.

Ans. *Distortion in amplifiers* : An ideal voltage amplifier should produce output whose wave form is identical to the wave form of the input signal. Any deviation present in the output wave form amounts distortion (noise) in the output. So, the change of output wave form from the input wave form is known as distortion in an amplifier.

Explanation

A transistor like other electronic instruments is essentially a non linear device. Therefore, whenever a signal is applied to the input of the transistor, the output signal is not exactly like the input signal. That means distortion is present. In case of voltage amplifiers, distortion does not become a problem for small signals. This is because transistor is a linear device for small variations about the operating point. In case of a power amplifier, large signals are employed. Therefore, the problem of distortion comes into play. Among two power amplifiers, the one which has the less distortion serves better. However, methods have been developed to reduce distortion in amplifiers. Three conditions must be satisfied for distortion to be absent :

(i) The output must contain only those frequencies that are contained in the input.

(ii) The output must contain all frequencies present in the input and the relative amplitude of various components must be the same as in the input.

(iii) If there is any phase shift between the output and the input signal corresponding to all frequency components it should have identical phase shift.

Types of distortion in amplifiers

As per the physical conditions noted above, three types of distortion occur, namely :

(a) Amplitude distortion or non linear distortion

(b) Frequency distortion

(c) Phase distortion

(a) Amplitude or non linear distortion

This type of distortion is caused when the amplitudes of various components in the output differ that from the input.

In any amplifier design the choice of operating point is vital. Let us take for example the operating point Q_1 on the load line curve shown in fig. 10.19.

For a base current swing of 30μA the collector current varies unequally over both halves. Similarly if we choose the operating point at Q over the load line and apply a strong signal to the base that causes a base current swing over a range 0-90-180 mamp, the

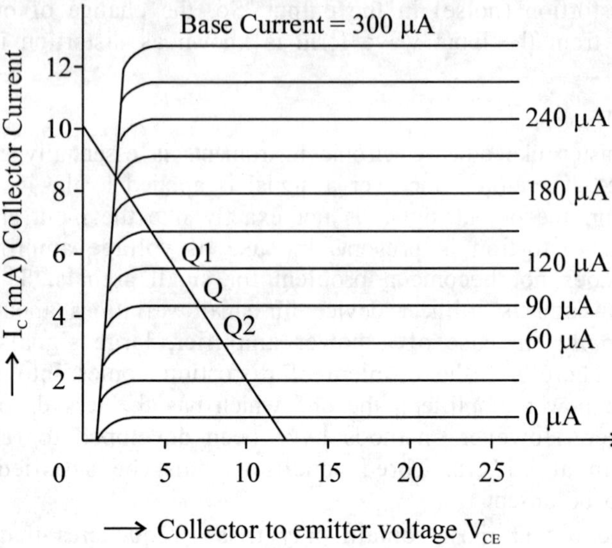

Fig. 10.19. Load line and operating point

collector current swing is between 0-5.85 mamp. Thus in both cases, the output voltage developed in +ve and –ve half of the input cycle is unequal, giving rise to nonlinear distortion.

When non-linear distortion occurs, the output, waveform is no longer a sine wave. The resulting complex wave consists of the original frequency f, and its harmonies multiples of the original frequency, 2*f*, 3*f*....etc.)

Amplitude distortion and hence harmonic generation is minimum in class-A amplifier and max^m in class-C amplifier.

(b) Frequency distortion

Frequency distortion occurs when all frequencies are not amplified equally. Frequency distortion is caused by the circuit associated with transistor that sets a lower and higher limit of frequencies which will be satisfactorily amplified. Moreover each transistor has a higher limiting frequency beyond which the transistor is not useful as an amplifier.

Thus an amplifier where the frequency response curve (fig. 10.21) is horizontal over a wide range is ideal.

(c) Phase distortion

Phase distortion is caused due to the presence of capacitive and inductive elements in the amplifier circuit, because the reactances of these elements change with frequency. This distortion isn't of importance in amplifier of the AF type since phase distortion is not perceptible to ear. But this type of distortion is very objectionable in systems with visual display such as in T.V. Radar system etc. Fig. 10.20 shows how phase distortion affects the wave shape.

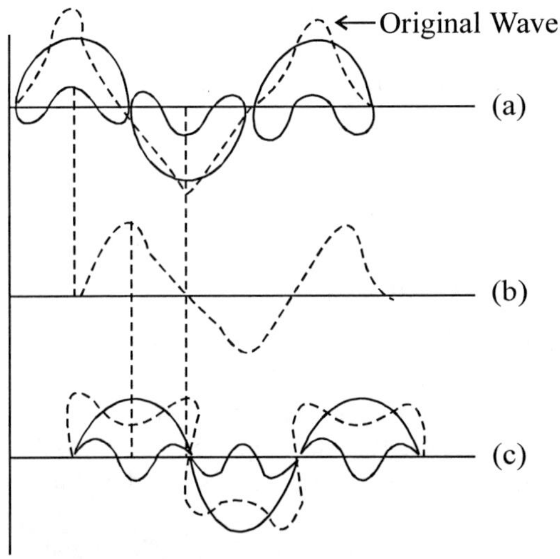

Fig. 10.20. (a) Original waveform (b) Distortionless output (c) Output with phase distortion

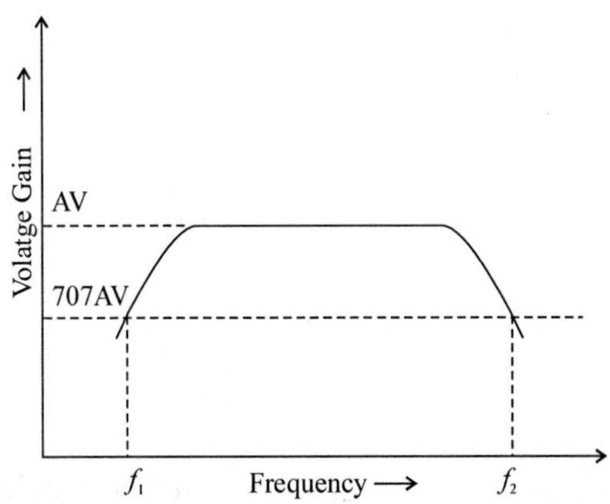

Fig. 10.21

EXERCISES

Numerical Problems with Solutions

1. In a CE amplifier, the load resistance in the collector circuit is 5 KΩ and V_{CC} = 15V. Locate the operating point if the zero signal base current is 20 μA and β = 100.

Solution

$$R_L = 5\ \text{K}\Omega = 5 \times 10^3\ \Omega$$
$$V_{CC} = 15\text{V},\ I_B = 20\ \mu\text{A} = 20 \times 10^{-6}\text{A}$$
$$\beta = 100,\ I_c = ?,\ V_{CE} = ?$$

We have,
$$I_C = \beta\, I_B = 100 \times 20 \times 10^{-6}\text{A}$$
$$= 2000 \times 10^{-6}\text{A}$$
$$= 2\ \text{mA}$$
$$V_{CE} = V_{CC} - I_C R_L$$
$$= 15\text{V} - 2 \times 10^{-3} \times 5 \times 10^3\ \text{V}$$
$$= 15 - 10 = 5\ \text{V}$$

So, the co-ordinates of the operating point are I_c = 2 mA and V_{CE} = 5 V.

2. A given transistor has a current gain factor β = 50. What is its voltage gain when used as an amplifier in CE configuration with assumed output impedance of 10 KΩ and input impedance of 2 KΩ?

Solution

$$\beta = 50,\ R_o = 10\ \text{K}\Omega = 10 \times 10^3\ \Omega$$
$$R_i = 2\ \text{K}\Omega = 2 \times 10^3\ \Omega$$

We have voltage gain
$$= \beta \times \frac{R_0}{R_i}$$
$$= 50 \times \frac{10 \times 10^3}{2 \times 10^3}$$
$$= 50 \times 5 = 250$$

3. A given transistor has a current gain factor β = 49. What is the voltage gain when used as an amplifier in CB configuration with assumed output impedance of 5 KΩ and input impedance of 10Ω?

Solution

$$\beta = 49,\quad R_o = 5\ \text{K}\Omega = 5 \times 10^3\ \Omega$$
$$R_i = 100\ \Omega,\ \text{Voltage gain } A_v = \ ?$$

Now, $\beta = \dfrac{\alpha}{1-\alpha} = 49$

or, $\alpha = 49 - 49\ \alpha$

or, $50\ \alpha = 49$

or, $\alpha = \dfrac{49}{50} = 0.98$

Voltage gain $A_v = \alpha \times \dfrac{R_o}{R_i}$

$$= \frac{0.98 \times 5 \times 10^3}{100}$$

$$= 9.8 \times 5 = 49$$

4. In a transistor amplifier, the collector current changes from 3 mA to 5 mA as the base current is changed from 5 μA to 15 μA. Find the current gain β.

Solution

$$\Delta I_C = 5 \text{ mA} - 2 \text{ mA} = 3 \text{ mA}$$
$$\Delta I_B = 15\ \mu\text{A} - 5\ \mu\text{A} = 10\ \mu\text{A}$$
$$\beta = ?$$

We have $\beta = \dfrac{\Delta I_C}{\Delta I_B} = \dfrac{3 \text{ mA}}{10\ \mu\text{A}}$

$$= \frac{3 \times 10^{-3}}{10 \times 10^{-6}} = 0.3 \times 10^3$$

$$= 300$$

5. In a transistor amplifier, if input signal is 1 mV (rms), R_{AC} = 2 KΩ, R_{in} = 1 KΩ and β = 50, then find the output voltage.

Solution

$$V_{in} = 1 \text{ mV},\ R_{AC} = 2 \text{ K}\Omega$$
$$R_{in} = 1 \text{ K}\Omega,\ \beta = 50,\ V_{out} = ?$$

We have, voltage gain $A_v = \beta \times \dfrac{R_{AC}}{R_{in}}$

$$= 50 \times \frac{2}{1}$$

$$= 100$$

Again, $A_v = \dfrac{V_{out}}{V_{in}}$

So, $V_{out} = A_v \times V_{in}$

$= 100 \times 1 \text{ mV}$

$= 100 \text{ mV}$

6. In a CB amplifier, the collector load resistance is 10 KΩ. If the input resistance is 100 Ω, find the voltage gain and power gain when $\alpha = 0.95$.

Solution

$R_L = 10 \text{ K}\Omega = 10 \times 10^3 \ \Omega$

$R_{in} = 100 \ \Omega, \quad \alpha = 0.95$

$A_v = ?$

We have, $A_v = \alpha \times \dfrac{R_L}{R_{in}}$

$= 0.95 \times \dfrac{10^4}{10^2}$

$= 95$

Power gain, $A_p = \alpha^2 \times \dfrac{R_L}{R_{in}}$

$= 90.25$

7. In a CE amplifier, the collector load resistance is 5 KΩ. If the input resistance is 500Ω, find the voltage gain and power gain when $\beta = 49$.

Solution

$R_L = 5 \text{ K}\Omega = 5 \times 10^3 \ \Omega$

$R_{in} = 500 \ \Omega, \quad \beta = 49$

$A_v = ?, \quad A_p = ?$

We have, $A_v = \beta \times \dfrac{R_L}{R_{in}}$

$= 49 \times \dfrac{5 \times 10^3}{5 \times 10^2}$

$= 490$

Power gain, $A_p = \beta^2 \times \frac{R_L}{R_{in}} = 24010$

8. In a transistor amplifier shown in Fig. 10.22, $R_c = 4\ K\Omega$, $R_E = 5\ K\Omega$ and $V_{CC} = 27V$. Draw the d.c. load line.

Solution

$R_C = 4\ K\Omega,\ R_E = 5\ K\Omega,\ V_{CC} = 27\ V$
$V_{CE} = ?,\ I_C = ?$

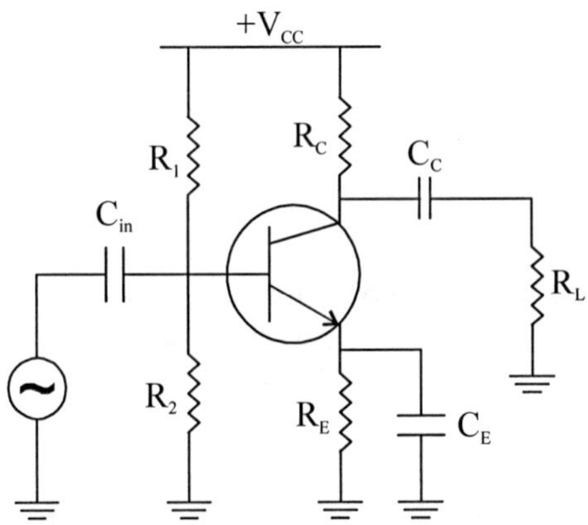

Fig. 10.22

In order to draw d.c. load line, two end points *i.e.*, maximum V_{CE} point and maximum I_C point are required.

Maximum $V_{CE} = V_{CC} = 27V$

This gives the point B (OB = 27V) of the d.c. load line.

$$\text{Maximum } I_C = \frac{V_{CC}}{R_C + R_E} = \frac{27\ V}{(4+5)\,K\Omega}$$

$$= \frac{27\ V}{9 \times 10^3\ \Omega} = 3 \times 10^{-3}\ A$$

$$= 3\ mA$$

This gives the point A (OA = 3mA) of the d.c. load line. Thus, the d.c. load line AB is shown in Fig. 10.23 below :

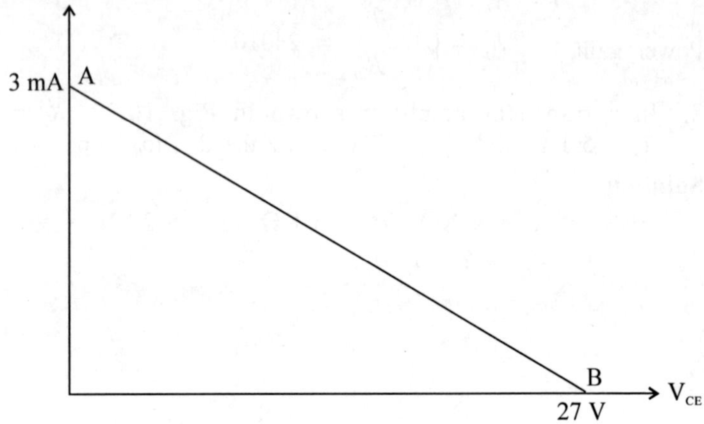

Fig. 10.23

9. Find the operating point for the transistor amplifier circuit of Fig. 10.22 of Q. No. 8, if V_{CC} = 30 V, R_1 = 20 KΩ, R_2 = 20 KΩ, R_C = 4 KΩ, R_E = 5 KΩ. Assume V_{BE} = 0.7 V.

Solution

$$V_2 = \text{The voltage across } R_2 = \frac{30}{2} = 15 \text{ V}$$

$$V_2 = V_{BE} + I_E R_E$$

or, $$I_E = \frac{V_2 - V_{BE}}{R_E} = \frac{(15-0.7)\text{ V}}{5 \text{ K}\Omega} = 2.86 \text{ mA}$$

$$I_C = I_E = 2.86 \text{ mA}$$

Now, $$\begin{aligned} V_{CE} &= V_{CC} - I_C (R_C + R_E) \\ &= 30 - 2.86 \text{ mA } (4+5) \text{ K}\Omega \\ &= 30 - 25.74 = 4.26 \text{ V} \end{aligned}$$

∴ Operating point Q is 4.26 V and 2.86 mA. This point Q can be located after plotting the d.c. load line.

10. For the circuit shown in Fig. 10.22 of Q No. 8, find the voltage gain if β = 100, R_c = 3 KΩ, R_L = 6 KΩ and R_{in} = 2 KΩ.

Solution

a.c. load $$R_{AC} = R_c \parallel R_L = \frac{R_C \times R_L}{R_C + R_L}$$

$$= \frac{3 \times 6}{3+6}$$

$$= \frac{18}{9} = 2\ \text{K}\Omega$$

$$\therefore \text{ Voltage gain } = \beta \times \frac{R_{AC}}{R_{in}} = 100 \times \frac{2}{2} = 100$$

11. In the circuit shown in Fig. 10.22 of Q. No. 8, V_{CC} = 30V, R_1 = 2 KΩ, R_2 = 1 KΩ, R_c = 2 KΩ, R_L = 2 KΩ, R_E = 1 KΩ. Draw the d.c. and a.c. load lines and the operating point Q.

Solution

(i) d.c. load line :

For drawing d.c. load line, we require two end points, *i.e.*, maximum V_{CE} point and maximum I_C point.

Maximum $V_{CE} = V_{CC}$ = 30 V

This locates the point B (OB = 30V) of the d.c. load line.

$$\text{Maximum } I_C = \frac{V_{CC}}{R_C + R_E} = \frac{30\ \text{V}}{(2+1)\ \text{K}\Omega}$$

$$= 10\ \text{mA}$$

This locates the point A (OA = 10 mA) of the d.c. load line. Fig. 10.24 below shows the d.c. load line AB.

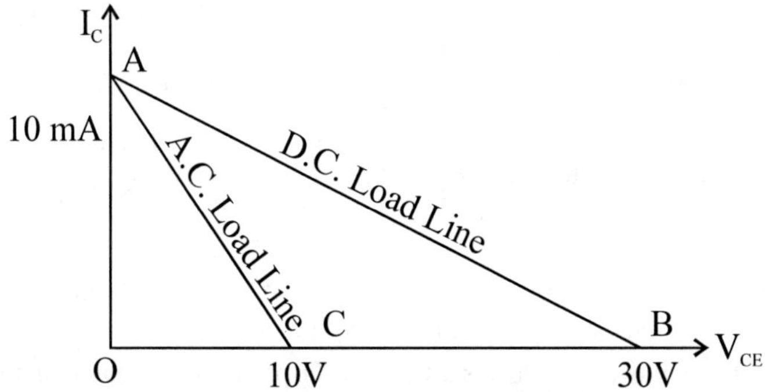

Fig. 10.24

(ii) Operating point Q :

$$V_2 = \text{The voltage across } R_2 = \frac{R_2}{R_1 + R_2} \times 30$$

$$30 \times \frac{1}{2+1} = \frac{30}{3} = 10 \text{ V}$$

Now, $V_2 = V_{BE} + I_E R_E$

or, $$I_E = \frac{V_2 - V_{BE}}{R_E} = \frac{(10-0)\,V}{1\,K\Omega} = 10 \text{ mA}$$

$$I_C = I_E = 10 \text{ mA}$$

Now, $$V_{CE} = V_{CC} - I_C (R_C + R_E)$$
$$= 30 - 10 \text{ mA } (2 + 1) \text{ K}\Omega$$
$$= 30 - 30 = 0.0 \text{ V}$$

Thus, the operating point Q is 0.0 V, 10 mA. So, the operating point Q concides with the point A of the d.c. load line AB drawn in Fig. 10.24 above.

(iii) a.c. load line :

In order to draw a.c. load line, we require two end points, *i.e.*, maximum V_{CE} point and maximum I_C point when signal is applied.

a.c. load, $$R_{AC} = R_C \parallel R_L = \frac{2 \times 2}{2+2} = \frac{4}{4}$$
$$= 1 \text{ K}\Omega$$

$$(V_{CE})_{max} = V_{CE} + I_C R_{AC}$$
$$= 0 + (10 \text{ mA} \times 1 \text{ K}\Omega)$$
$$= 10 \text{ V}$$

This locates the point 'C' (OC = 10V) on the V_{CE} Axis.

$$\text{Maximum collector current} = I_C + \frac{V_{CE}}{R_{AC}} = I_C + \frac{0.0 \text{ V}}{1 \text{ K}\Omega}$$
$$= I_C = 10 \text{ mA}$$

This locates the point A (OA = 10 mA) on the I_C axis. We get the a.c. load line AC by joining the points A and C.

It should be noted here that the Q-point is the point of intersection of d.c. and a.c. load lines. In this problem the point 'A' is the operating point Q.

12. A transistor amplifier circuit shown in Fig. 10.25 below consists of R_1 = 10 KΩ, R_2 = 5 KΩ, R_C = 1 KΩ, R_E = 2 KΩ and R_L = 1 KΩ. Draw d.c. and a.c. load lines taking V_{BE} to be 0.7 V.

Solution

In order to draw d.c. load line, we require two end points, *i.e.*, maximum V_{CE} point and maximum I_C point.

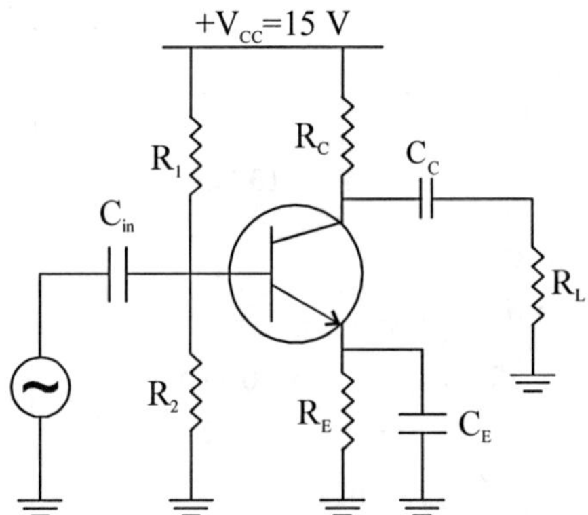

Fig. 10.25

Maximum $V_{CE} = V_{CC}$ = 15 V

This locates the point B (OB = 15 V) of the d.c. load line.

$$\text{Maximum} \quad I_C = \frac{V_{CC}}{R_C + R_E}$$

$$= \frac{15\,V}{(1+2)\,K\Omega} = 5\,mA$$

This locates the point A (OA = 5 mA) of the d.c. load line. Fig. 10.26 shows the d.c. load line AB.

(ii) Operating point Q

$$V_2 = \text{The voltage across} \quad R_2 = 15 \times \frac{R_2}{R_1 + R_2}$$

$$= 15 \times \frac{5}{(10+5)}$$

$$= 5\text{ V}$$

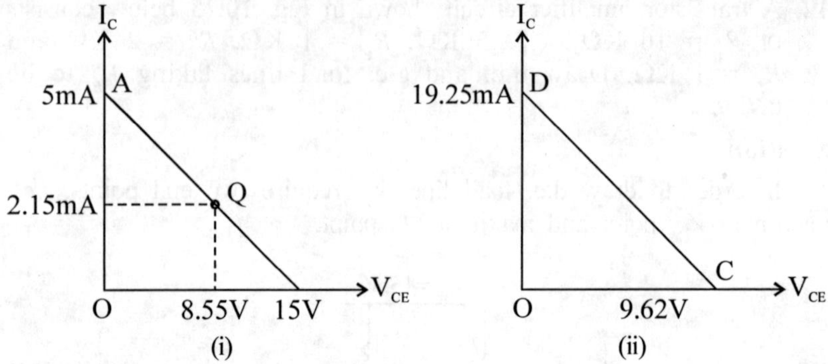

Fig. 10.26

Now, $V_2 = V_{BE} + I_E R_E$

or, $I_E = \dfrac{V_2 - V_{BE}}{R_E} = \dfrac{(5-0.7)\text{ V}}{2\text{ K}\Omega} = 2.15\text{ mA}$

$I_C = I_E = 2.15\text{ mA}$

Hence $V_{CE} = V_{CC} - I_C\,(R_C + R_E)$

$= 15 - 2.15\text{ mA} \times (1 + 2)\text{ K}\Omega$

$= 15 - (2.15 \times 3)$

$= 15 - 6.45 = 8.55\text{ V}$

∴ Operating point Q is 8.55 V, 2.15 mA. This is shown on the d.c. load line.

(iii) a.c. load line :

To draw a.c. load line, we require two end points, *i.e.*, maximum V_{CE} point and maximum I_C point when the signal is applied.

a.c. load, $R_{AC} = R_C \,||\, R_L$

$$= \frac{R_C \times R_L}{R_C + R_L}$$

$$= \frac{1 \times 1}{1+1} = \frac{1}{2}\text{ K}\Omega$$

$= 0.5\text{ K}\Omega$

$(V_{CE})_{max} = V_{CE} + I_C\, R_{AC}$

$= 8.55 + (2.15\text{ mA} \times 0.5\text{ K}\Omega)$

$= 9.62$ Volts

This locates the point C (OC = 9.62 V) of the V_{CE} axis.
Maximum collector current,

$$(I_C)_{\max} = I_C + \frac{V_{CE}}{R_{AC}}$$

$$= 2.15 + \frac{8.55}{0.5}$$

$$= 2.15 + 17.10$$

$$= 19.25 \text{ mA}$$

This locates the point D (OD = 19.25 mA) on the I_C axis. By joining points C and D, a.c. load line CD is drawn.

13. In a transistor amplifier, when the signal changes by 0.02 V, the base current changes by 10 μA and collector current by 1 mA. If collector load R_C = 5 KΩ and R_L = 10 KΩ, find (i) current gain (ii) input impedance (iii) a.c. load (iv) voltage gain (v) power gain.

Solution

ΔI_B = 10 μA, ΔI_C = 1 mA, ΔV_{BE} = 0.02 V
R_C = 5 KΩ, R_L = 10 KΩ

(i) Current gain, $\beta = \frac{\Delta I_C}{\Delta I_B} = \frac{1 \text{ mA}}{10\ \pi\text{A}} = 100$

(ii) Input impedance, $R_{in} = \frac{\Delta V_{BE}}{\Delta I_B} = \frac{0.02\ V}{10\ \mu A}$

$$= 2 \text{ K}\Omega$$

(iii) a.c. load, $R_{AC} = R_C \parallel R_L = \frac{R_C \times R_L}{R_C + R_L}$

$$= \frac{5 \times 10}{5 + 10} = \frac{50}{15} = 3.33 \text{ K}\Omega$$

(iv) Voltage gain, $A_v = \beta \times \frac{R_{AC}}{R_{in}} = 100 \times \frac{3.33}{2} = \frac{333}{2} = 166.6$

(v) Power gain, A_P = Current gain × Voltage gain

$$= \beta \times A_V$$

$$= 100 \times 166.6$$

$$= 16660$$

14. In a transistor amplifier circuit, $\beta = 50$, $R_1 = 10$ KΩ, $R_2 = 5$ KΩ, $R_C = 3$ KΩ, $R_E = 3$ KΩ, $R_L = 6$ KΩ, $R_{in} = 0.5$ KΩ. Find output voltage if input voltage is 1 mV (rms).

Solution

a.c. load resistance, $R_{AC} = R_C \parallel R_L$

$$= \frac{R_C \times R_L}{R_C + R_L}$$

$$= \frac{3 \times 6}{3+6}$$

$$= \frac{18}{9} = 2 \text{ K}\Omega$$

∴ Voltage gain, $A_V = \beta \times \frac{R_{AC}}{R_{in}}$

$$= 50 \times \frac{2}{0.5}$$

$$= 200$$

Hence, $\frac{V_{out}}{V_{in}} = 200$

$$V_{out} = 200 \times V_{in} = 200 \times 1 \text{ mA}$$
$$= 200 \text{ mV}$$

15. A silicon NPN transistor has the operating point at $I_C = 2$ mA in common emitter configuration. If $V_{CC} = 9$ V, find the value of R_B in the base circuit. Given, $\beta = 50$.

Solution

$I_C = 2$ mA, $V_{CC} = 9$ V, $\beta = 50$,

$R_B = ?$, $I_C = \beta I_B$ or $I_B = \frac{I_C}{\beta} = \frac{2\ mA}{50} = 40\ \mu\text{A}$

For silicon transistor, $V_{BE} = 0.7$ V

Using the relation,

$$V_{CC} = I_B R_B + V_{BE}$$

or, $R_B = \frac{V_{CC} - V_{BE}}{I_B}$

$$= \frac{9 - 0.7}{40 \times 10^{-6}} \text{ ohm}$$

$$= \frac{8.3 \times 10^{+6}}{40} = \frac{8.3}{4} \times 10^{5} \text{ ohm}$$

$$= \frac{830}{4} \text{ Kilo-ohm}$$

$$= 207.5 \text{ K}\Omega$$

16. An amplifier, when loaded by 2 KΩ resistor, has a voltage gain of 80 and a current gain of 120. Determine the necessary signal voltage and current to give an output voltage of 1 V. Find the power gain of the amplifier.

Solution

$R_L = 2 \text{ K}\Omega,\ A_V = 80,\ A_i = 120,$
$V_2 = 1 \text{ V},\ V_1 = ?,\ I_I = ?$

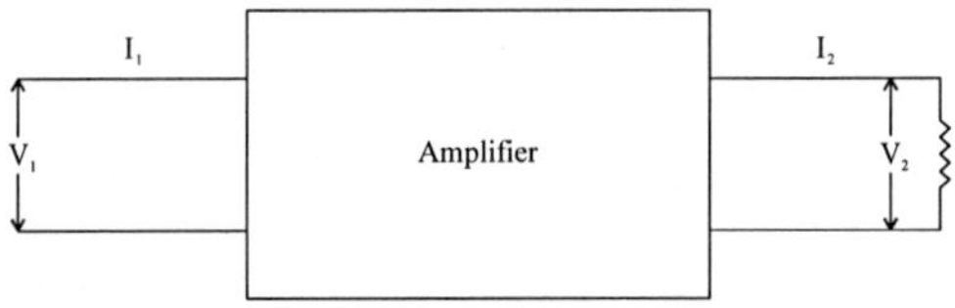

Fig. 10.27

$$A_v = \frac{V_2}{V_1} = 80$$

$$\therefore \quad V_1 = \frac{V_2}{A_V} = \frac{1}{80} = 0.0125 \text{ V} = 12.5 \text{ mV}$$

Again, $A_v = \frac{A_0 R_L}{R_0 + R_L}$ where A_o is the open circuit voltage gain.

$$A_i = \frac{A_0 R_{in}}{R_0 + R_L}$$

$$\therefore \quad \frac{A_v}{A_i} = \frac{R_L}{R_{in}}$$

or, $$\frac{80}{120} = \frac{2}{R_{in}}$$

or, $R_{in} = \dfrac{2 \times 120}{80} = 3 \text{ K}\Omega$

Now, $I_1 = \dfrac{V_{in}}{R_{in}} = \dfrac{V_1}{R_{in}} = \dfrac{12.5 \text{ mV}}{3 \text{ K}\Omega}$

$= 4.17\ \mu\text{A}$

Power gain $= A_v \times A_i = 80 \times 120 = 9600$

17. In a CE circuit employing base resistor bias method, $R_E = 500\ \Omega$, $R_B = 20{,}000\ \Omega$ and $\alpha = 0.95$. Calculate the stability factor.

Solution

$R_E = 5 \times 10^2$ ohms, $R_B = 2 \times 10^4$ ohms

$\alpha = 0.95$, $S = ?$

Here, $$S = \frac{1 + R_E/R_B}{1 - \alpha + R_E/R_B}$$

$$= \frac{1 + \frac{5\times10^2}{2\times10^4}}{1 - 0.95 + \frac{5\times10^2}{2\times10^4}}$$

$$= \frac{1 + 2.5 \times 10^{-2}}{0.05 + 2.5 \times 10^{-2}}$$

$$= \frac{1 + 0.025}{0.05 + 0.025}$$

$$= \frac{1.025}{0.075}$$

$$= \frac{1025}{75} = 13.666....$$

$$\approx 13.7$$

18. In the circuit given in Q. No. 17, if R_E is shorted, what is the value of stability factor?

Solution

$R_E = 0$

So, $$S = \frac{1}{1-\alpha} = \frac{1}{1-0.95}$$

$$= \frac{51 \times 10^{-3}}{2 \times 3.142 \times 50 \left[1 + \frac{20}{9}\right]}$$

$$= \frac{51 \times 10^{-3} \times 9}{314.2 \times 29} = 50 \times 10^{-6} F = 50\ \mu F$$

26. In an RC coupled amplifier shown in Fig. 10.28 of Q. No. 24, calculate the lower cut-off frequency assuming $R_L = 5$ KΩ, $R_1 = 6$ KΩ, $R_2 = 12$ KΩ, $R_{in} = 100$ Ω and $C_C = 10$ μF.

Solution

$$R_B = \frac{R_1 R_2}{R_1 + R_2}$$

$$= \frac{6 \times 12}{6 + 12} = \frac{72}{18} = 4\ K\Omega$$

$$f_1 = \frac{1}{2\pi C_C \left(R_L + \frac{R_B R_{in}}{R_B + R_{in}}\right)}$$

$$= \frac{1}{2\pi \times 10 \times 10^{-6}\left(50 \times 10^3 + \frac{4 \times 10^3 \times 100}{4000 + 100}\right)}$$

$$= \frac{1}{20\pi \times 10^{-6}\left(50 \times 10^3 + \frac{4 \times 10^5}{4100}\right)}$$

$$= \frac{1}{20\pi \times 10^{-6}\left(50 \times 10^3 + \frac{4 \times 10^3}{41}\right)}$$

$$= \frac{1}{20\pi \times 10^{-3}\left(5 + \frac{4}{41}\right)}$$

$$= \frac{10^3}{20 \times 3.142 \times \frac{209}{41}}$$

$$= \frac{41 \times 10^3}{3.142 \times 2 \times 209} \approx 3\ Hz$$

27. A single stage amplifier has gain of 50. The collector load R_C = 500 Ω and the input impedance is 1 KΩ. Calculate the overall gain when two such stages are cascaded through R-C coupling. Discuss on the result.

Solution

The gain of second stage remains 50 because it has not loading effect of any stage. However, the gain of first stage is less than 50 due to the loading effect of the input impedance of second stage.

$\therefore$ Gain of second stage = 50

Effective load of first stage = R_C // R_{in}

$$R_{AC} = \frac{500 \times 1000}{500 + 1000} = 333 \ \Omega$$

$$\text{Gain of first stage} = \frac{50 \times 333}{2 \times 500} = 16.6$$

Total gain = 50 × 33.3/2 = 1665/2 = 832.5

Discussion. The gain of individual stage is 50. But, coupling the two stages, he gain is not 50 × 50 = 2500 as might be expected rather it is less and is equal to 1665 in this case. It is because the first stage has a loading effect of the input impedance of second stage and as a result of which its gain is reduced. However, the second stage has no loading effect of any subsequent stage. Hence, the gain of second stage remains 50.

28. A two-stage RC coupled amplifier is shown in Fig. 10.29. If the input resistance R_{in} of each stage is 1 KΩ, find (i) voltage gain of first stage (ii) voltage gain of second stage and (iii) total voltage gain.

Solution

R_{in} = 1 KΩ, β = 100, R_C = 2 KΩ

(i) The first stage has a loading of input resistance of second stage.

$\therefore$ Effective load of first stage,

$$R_{AC} = R_C \,//\, R_{in} = \frac{2 \times 1}{2 + 1}$$

$$= \frac{2}{3} = 0.66 \ \text{K}\Omega$$

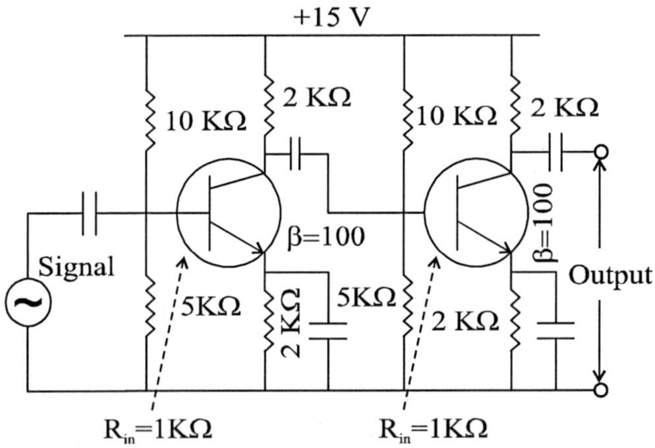

Fig. 10.29

∴ Voltage gain of first stage

$$= \beta \times \frac{R_{AC}}{R_{in}} = \frac{100 \times 0.66}{1} = 66$$

(ii) The collector of the second stage has a load of only R_C (= 2 KΩ) as there is no loading effect of any subsequent stage.

∴ Voltage gain of second stage

$$= \beta \times \frac{R_C}{R_{in}} = \frac{100 \times 2}{1} = 200$$

(iii) Total voltage gain = 66 × 200 = 13200

29. A multistage amplifier consists of four identical stages of RC coupled amplifiers. If the bandwidth of individual stage is 16 KHz, what is the bandwidth of the amplifier?

Solution

$$\Delta f = f_2 - f_1 = 16 \text{ KH}_z$$

$$B \approx \Delta f / \sqrt{n} = \frac{16\text{KHz}}{\sqrt{4}} = \frac{16}{2} \text{ KHz}$$

$$= 8 \text{ KHz}$$

30. A multistage amplifier consists of two identical stages of RC coupled amplifiers. If the voltage gain for individual stage is 20, calculate the voltage gain of the amplifier assuming no loss in interstage transfer and no loading effect.

Solution

$$n = 2,\ A = 20$$

$$A_V = A^n = (20)^2 = 400$$

31. A single stage RC coupled amplifier has a voltage gain of 20. The collector load R_L = 5 KΩ and the input impedance R_{in} = 500 Ω. Calculate the overall gain when two such stages are cascaded.

Solution

$$R_L = 5\ \text{K}\Omega,\quad R_{in} = 500\ \Omega$$

The gain of the second stage remains 20 because it has no loading effect of any stage. However, the gain of first stage is less than 20 due to loading effect of the input impedance of second stage.

∴ Gain of second stage = 20

Effective load of first stage

$$R_{AC} = R_L \,//\, R_{in}$$

$$= \frac{R_L \times R_{in}}{R_L + R_{in}}$$

$$= \frac{5 \times 10^3 \times 500}{5000 + 500}$$

$$= \frac{25 \times 10^5}{5500} = 454\ \Omega$$

Gain of first stage

$$= 20 \times \frac{R_{AC}}{R_{in}} = \frac{20 \times 454}{500}$$

$$= 18.16$$

∴ Overall gain of the amplifier = 20 × 18.16

= 363.2

32. In a cascaded amplifier, transformer coupling is used in the last stage to deliver the output to a speaker of resistance 16 Ω. If the output resistance of the final stage is 3200 Ω, find the turn ratio of the output transformer for maximum power transfer.

Solution

For maximum power transfer, the impedance of the primary should be equal to the output impedance of the 1st stage and impedance of the secondary should be equal to the load impedance.

Primary impedance = 3200 Ω

Turn ratio of the transformer, $n = \dfrac{N_S}{N_P}$

Primary impedance $\left(\dfrac{N_P}{N_S}\right)^2 \times$ Load impedance

or, $$\left(\frac{N_P}{N_S}\right)^2 = \frac{\text{Primary impedance}}{\text{Load impedance}}$$

$$= \frac{3200}{16} = \frac{200}{1}$$

or, $$\frac{N_P}{N_S} = \frac{10\sqrt{2}}{1} = \frac{10 \times 1.414}{1} = \frac{14.14}{1}$$

or, $$\frac{N_S}{N_P} = n = 1:14.14$$

33. Transformer coupling is used in the final stage of a cascaded amplifier. If the output resistance of the final stage is 800 Ω and the output is delivered to a speaker of resistance 8 Ω, find the turn ratio of the transformer, so that maximum power is transferred to the load.

Solution

Primary impedance = 800 Ω

Load impedance = 8 Ω

$$\left(\frac{N_P}{N_S}\right)^2 = \frac{\text{Primary impedance}}{\text{Load impedance}}$$

$$= \frac{800}{8} = \frac{100}{1}$$

or, $$\frac{N_P}{N_S} = \frac{10}{1}$$

$\therefore$ Turn ratio of the transformer,

$$n = \frac{N_S}{N_P} = 1 : 10$$

Thus, a step down transformer with the turn ratio 1 : 10 is needed.

34. We are to match a 16 Ω speaker load to an amplifier so that the effective load resistance is 10 KΩ. What should be the transformer turn ratio?

Solution

$$R_L = 16\ \Omega,\ R_L^1 = 10\ \text{K}\Omega = 10 \times 10^3\ \Omega$$

$$\text{Transformer turn ratio} = n = \frac{N_s}{N_p}$$

$$\therefore \left(\frac{N_s}{N_p}\right)^2 = \frac{R_L}{R_L^1} = \frac{16\Omega}{10 \times 10^3\ \Omega}$$

$$= \frac{16}{10^4}$$

or, $$\frac{N_S}{N_P} = \sqrt{\frac{16}{10^4}} = \frac{4}{10^2} = \frac{1}{25}$$

or, $n = 1 : 25$

35. Determine the necessary transformer turn ratio for transferring maximum power to a 16 Ω load from a source that has an output impedance of 10 KΩ. Also, calculate the voltage across the external load if the terminal voltage of the source is 10 V rms.

Solution

$$R_L = 16\Omega,\quad R_L^1 = 10\ \text{K}\Omega = 10 \times 10^3\ \Omega$$

$$V_P = 16\text{V},\quad V_S = ?$$

$$\text{Transformer turn ratio, } n = \frac{N_S}{N_P}$$

$$\left(\frac{N_S}{N_P}\right)^2 = \frac{R_L}{R_L^1} = \frac{16}{10^4}$$

or, $$n = \frac{N_S}{N_P} = \frac{4}{10^2} = \frac{4}{100} = \frac{1}{25}$$

Again, $$\frac{V_S}{V_P} = \frac{N_S}{N_P}$$

or, $$V_S = \left(\frac{N_S}{N_P}\right) \times V_P$$

$$= \frac{1}{25} \times 10 = \frac{10}{25} = 0.4 \text{ V}$$

36. A transistor uses transformer coupling for amplification. The impedance of transistor is 20 KΩ while the input impedance of next stage is 5 KΩ. Determine the inductance of primary and secondary of the transformer for perfect impedance matching at a frequency of 200 Hz.

Solution

$f = 200$ Hz,

$R_o = 20$ KΩ, $R_i = 5$ KΩ

$L_P = ?$, $L_s = ?$

Output impedance of transistor = primary impedance

or, $20 \times 10^3 = 2\pi f L_p$

or, $$L_P = \frac{20 \times 10^3}{2\pi \times 200} = \frac{1000}{20 \times 3.142}$$

$$= \frac{50}{3.142} = 15.9 \approx 16 \text{ H}$$

Again,

Input impedance of next stage = Impedance of Secondary

or, $5 \times 10^3 = 2\ \pi f L_s$

or, $$L_S = \frac{5 \times 10^3}{2\pi \times 200} = \frac{50}{4\pi} = \frac{25}{2 \times 3.142}$$

$$= 3.978 \text{ H} \approx 4 \text{ H}$$

37. A three stage amplifier has a first stage voltage gain of 100, second stage voltage gain of 200 and third stage gain of 400. Find the total voltage gain in db.

Solution

First stage voltage gain in db $= 20 \log_{10} 100$
$= 40$

Second stage voltage gain in db $= 20 \log_{10} 200$
$= 20 \times 2.3 = 46$

Third stage voltage gain in db $= 20 \log_{10} 400$
$= 20 \times 2.6 = 52$

Total voltage gain $= 40 + 46 + 52 = 138$ db.

❑ ❑ ❑

11

Feedback Amplifiers

Q. 1. What do you understand by feedback?

Ans. The process of injecting a fraction of output energy of some device back to the input is known as feedback.

Q. 2. What are different types of feedback?

Ans. There are two types of feedback :

(a) Positive feedback

(b) Negative feedback

(a) Positive feedback

When the feedback energy is in phase with the input signal, and thus increases the input signal, the feedback is called +ve feedback.

Positive feedback increases gain of the amplifier. However it has the disadvantage of increased distortion and instability. Therefore +ve feedback isn't used in amplifiers. But it is employed in oscillators. If +ve feedback is very large, it leads to oscillations.

(b) Negative feedback

When the feedback energy is out of phase with the input signal, and thus decreases the strength of the input signal, the feedback is called –ve feedback.

Negative feedback reduces gain of the amplifier. Its advantages are reduction in distortion, stability in gain, increased bandwidth and improved input and output impedances. Therefore it is frequently used in amplifiers.

Q. 3. Why –ve feedback is necessary in amplifier?

Ans. A practical amplifier has a very high gain. That means the output is one million times the input. As a result of which, even a small disturbance in the input will appear in the amplified form in the output. This introduces noise in the output. Therefore every high gain amplifier produces noise in the output of the amplifier. This is not desirable. Negative feedback in amplifiers reduces this noise considerable. By the use of –ve feedback, a fraction of output is fed-back to the input in a phase opposite to the input signal. So distortion is reduced, gain of the amplifier becomes stable. Due to these advantages the –ve feedback is frequently employed in amplifiers.

Q. 4. Explain the principles of –ve feedback in amplifiers.

Ans. There are two sections in a feedback amplifier, such as an amplifier and a feedback circuit. The feedback circuit usually consists of resistors. It returns a fraction of output energy back to the input. Fig. 11.1 shows the principles of –ve feedback in an amplifier.

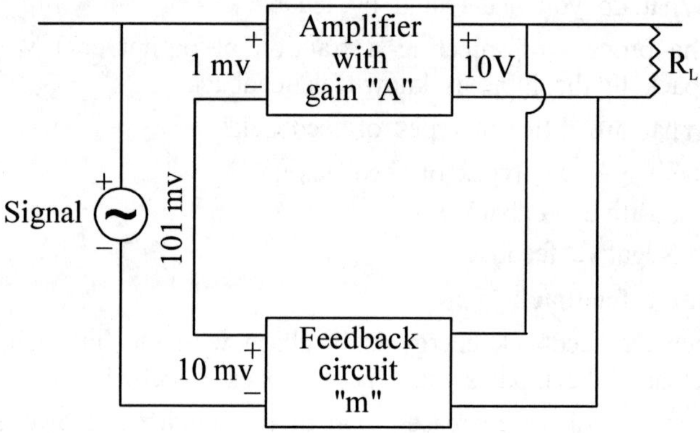

Fig. 11.1

Let us assume typical values of input and output voltage of the amplifier. Let the output of the amplifier is 10 V. The fraction "*m*" of this output is fed back to the input, where it is applied in series with the input signal of 101 mV. Let the fraction of the output fed back to the input be 100 mV. As the feedback is –ve the resultant input signal becomes (101 mV – 100 mV = 1 mV)

Gain of amplifier without feedback

$$A = \frac{10\text{ V}}{1\text{ mV}} = \frac{10\text{ V}}{10^{-3}\text{ V}} = 10^4$$

Fraction of output fed back,

$$m = \frac{100\text{ V}}{10\text{V}} = 100\times10^{-3}\text{ V}\times10^{-1} = 10^{-2} = 0.01$$

Gain of the amplifier with feedback,

$$A_{fb} = \frac{10\ V}{101\ mV} = \frac{10}{101} \times 10^3 \approx 100$$

Thus –ve feedback reduces the gain of the amplifier.

Points to remember

(i) When –ve feedback is applied, the gain of the amplifier is reduced.

(ii) When –ve feedback is applied, the voltage actually applied to the amplifier becomes extremely small.

(iii) In the –ve feedback circuit, the feedback fraction '*m*' is always between zero and 1.

(iv) The gain with feedback is sometimes called closed-loop gain. The gain without feedback is called open-loop gain.

Q. 5. Find an expression for the voltage gain with –ve feedback in an amplifier.

Ans. Let us consider the –ve feedback amplifier shown in fig. 11.2 below. The gain of the amplifier without feedback is A. Negative feedback is then applied by feeding a fraction '*m*' of the output voltage e_0 back to amplifier input. Therefore, the actual input to the amplifier is the signal voltage *e.g.* minus feedback voltage me_0 *i.e.*,

Actual input to amplifier = $e_g - me_0$

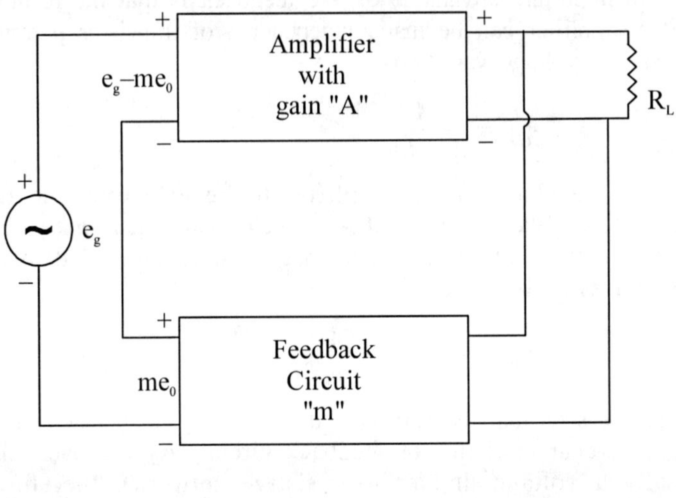

Fig. 11.2

The output e_o must be equal to the input voltage $e_g - me_o$ multiplied by gain A of the amplifier as $A = \frac{e_o}{e_g - me_o}$ ...(1)

So, $(e_g - me_o)A = e_o$

or, $Ae_g - Ame_o = e_o$

or, $e_o(1 + Am) = Ae_g$

or, $\frac{e_o}{e_g} = \frac{A}{1+Am}$

But e_o/e_g is the voltage gain of the amplifier with feedback

∴ Voltage gain with –ve feedback,

$$A_{fb} = \frac{A}{1+Am} \quad ...(2)$$

It may be seen that the gain of the amplifier without feedback is A. However, when –ve feedback is applied, the gain is reduced by a factor $1 + Am$.

Q. 6. Discuss the advantages of –ve feedback.

Ans. *Advantages of –ve feedback :* The following are the advantages of –ve feedback in amplifiers :

(i) Gain stability

An important advantage of –ve feedback is that the resultant gain of the amplifier can be made independent of transistor parameters or the supply voltage variations.

$$A_{fb} = \frac{A}{1+Am}$$

For –ve feedback in an amplifier to be effective, the designer deliberately makes the product Am much greater than unity. Therefore, in the above relation, 1 can be neglected as compared to *Am* and the expression becomes :

$$A_{fb} = \frac{1}{1+Am} = \frac{1}{m}$$

It may be seen that the gain now depends only upon feedback fraction '*m*' *i.e.*, characteristics of feedback circuit. As feedback circuit is usually a voltage divider (a resistive network), therefore, it is unaffected by changes in temperature, variations in transistor parameters and frequency. Hence the gain of the amplifier is extremely stable.

(ii) Reduces non-linear distortion

A large signal stage has non-linear distortion because its voltage gain changes at various points in the cycle. The –ve feedback reduces the non-linear distortion in large signal amplifiers. It can be proved that :

$$D_{fb} = \frac{D}{1 + Am}$$

where D = distortion in amplifier without feedback

D_{fb} = distortion in amplifier with the feedback.

It is clear that by applying –ve feedback to an amplifier, distortion is reduced by a factor $1 + Am$.

(iii) Improves frequency response

As feedback is usually obtained through a resistive network, therefore, voltage gain of the amplifier is independent of signal frequency. The result is that voltage gain of amplifier will be substantially constant over a wide range of signal frequency. The –ve feedback, therefore, improves the frequency response of the amplifier.

(iv) Increases circuit stability

The output of an ordinary amplifier is easily changed due to variations in ambient temperature, frequency and signal amplitude. This changes the gain of the amplifier, resulting in distortion. However, by applying –ve feedback, voltage gain of the amplifier is stabilised or accurately fixed in value. This can be easily explained. Suppose the output of a –ve feedback amplifier has increased because of temperature change or due to some other reason. This means more –ve feedback since feedback is being given from output. This tends to oppose the increase in an amplification and maintain it stable. The same is true should the output voltage decrease. Consequently, the circuit stability is considerably increased.

(v) Increases input impedance and decreases output impedance

The –ve feedback increases the input impedance and decreases the output impedance of amplifier. Such a change is profitable in practice as the amplifier can then serve the purpose of impedance matching.

(a) Input impedance

The increase in input impedance with negative feedback can be explained by referring to fig. 11.3. Suppose the input impedance of the amplifier is Z_{in} without feedback and Z'_{in} with –ve feedback. Let us further assume that input current is i_1.

Referring to fig. 11.3,

$$e_g - me_o = i_1 Z_{in}$$

Now,
$$e_g = (e_g - me_o) + me_o$$
$$= (e_g - me_o) + Am\ (eg - me_o)$$
$$[\because e_o = A\ (e_g - me_o)]$$
$$= (e_g - me_o)\ (1 + Am)$$
$$= i_1 Z_{in}\ (1 + Am) \qquad (\because e_g - me_o = i_1 Z_{in})$$

or,
$$\frac{e_g}{i_1} = Z_{in}\ (1 + Am)$$

But $e_g/i_1 = Z'_{in}$, the input impedance of amplifier with –ve feedback.

$$\therefore \quad Z'_{in} = Z_{in}\ (1 + Am)$$

It is clear that by applying –ve feedback, the input impedance of the amplifier is increased by a factor 1 + *Am*. As '*Am*' is much greater than unity, therefore, input impedance is increased considerably.

(b) Output impedance

Following similar line, we can show that output impedance with –ve feedback is given by,

$$Z'_{out} = \frac{Z_{out}}{1 + Am}$$

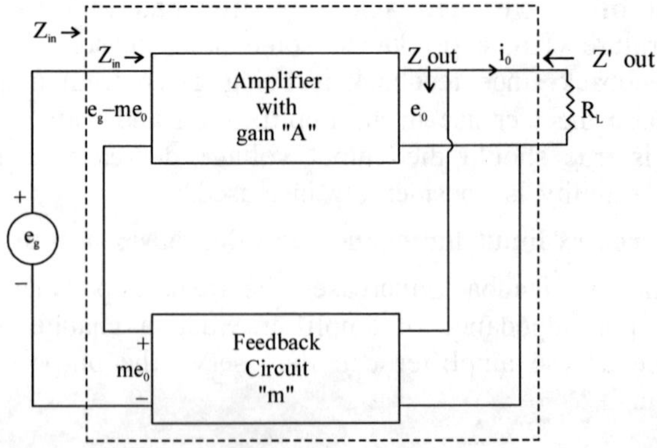

Fig. 11.3

where Z'_{out} = Output impedance with –ve feedback

Z_{out} = Output impedance without feedback

It is clear that by applying –ve feedback, the output impedance of the amplifier is decreased by a factor $1 + Am$.

Q. 7. What is a feedback circuit? Explain how it provides feedback in amplifiers.

Ans. *Feedback circuit :* The function of the feedback circuit is to return a fraction of the output voltage to the input of the amplifier. Fig. 11.4 shows the feedback circuit of –ve feedback amplifier. It is essentially a potential divider consisting of resistances R_1 & R_2. The output voltage of the amplifier is fed to this potential divider which gives the feedback voltage to the input. Referring to fig. 11.4, it is clear that :

$$\text{Voltage across } R_1 = \left(\frac{R_1}{R_1 + R_2}\right) e_o$$

Feedback fraction,

$$m = \frac{\textit{Voltage across } R_L}{e_o}$$

$$\therefore \quad m = \frac{R_1}{R_1 + R_2}$$

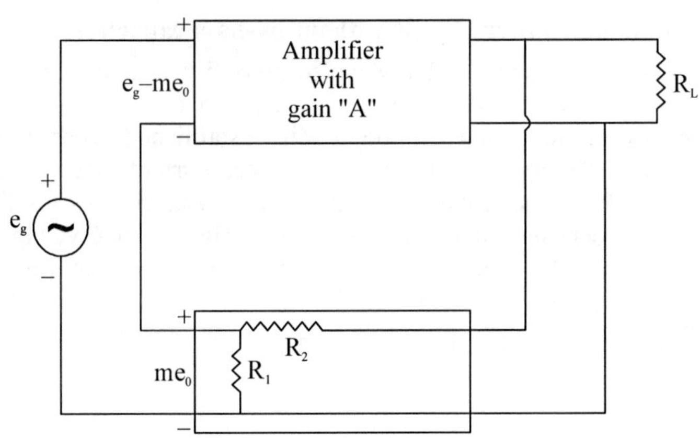

Fig. 11.4. Feedback Circuit

Q. 8. What are different types of –ve feedback circuits.

Ans. Two types of feedback are used, *viz.*,

(1) Negative voltage feedback
(2) Negative current feedback

(1) Negative voltage feedback

In this method, the voltage feedback to the input of the amplifier is proportional to the output voltage. Negative voltage feedback reduces the output impedance and also decreases the gain of the amplifier. However, it reduces distortion and increases bandwidth. Therefore, it is extensively used in electronic circuits. In all modern high gain transistor amplifiers (*e.g.* public address system, transistor radio), negative voltage feedback is always employed.

(2) Negative current feedback

In this method, the voltage fedback to the input of the amplifier is proportional to the output current. Negative current feedback improves the high frequency response of an amplifier. It increases the input impedance and decreases the output impedance of the amplifier. For thin reason, it is often employed for impedance matching.

Q. 9. Describe the action of negative current feedback transistor circuits.

Ans. *Negative current feedback transistor circuits* : There are two important negative current feedback circuits *viz.*

(i) Common emitter circuit without by-pass capacitor.
(ii) Common emitter follower.

(i) Common emitter circuit without by-pass capacitor

Fig. 11.5 shows an R-C coupled amplifier without emitter by-pass capacitor across emitter resistance R_E. This is a negative current feedback circuit. This can be readily established when a signal is applied at the input, resulting a.c. emitter current *i.e.*, i_e produces a voltage $i_e R_E$ across emitter resistance. This voltage opposes the signal as it is in opposite direction to the signal. The negative voltage fedback is proportional to the emitter current i_e. It is in this way that the circuit provides –ve current feedback.

Referring to fig. 11.5, it is clear that

$$i_e = \frac{-e_o}{R_c}$$

The –ve sign comes with e_o, because input and output are 180° out of phase with each other.

∴ Base emitter voltage,

$$V_{be} = e_s - i_e R_E$$
$$= e_s - i_c R_E$$
$$= e_s + \frac{e_o R_E}{R_c}$$
$$= e_s + me_o \qquad \left[\because i_c = \frac{-e_o}{R_c}\right]$$

where m is the feedback fraction and is equal to (R_E/R_c). As already seen, voltage gain with –ve feedback is given by

$$A_{fb} = \frac{1}{m} = \frac{R_c}{R_E} \qquad [\because A_M >> 1] \qquad ...(1)$$

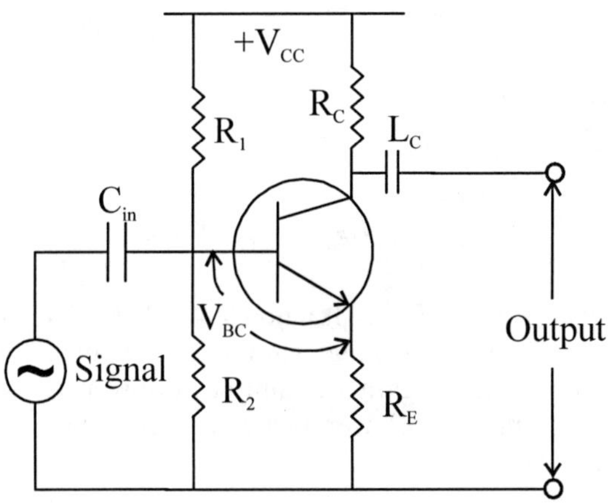

Fig. 11.5

Expression (1) gives the effective voltage gain of this circuit. It is clear that gain also depends upon the collector load R_c. If R_c is frequency dependent (*e.g.*, reactive load), then there will be distortion in the output.

(ii) Emitter follower

It is also a negative current feedback circuit. Its most important characteristic is that it has high input impedance and low output impedance. This makes it an ideal circuit for impedance matching. Fig. 11.6 shows the circuit of an emitter follower. It differs from the circuitry of conventional amplifiers by the absence of collector load and emitter by-pass capacitor. The emitter resistance R_E itself acts as the load and the a.c. output voltage e_o is taken across it. The biasing is generally provided by base resistor method. When signal e_s is applied, the resulting a.c. emitter current *i.e.* produces an output voltage $i_e R_E$ across emitter resistance. This voltage opposes the signal voltage, thus providing –ve feedback. Obviously, it is a current feedback circuit since the voltage feedback is proportional to the emitter current *i.e.*, output current. It is called emitter follower because voltage variations across base emitter junction follow the emitter.

Referring to fig. 11.6, it is clear that ;

$$i_e = \frac{-e_o}{R_E} \qquad (\because \text{ output is } -e_o)$$

Base emitter voltage,

$$V_{be} = e_s - i_e R_E$$

$$= e_s - \left(\frac{e_o}{R_E}\right) R_E$$

It is clear that feedback fraction $m = 1$ *i.e.*, feedback is 100%. Consequently, feedback voltage gain is nearly one. Similarly, it can be shown that current gain of emitter follower is, less than 1. In this sense, an emitter follower is not an amplifier in the real sense of the word

$$\text{Output impedance} = \frac{R_{out}}{\beta + 1}$$

$$\text{Input impedance} = \beta R_E + R_{in}$$

Thus, output impedance of an emitter follower is considerably reduced whereas input impedance is greatly increased. Consequently instead of transformer, we can use emitter follower for impedance matching. Emitter follower isn't only more convenient than a transformer but also it has much better frequency response *i.e.*, it works over a large frequency range.

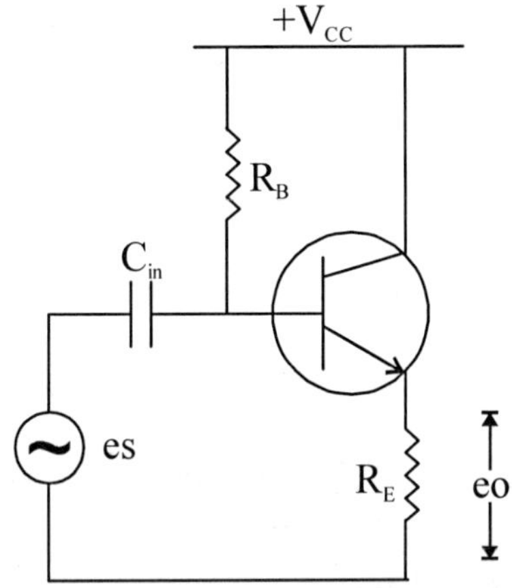

Fig. 11.6

Q. 10. Discuss on the frequency response of a feedback amplifier.

Ans. *Frequency Response of a Feedback Amplifier :* The negative feedback in an amplifier affects both the lower cut-off and upper cut-off frequencies.

High frequency

If the upper cut-off frequency of the amplifier without feedback, be f_2 and that with feedback be f_2^1, then

$$f_2^1 = f_2\,(1+\alpha A), \qquad ...(1)$$

Where A is the midfrequency gain of the amplifier without feedback. Thus, the upper cut-off frequency is increased by a factor $(1 + \alpha A)$ which is greater than 1.

Low frequency response

If the lower cut-off frequency of the amplifier without feedback be f_1 and that with feedback be f_1^1, then

$$f_1^1 = f_1 / (1+\alpha A) \qquad ...(2)$$

Thus, the lower cut-off frequency is decreased by a factor $(1 + \alpha A)$.

Bandwidth

Bandwidth of the amplifier without feedback is

$$B = f_2 - f_1, \qquad ...(3)$$

Bandwidth of the amplifier with feedback,

$$B^1 = f_2^1 - f_1^1, \qquad ...(4)$$

$$= f_2\,(1+\alpha A) - \frac{f_1}{(1+\alpha A)} \qquad ...(5)$$

Thus, it is obvious from (4) and (5) that $B^1 > B$, *i.e.*, the bandwidth of the amplifier with feedback is greater than that without feedback.

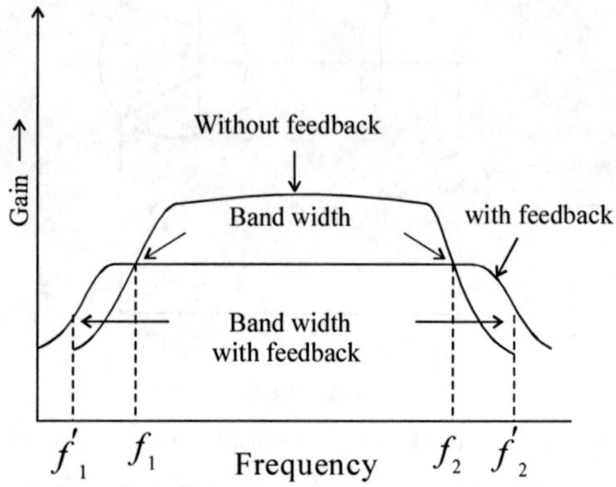

Fig. 11.7

EXERCISES

Numerical Problems with Solutions

1. An amplifier has a voltage gain of 50. The input resistance of the amplifier is 1 KΩ and the output resistance is 50 KΩ. The amplifier is now provided with 5% negative voltage feedback. Calculate (i) the voltage gain (ii) input and output resistances with feedback.

Solution

$A = 50$

$m = 0.05,\ R_{in} = 1\ \text{K}\Omega,\ R_{out} = 50\ \text{K}\Omega$

$A_{fb} = ?,\ (R_{in})_{fb} = ?,\ (R_{out})_{fb} = ?$

(i) $$A_{fb} = \frac{A}{1+mA} = \frac{50}{1+0.05\times 50} = \frac{50}{1+2.5}$$

$$= \frac{50}{3.5} = 14.28$$

(ii) $(R_{in})_{fb} = (1 + \text{mA})\ R_{in}$

$= (1 + 0.05 \times 50)\ 1\ \text{K}\Omega$

$= 3.5\ \text{K}\Omega$

(iii) $$(R_{out})_{fb} = \frac{R_{out}}{1+mA} = \frac{50}{3.5} = 14.28 \quad K\Omega$$

2. The voltage gain of an amplifier is 100. Calculate its voltage gain when negative feedback is introduced in the circuit for (i) m = 0.01 and (ii) m = 0.04.

Solution

(i) $$A = 100, \quad m = 0.01$$

$$A_{fb} = ?$$

$$A_{fb} = \frac{A}{1+mA} = \frac{100}{1+0.01\times 100} = \frac{100}{2}$$

$$= 50$$

(ii) $$m = 0.04$$

$$A_{fb} = \frac{A}{1+0.04\times 100} = \frac{100}{1+4}$$

$$= \frac{100}{5} = 20$$

3. The voltage gain of an amplifier reduces from 100 to 10 with introduction of negative feedback. Calculate the feedback ratio, m.

Solution

$$A = 100, \quad A_{fb} = 10, \quad m = ?$$

$$A_{fb} = \frac{A}{1+mA} \quad or, \quad 10 = \frac{100}{1+100m}$$

or, $10 + 1000\,m = 100$

or, $1000\,m = 90$

or, $$m = \frac{90}{1000} = \frac{9}{100} = 0.09$$

4. The bandwidth of an amplifier is 10 KHz. If negative voltage feedback is introduced in the circuit and m = 0.05, what is the altered bandwidth if the voltage gain of the amplifier without feedback is 100.

Solution

$$f_2 - f_1 = \Delta f = 10 \text{ KHz}$$

$$m = 0.05, \quad A = 100$$

$$f_2^1 - f_1^1 = ?$$

$$\Delta f^1 = f_2^1 - f_1^1 = \Delta f\ (1 + mA)$$
$$= (f_2 - f_1)\ (1 + 0.05 \times 100)$$
$$= 10 \times (1 + 5)$$
$$= 10 \times 6 = 60 \text{ KHz}$$

5. In a feedback amplifier circuit shown in Fig. 11.8 below, $R_1 = 100\ \Omega$, $R_2 = 10\ \text{K}\Omega$, $R_E = 2\ \text{K}\Omega$, $R_L = 10\ \text{K}\Omega$. The gain of the amplifier without feedback is 1000. Calculate (i) the feedback ratio (ii) gain with feedback (iii) Bandwidth ratio with and without feedback.

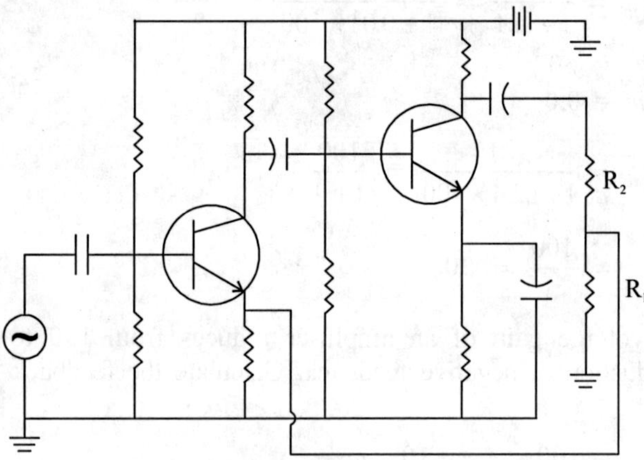

Fig. 11.8. Feedback amplifier employing both voltage and current feedback

(i) $$m = \frac{R_1}{R_1 + R_2} + \frac{R_E}{R_L}$$

$$= \frac{100}{100 + 10000} + \frac{2}{10}$$

$$= \frac{100}{10100} + \frac{2}{10} = 0.2099$$

(ii) $$A_{fb} = \frac{A}{1 + mA} = \frac{1000}{1 + 0.2099 \times 1000}$$

$$= \frac{1000}{1 + 209.9}$$

$$= \frac{1000}{210.9} = 4.74$$

(iii) Bandwidth with feedback

$$\Delta f^1 = \Delta f\ (1 + Am)$$

$$\frac{\Delta f^1}{\Delta f} = 1 + mA$$

$$= 1 + 0.2099 \times 1000$$

$$= 1 + 209.9$$

$$= 210.9$$

6. The overall gain of a multistage amplifier is 140. When negative feedback is applied, the gain is reduced to 17.5. Find the fraction of the output that is fedback to the input.

Solution

$$A = 140,\ A_{fb} = 17.5$$

$$m = ?$$

$$A_{fb} = \frac{A}{1 + mA}$$

or $$17.5 = \frac{140}{1 + 140m} \quad or \quad 17.5 + 2450m = 140$$

or, $$m = \frac{140 - 17.5}{2450}$$

$$= \frac{1}{20}$$

7. An amplifier has a voltage amplification A and a fraction m of its output is fedback in opposition to the input. If $m = 0.1$ and $A = 100$, calculate the percentage change in the gain of the system if A falls 6 db due to ageing.

Solution

$$A = 100, \quad m = 0.1, \quad A_{fb} = ?$$

$$A_{fb} = \frac{A}{1 + mA} = \frac{100}{1 + 0.1 \times 100} = \frac{100}{11}$$

$$= 9.09$$

Fall in gain $= 6$ db

Let A_1 be the new absolute voltage gain without feedback.

Then, $20 \log_{10}\dfrac{A}{A_1} = 6$

or, $\log_{10} \dfrac{A}{A_1} = \dfrac{6}{20} = 0.3$

or, $\dfrac{A}{A_1} = \text{Antilog } (0.3) = 2$

or, $A_1 = \dfrac{A}{2} = \dfrac{100}{2} = 50$

New $A_{fb}^1 = \dfrac{A_1}{1+mA_1} = \dfrac{50}{1+0.1\times 50} = \dfrac{50}{6}$

$= 8.33$

%age change in system gain

$$= \frac{9.09-8.33}{9.09}\times 100$$

$$= 8.36\%$$

8. The total harmonic distortion of an amplifier is reduced from 10% to 2% when an amplifier is provided with 5% negative feedback voltage in series with the input. Find (i) the voltage gain without feedback and (ii) the voltage gain with feedback.

Solution

$D = 10\%,\ D^1 = 2\%$

$m = 0.05,\ A = ?,\ A_{fb} = ?$

(i) $D^1 = \dfrac{D}{1+mA}$ *or* $1+mA = \dfrac{D}{D^1}$

or, $mA = \dfrac{D}{D^1} - 1$

$$A = \frac{1}{m}\left(\frac{D}{D^1}-1\right)$$

$$\therefore\ A = \frac{1}{0.05}\left(\frac{10}{2}-1\right)$$

$$= \frac{4}{0.05} = 80$$

(ii) $$A_{fb} = \frac{A}{1+mA} = \frac{80}{1+0.05\times 80} = \frac{80}{5} = 16$$

9. The circuit of a negative feedback amplifier is shown in Fig. 11.9 below. If the gain is 10,000 without feedback, Z_{in} = 10 KΩ, Z_{out} = 100 Ω, then calculate (i) feedback fraction (ii) gain with feedback (iii) input impedance with feedback and (iv) output impedance.

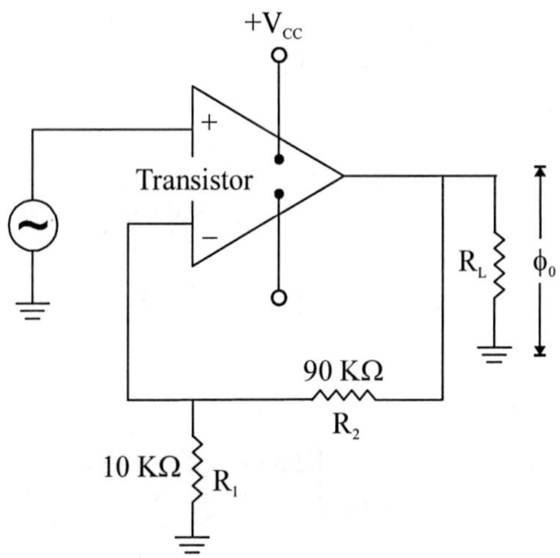

Fig. 11.9

Solution

(i) $$m = \frac{R_1}{R_1+R_2} = \frac{10}{10+90} = \frac{10}{100} = 0.1$$

(ii) $$A_{fb} = \frac{A}{1+mA} = \frac{10{,}000}{1+0.1\times 10000}$$

$$= \frac{10{,}000}{1001} = 10$$

(iii) $$Z'_{in} = (1 + mA)\ Z_{in}$$
$$= (1+0.1 \times 1000)\ 10\ \text{K}\Omega$$

$$= 1001 \times 10 \text{ K}\Omega$$
$$= 10010 \text{ K}\Omega = 10.01 \text{ M}\Omega$$

(iv) $$Z'_{out} = \frac{Z_{out}}{1+mA} = \frac{100}{1+0.1\times 10000}$$

$$= \frac{100}{1001} = 0.1\ \Omega$$

10. A negative feedback amplifier is provided with a voltage feedback factor of 0.1. The lower cut-off frequency and upper cut-off frequency value of the amplifier without feedback is 50 Hz and 10 KHz, respectively and its voltage gain is 50. What are the new values of lower and upper cut-off frequencies with feedback?

Solution

$$f_1^1 = \frac{f_1}{1+mA} = \frac{50}{1+0.1\times 50} = \frac{50}{6}$$
$$= 8.3 \text{ Hz}$$

Where, $f^1{}_1$ and f_1 are the lower cut-off frequencies with and without feedback.

But, $f_2^1 = f_2\ (1 + Am)$
$$= 10 \text{ K}\Omega\ (1 + 50 \times 0.1)$$
$$= 60 \text{ K}\Omega$$

where, f_2^1 and f_2 are the upper cut-off frequencies with and without feedback.

❑ ❑ ❑

12
Oscillators

Q. 1. What is an oscillator? What is its need? Discuss the advantages of oscillators :

Ans. An electronic device that generates oscillations of desired frequency is known as an oscillator. Oscillators can produce sinusoidal or non-sinusoidal waves.

Many electronic devices require a source of energy at a specific frequency which may range from a few Hz to several mega Hz. This is achieved by an oscillator. Oscillators are extensively used in electronic equipment. For example in radio and television receivers oscillators are used to generate high frequency wave (called carrier wave) in the tuning stages. Audio frequency and radio frequency signals are required for the repair of radio, television and other electronic equipment. Oscillators are also widely used in radar, electronic computers and other electronic devices.

Advantages

Although oscillations can be produced by mechanical devices (*e.g.* alternators), but electronic oscillators have the following advantages :

(i) An oscillator is a non-rotating device. Consequently there is little wear and tear and hence longer life.

(ii) Due to the absence of moving parts, the operation of an oscillator is quite silent.

(iii) An oscillator can produce from small (20 Hz) to extremely high frequencies (>100 mHz).

(iv) The frequency of oscillators can be easily changed when desired.

(v) It has good frequency stability *i.e.*, frequency once set remains constant for a considerable period of time.

(vi) It has very high efficiency.

Q. 2. What are different types of wave form?

Ans. Oscillators can generate a.c. with different wave forms, namely :

(i) Non-Sinusoidal wave form

(ii) Sinusoidal wave form.

Q. 3. What do you mean by Sinusoidal wave form?

Ans. *Sinusoidal oscillation :* Sinusoidal oscillations can be of two types, such as,

(i) damped (Fig. 12.1)

(ii) undamped (Fig. 12.2)

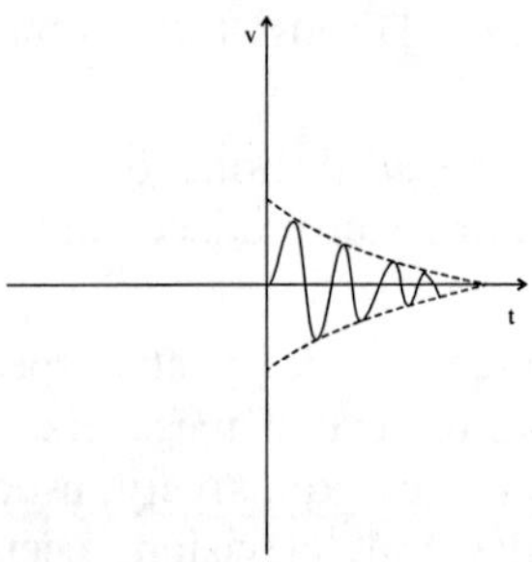

Fig. 12.1. Damped Oscillation

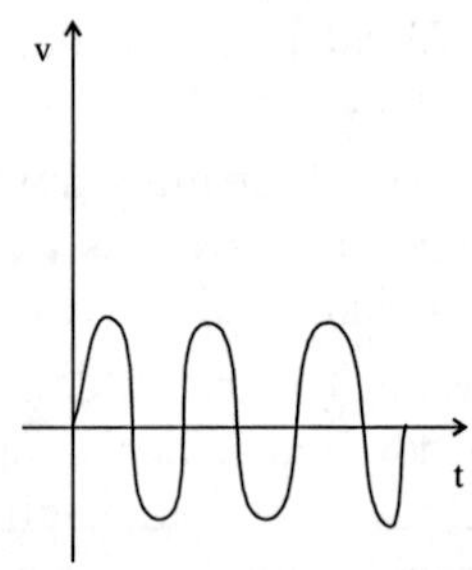

Fig. 12.2. Undamped Oscillation

In damped oscillation, the amplitude of electrical oscillation goes on decreasing due to losses inherent in the electrical system in which oscillation is generated and ultimately oscillation ceases. However, if these losses could be compensated, amplitude of oscillation would remain constant and as such the oscillation would continue indefinitely. This type of oscillation is called undamped oscillation. The electrical oscillation produced by an oscillator, therefore, is of undamped type.

Q. 4. What is a tank circuit? How does it work?

Ans. *Oscillatory circuit or tank circuit :* An oscillatory circuit or tank circuit is that circuit which produces electrical oscillations of any desired frequency. A parallel combination of an inductor (L) and a capacitor (C) is the example of the most commonly used tank circuit.

Frequency of oscillation

The resonant frequency of an oscillatory circuit is given by

$$f_r = \frac{1}{2\pi\sqrt{LC}} \qquad \text{...(1)}$$

With this frequency the tank circuit actually oscillates. This frequency of oscillation depends on the values of L and C of the circuit.

Working

(a) If the capacitor is charged from a d.c. source by closing the switch S_1 [Fig. 12.3(i)] while the switch S_2 is open, the following situation arises. The upper plate of the capacitor is +vely charged and the lower plate is –vely charged. As a result of which, there is a P.D. across the capacitor and thus electrostatic energy is stored in it.

(b) Next S_1 is opened and S_2 is closed as shown in Fig. 12.3(ii). The capacitor gets discharged and the electron flow takes place along the direction indicated. This electron flow sets up magnetic field around the coil and hence magnetic energy gradually builds up in the coil. There is slow growth of the circuit current due to inductance L till the current attains a transient maximum value. At this stage, the capacitor is completely discharged and electrostatic energy in it is zero. On the other hand, current in the coil being maximum, electro-magnetic energy stored in the coil is maximum. That means the electrostatic energy stored in the capacitor is fully converted into electromagnetic energy in the field of the inductor.

(c) After the capacitor being discharged, the magnetic field will start collapsing. Consequently, back emf will be produced. According to Lenz's law, the back emf will tend to oppose the collapse of magnetic field, *i.e.*, the current will continue to flow in the same direction as it was during the discharge of the capacitor. Due to this, the upper plate of the capacitor gets negatively charged and lower plate becomes positively charged as shown in Fig. 12.3(iv).

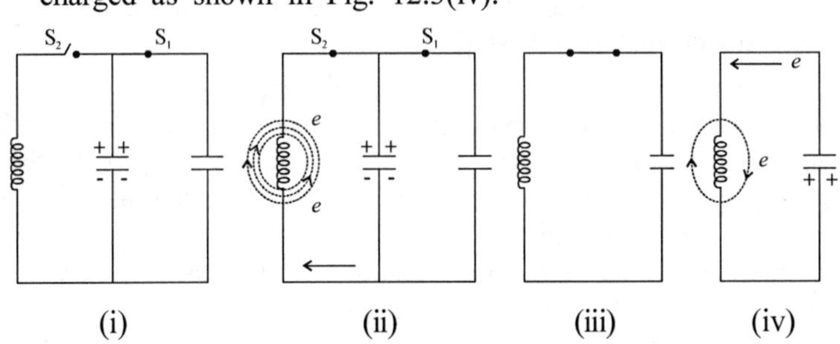

Fig. 12.3. Mechanism of oscillation in a tank circuit

The field will continue to collapse till the capacitor is completely charged with reversal of polarity. Thus, electromagnetic energy is converted back to electrostatic energy.

(d) Since the capacitor is now fully charged, the discharge process starts and again magnetic field around the coil builds up.

The alternating motion of electrons is resulted due to the sequence of charging and discharging and this produces alternating current (a.c.). Obviously, interchange of energy between L and C is repeated continuously to produce oscillation.

Q. 5. What do you mean by damped and undamped oscillations?

Ans. *Damped oscillation :* The oscillation in which the amplitude gradually decreases with time is called the damped oscillation. Such type of oscillations does not continue for a longer time and finally, it ceases. Where there is loss of energy, the motion becomes damped.

Undamped oscillation : The oscillation is said to be undamped if the amplitude does not fall with time. That means there is no loss of energy.

Since all types of vibrations are more or less damped, it is required to compensate the energy losses by supplying additional energy from an external agency to make the oscillations undamped.

Q. 6. Explain how undamped oscillations are produced in an oscillator circuit.

Ans. The interchange of electrostatic energy to electromagnetic energy and *vice versa* once started in a tank circuit would continue indefinitely if there were no losses in the circuit. In a practical tank circuit, there are resistive and radiation losses in the coil and also dielectric loss in the capacitor. Hence, during each cycle of oscillation, a small part of energy originally supplied from the battery (d.c. source) is used up to overcome these losses. As a result of which, the oscillation gets damped and finally stops.

In order to produce a.c. output, we require continuous undamped oscillations in an electronic oscillator. Hence, for producing undamped oscillations in the tank circuit, it is necessary to compensate the losses by supplying additional energy at proper time intervals. Any energy supplied to the circuit from outside should be in phase with the oscillations set up. Thus, the condition of resonance persists and the oscillation is sustained without damping.

Q. 7. Write down the conditions for producing sustained oscillations.

Ans. *Conditions for sustained or undamped oscillation :* The following conditions are to be fulfilled for producing undamped oscillations :

(i) There should be d.c. source for supplying energy which is to be converted to a.c. oscillations. The amount of energy supplied should be such as to compensate the losses in the tank circuit as well as the a.c. energy removed from the circuit by the load.

(ii) The frequency of additional energy supplied should be same as that of oscillation in the tank circuit.

(iii) The energy supplied should be in phase with the oscillations set up in the tank circuit. This is usually achieved by providing positive feedback.

Q. 8. What are different types of oscillators?

Ans. (a) Classification according to frequency. According to frequency range of oscillations produced, there are :

(i) RF oscillators,

(ii) AF oscillators,

(iii) Microwave oscillators, etc.

(b) Classification according to waveshape. According to waveshape, generally two types of oscillators are there, such as :

(i) Non-sinusoidal oscillators,

(ii) Sinusoidal oscillators.

Examples of non-sinusoidal oscillators are Multivibrators, sawtooth wave generators, squarewave generators and blocking oscillators. These are also called relaxation oscillators.

Sinusoidal oscillators are divided into two broad categories, namely, Feedback oscillators and negative resistance oscillators.

Examples of Feedback oscillators are Tuned Collector, Tuned base, Tuned Collector-Tuned base, Hartley, Collpitt, R-C phase-shift and Weinbridge oscillators.

The negative resistance oscillators are Dynatron, Transitron, Tunnel diode and push-pull, etc.

Crystal controlled oscillators are used when the frequency of oscillation is to be maintained at a fixed value. The crystal is not used to produce oscillations. It only controls the output frequency of the oscillator with which it is used.

Q. 9. Discuss the criteria for oscillations in a feedback oscillator.

Ans. The voltage gain of a feedback amplifier is given by :

$$A_f = \frac{A}{1-\alpha A} = \frac{V_o}{V_i} \qquad ...(1)$$

where, A is the voltage gain without feedback

α is the feedback factor,

V_o and V_i are the output and input voltages respectively.

When the loop gain $|\alpha A|$. $\to$ 1, $A_f \to \infty$. That means output voltage is present even if there is no input voltage. At this condition, the circuit becomes an oscillator. Thus, there are two basic conditions for oscillations to occur in a feedback oscillator. These are :

(i) the loop gain $|\alpha A| = 1$

(ii) the feedback is +ve

However, in all practical oscillators, $|\alpha A| > 1$. Because, when $|\alpha A| = 1$, oscillation may stop due to reduction in feedback factor caused by ageing of circuit components and other factors. It is obvious that oscillations can be set up in an L-C circuit by shock excitation method. Besides, in a transistor or in a valve electrical noise of wide range of frequencies are present when d.c. voltage is applied to them. Therefore, in a high gain amplifier, even without application of input voltage, a small amount of voltage may develop at the output. The output available in both cases are usually small. But in case of a +ve feedback (Fig. 12.4), a part of the output is applied to the input at the same phase. Consequently, the input voltage increases causing higher output voltage. This in turn increases the input voltage. This process is mutual. The collector current increases with the rise of base (input) current. This process continues till saturation is reached. After this position, there is no increase in the collector current in spite of increase in base current. At the state of saturation, the amplitude of oscillation (output) remains steady. This condition is achieved when energy lost in the circuit is exactly compensated by the feedback energy.

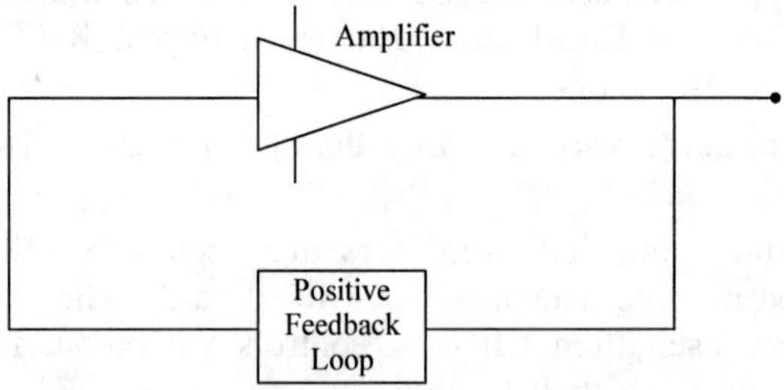

Fig. 12.4. Positive Feedback for oscillator

In an oscillator, no input voltage is applied. The initial input voltage is derived from within the circuit and device at the desired frequency. The frequency of oscillation is determined by the tuned frequency of the external tank circuit or the characteristic of the feedback loop. Therefore, an oscillator can also be named as a frequency selective amplifier.

Q. 10. With a neat circuit diagram, describe the working of tuned-collector oscillator.

Ans. Tuned-Collector Oscillator

Principle. The circuit arrangement of a tuned-collector oscillator is shown in Fig. 12.5. It consists of tuned circuit *Lp*–*Cp* in the collector and hence the name. The frequency of oscillations depends upon the values if *Lp* and *Cp* and is given by,

$$f_r = \frac{1}{2\pi\sqrt{L_p C_p}} \quad ...(1)$$

The feedback coil L_s in the base circuit is magnetically coupled to the tank circuit coil L_p. Practically, L_p and L_s are the primary and the secondary coils of the transformer. The biasing is provided by potential divider arrangement. The capacitor *C* connected in the base circuit provides low reactance path to the oscillations.

Circuit Operation. When the circuit is on by closing the switch *S*, collector current starts growing and charges the capacitor C_p. After being fully charged, the capacitor C_p discharges through coil L_p and produces oscillations of frequency given in equn. (1). These oscillations induce some voltage in coil L_s by mutual induction. The frequency of voltage in coil L_s is the same as that of tank circuit but its magnitude depends upon the number of turns of L_s and coupling between L_p and L_s. The voltage across L_s is applied between base and emitter and appears in the amplified form in the collector circuit. Consequently, the losses occurring in the tank circuit is overcome.

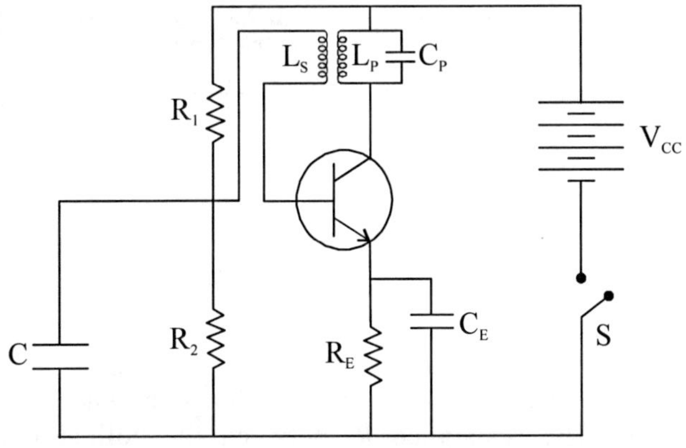

Fig. 12.5. Tuned-collector oscillator

The number of turns of L_s and coupling between L_p and L_s are so adjusted that oscillations across L_s are amplified to a level just sufficient to supply losses to the tank circuit.

It is observed that the phase of feedback is correct, *i.e.*, energy supplied to the tank circuit is in phase with the general oscillations. A phase shift of 180° is created between the voltages of L_p and L_s due to transformer action. It is a fact that all transformers introduce a phase shift of 180° between primary and secondary.

Q. 11. With a neat diagram, explain the action of Hartley oscillators.

Ans. *Hartley oscillator :* Hartley oscillator is very popular and is commonly used as a local oscillator in radio receivers. It has two main advantages, *viz.* adaptability and is easy to tune. Fig. 12.6 shows the circuit of Hartley oscillator.

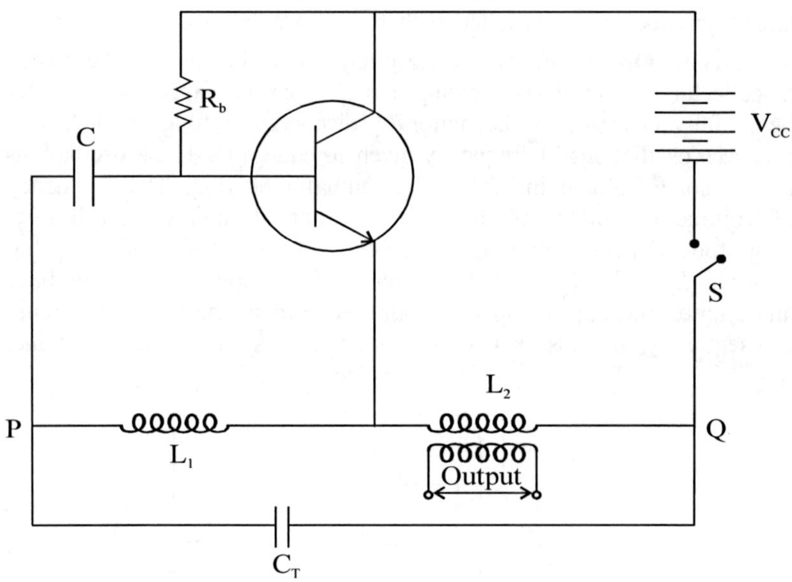

Fig. 12.6

Circuit Connections

This tank circuit is made up of C_T, L_1 and L_2. The coil L_1 is inductively coupled to coil L_2, the combination functions as auto-transformer. The resistance R_b between the collector and base provides the necessary biasing. The capacitor C blocks the d.c.

component. The frequency of oscillations is determined by the values of L_1, L_2 and C_T and is given by,

$$f = \frac{1}{2\pi\sqrt{C_T\left(L_1 + L_2\right)}} \quad ...(1)$$

Circuit operation

When switch 'S' is closed, collector current starts rising and charges the capacitor C_T. When this capacitor is fully charged, it discharges through coils L_1 and L_2, setting up oscillations of frequency determined by exp. (1). The oscillations across L_1 are applied to the base-emitter junction and appear in the amplified form in the collector circuit. The coil L_2 couples the collector circuit energy back into the tank circuit by means of mutual inductance between L_1 and L_2. In this way, energy is being continuously supplied to the tank circuit to overcome the losses occurring in it. Consequently, continuous undamped output is obtained.

It may be seen that energy supplied to the tank circuit is of correct phase. The ends P and Q of the auto transformer $L_1 - L_2$ are 180° out of phase. A further phase shift of 180° is produced by base and collector circuit of transistor. In this way, energy feedback to the tank circuit is in phase with the generated oscillations.

Q. 12. With a neat diagram, explain the action of Collpitt's oscillator.

Ans. *Circuit connections :* The Collpitt's oscillator is similar to Hartley oscillator with minor modifications. Instead of using a tapped inductance, two capacitors C_1 & C_2 are placed across a common inductance L and the center of the two capacitors is tapped as shown in Fig. 12.7. The tank circuit is made up of C_1, C_2 & L. The frequency of oscillations is determined by the values of C_1, C_2 & L and is given by,

$$f = \frac{1}{2\pi\sqrt{LC_T}} \quad ...(2)$$

where $$C_T = \frac{C_1 C_2}{C_1 + C_2}$$

Circuit operation

When switch S is closed, the capacitors C_1 & C_2 are charged. These capacitors discharge through coil L, setting up oscillations of frequency determined by exp. (1). The oscillations across C_2 are applied to the base-emitter junction and appear in the amplified form in the collector circuit and supply losses to the tank circuit. The

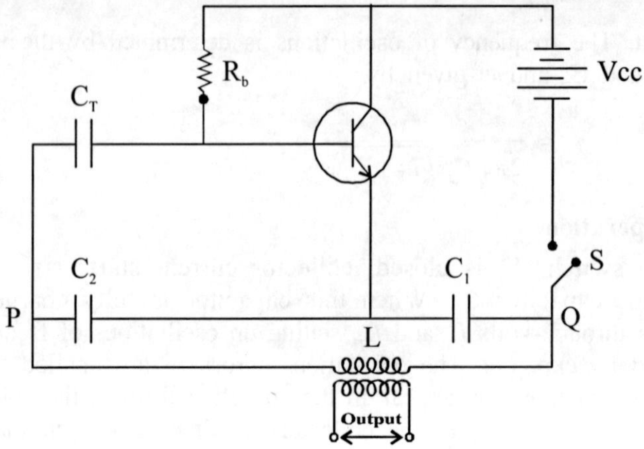

Fig. 12.7

amount of feedback depends upon the relative capacitance value of C_1 & C_2. The smaller the capacitance C_1, the greater the feedback. It is easy to ascertain that energy feedback to tank circuit is of correct phase. The capacitors C_1 & C_2 act as a simple alternating voltage divider. Therefore, points *P* and *Q* are 180° out of phase. A further phase shift of 180° is produced by the transistor. In this way, feedback is properly phased to produce continuous undamped oscillations.

Q. 13. Focuss on the principle of phase shift oscillators.

Ans. The drawbacks of L–C oscillators are eliminated by employing resistive (*R*) and capacitive (*C*) elements. Such circuits are called R-C or phase shift oscillators and have the additional merit that they can be used for very low frequencies. In a phase shift oscillator, a phase shift of 180° is obtained with a phase shift circuit instead of inductive or capacitive coupling. A further phase shift of 180° is introduced due to the transistor properties. Thus, energy supplied back to the tank circuit is assured of correct phase.

Phase-shift circuit. A phase-shift circuit essentially consists of an R–C network. Fig. 12.8(i) shows a single section of *R–C* network. From the elementary theory of electrical technology, it can be shown that alternating voltage V_1^1 across *R* leads the applied voltage voltage V_1 by ϕ°. The value of ϕ depends upon the value of *R* and *C*. If resistance R is varied, the value ϕ also changes. If R were reduced to zero, V_1^1 will lead V_1 by 90° *i.e.*, $\phi = 90°$. However, adjusting R to zero would be impracticable because it would lead to no voltage across R. Therefore, in practice, R is varied to such a value that makes V_1^1 to lead V_1 by 60°.

Fig. 12.8(ii) shows the three sections of R-C network. Each section produces a phase shift of 60°. Consequently, a total phase shift of 180° is produced, *i.e.*, voltage V_2 leads the voltage V_1 by 180°.

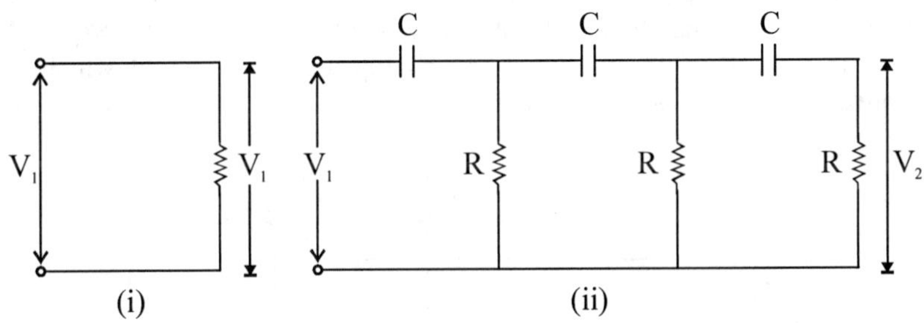

Fig. 12.8

Q. 14. What are the drawbacks of L–C oscillators? Explain.

Ans. *Drawbacks of L–C oscillators :* The main objective of an oscillator is that it should feedback energy of correct phase to the tank circuit to overcome the losses occurring in it. In L–C oscillators, the tank circuit consists of inductive (L) and capacitive (C) elements. In such circuits, a phase shift of 180° is obtained due to inductive or capacitive coupling and a further phase shift of 180° is obtained due to transistor properties. Consequently, energy supplied to the tank circuit is in phase with the generated oscillations. The oscillator circuit using L–C elements have two general drawbacks.

(i) They suffer from frequency instability and poor wave form.

(ii) They cannot be used for very low frequencies, because they become too much bulky and expensive.

Q. 15. With a neat circuit diagram, describe the operation of a *RC* phase shift oscillator.

Ans. *Circuit arrangements of a RC phase shift oscillator :* The circuit arrangements of a phase shift network is shown in Fig. 12.9. It consists of a conventional single transistor amplifier and a RC phase shift network. The phase shift network consists of three sections R_1

C_1, R_2 C_2 and R_3 C_3. At some particular frequency f_o, the phase shift in each RC section is 60°, so that the total phase-shift produced by the RC network is 180°. The frequency of oscillation is given by

$$f_0 = \frac{1}{2\pi RC\sqrt{6}} \qquad ...(1)$$

where, $R_1 = R_2 = R_3 = R$
$C_1 = C_2 = C_3 = C$

Circuit Operation. As soon as the circuit is on, it produces oscillations of frequency given by exp. (1). The output E_o of the amplifier is fedback to RC feedback network. This network produces a phase shift of 180° and a voltage E_i appears at its output which is applied to the transistor amplifier.

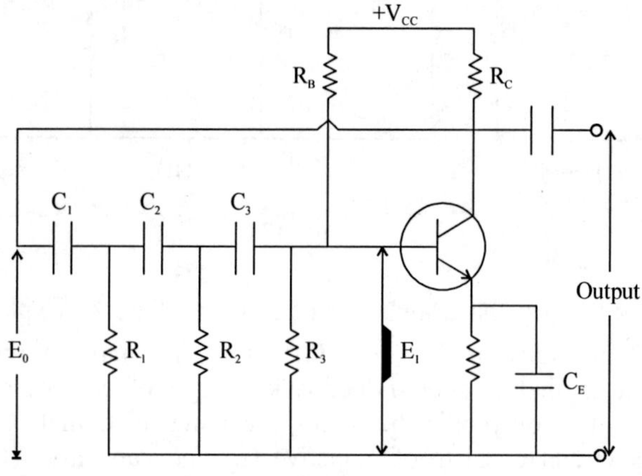

Fig. 12.9

Obviously, the feedback fraction $m = E_i/E_o$. The feedback phase is correct. A phase shift of 180° is produced by the transistor amplifier. A further phase shift of 180° is produced by the RC network. As a result, the phase shift around the entire loop is 360°.

Merits. A phase shift RC network has the following advantages :

(i) It does not require transformers or inductors.

(ii) It can be used to produce very low frequencies.

(iii) The circuit provides good frequency stability.

Demerits.

(i) It is difficult for the circuit to start oscillations as the feedback is generally weak.

(ii) The output of the circuit is not large.

Q. 16. With a neat circuit diagram, describe the functions of a Wien Bridge Oscillator.

Ans. *Circuit arrangements of a Wien Bridge oscillator :* The Wien Bridge oscillator is the standard oscillator circuit for all low frequencies in the range of 10 Hz to about 1 MHz. It is the most frequently used type of audio oscillator as the output is free from circuit fluctuations and ambient temperature.

Fig. 12.10 is the circuit of Wien bridge oscillator. It is essentially a two-stage amplifier with R–C bridge circuit. The bridge circuit has the arms R_1C_1, R_3, R_2C_2 and tungsten lamp L_p of resistance R_p. Resistances R_3 and R_p are used to stabilise the amplitude of the output. The transistor T_1 serves as an oscillator and amplifier while the other transistor T_2 serves as an inverter producing a phase shift of 180°. The circuit uses positive and negative feedbacks. The positive feedback is through R_1C_1, R_2C_2 to the transistor T_1. The negative feedback is through the voltage divider to the input of transistor T_2. The frequency of oscillations is determined by the series element R_1C_1 and parallel element R_2C_2 of the bridge. The frequency of oscillation is given by

$$f_0 = \frac{1}{2\pi\sqrt{R_1C_1R_2C_2}} \quad \text{...(1)}$$

If $R_1 = R_2 = R$ and $C_1 = C_2 = C$,

then $$f_0 = \frac{1}{2\pi RC} \quad \text{...(2)}$$

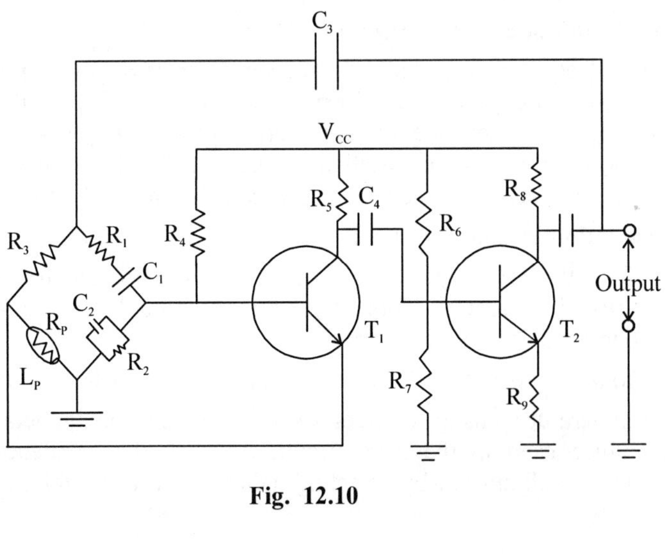

Fig. 12.10

Circuit Operation. When the circuit is switched on, oscillations of frequency determined by equn. (2) are produced. The two transistors produce a total phase shift of 360°, so that proper positive feedback is ensured. The negative feedback in the circuit ensures constant output. This is achieved by the temperature sensitive tungsten lamp L_p. Its resistance increases with current. If the amplitude of output tends to increase, more current will provide more negative feedback. The result is that the output will return to original value. If the output tends to fall, reverse will happen.

Merits

(i) It gives constant output.

(ii) The circuit works quite easily.

(iii) The overall gain is high because of two transistors.

(iv) The frequency of oscillations can be easily changed by using a potentiometer.

Demerits

(i) The circuit requires two transistors and a large number of components.

(ii) It cannot generate very high frequencies.

Q. 17. Discuss on frequency drift in oscillators.

Ans. *Frequency drift in oscillators :* Frequency drift in oscillators is caused due to variation in the operating characteristics of transistor/valve or circuit parameters. The frequency of oscillation become unstable, *i.e.*, it drifts from its designed valve. Various causes of such frequency drift in oscillators are :

(i) Variation in collector supply voltage

The frequency of oscillation becomes unstable due to variation in collector supply voltage. When the supply voltage changes beyond certain limit, either due to a defective power supply or rundown batteries, the frequency of oscillation changes. This is avoided by using either a good regulated power supply or cells in good condition.

(ii) Variation in collector load

When there is any variation in collector load, the frequency of oscillation is affected. This can be eliminated by using a buffer amplifier to isolate the load from the oscillator.

(iii) Variation in the tank circuit due to change in temperature

Temperature changes affect to some extent the inductance and capacitance used in the tank circuit. As such the frequency would drift. This is eliminated by selecting inductor and capacitor in the tank circuit with very low temperature co-efficients or with such temperature variations that changes in one parameter are countered by opposite changes in the other parameter.

(iv) Variations on transistor characteristics

Due to excessive temperature variation or even due to long use, the transistor characteristic may undergo certain variations. This may produce frequency drift.

(v) Antiresonant frequency of the output circuit

The frequency drift is also caused due to antiresonant frequency of the output circuit. This occurs if the phase-shift in the feedback loop is not exactly 180°. The maximum frequency stability occurs when the phase-shift is exactly 180°.

Q. 18. Why stabilization of frequency is necessary in oscillators?

Ans. *Stabilization of frequency in oscillators* : The major problem in LC and RC oscillators is that there is no frequency stability. It is because of the values of resistors and inductors, which are the frequency determining factors in these circuits, change with temperature. This, in turn, causes the variation in the frequency of the oscillator. However, in most of the practical oscillators, it is required to keep frequency at constant value with extreme low tolerances. For example, the frequency tolerance for a broadcasting station should not exceed 0.002%, *i.e.*, change in frequency due to any reason should not be greater than 0.002% of the specified frequency. The broadcasting stations have frequencies which are quite close to each other. Actually, the frequency difference between two broadcasting stations is less than 1%. It is apparent that any frequency drift higher than this will interfere with the working of the transmitters. This will result in overlaping the frequencies of adjacent radio stations. Therefore, the need of frequency stability in oscillators is obvious.

Q. 19. Describe the functions of a *piezoelectric crystal* in stabilizing the frequency of oscillation. Find expressions for series and parallel resonant frequency.

Ans. *Piezoelectric crystals* are employed in place of LC or RC circuits in order to stabilize the frequency of oscillation. Oscillators of this type are known as *Crystal oscillators*. The frequency of a crystal oscillator changes by less than 0.1% due to temperature and other variations. Therefore, such oscillators are employed in wide range of electronic applications and they offer the most suitable method of stabilizing the frequency.

Piezoelectric Crystal

Rochelle salt, quartz and tourmaline are some examples of crystalline materials which exhibit the *piezoelectric effect*, *i.e.*, when we apply an a.c. voltage across them, they vibrate at the frequency of the applied voltage. Conversely, when they are compressed or placed under mechanical strain to vibrate, they produce an a.c. voltage. Such crystals which exhibit piezoelectric effect are called *piezoelectric crystals*. Though there are various piezoelectric crystals, quartz crystals are commonly used because of their easy availability, cheapness, better mechanical strength and simplicity in manufacture.

Quartz crystal

Generally, quartz crystals are used in crystal oscillators owing to their high mechanical strength and simplicity of manufacture. The natural shape of quartz crystal is hexagonal as shown in Fig. 12.11. The three axes are shown : the Z-axis is called the optical axis, the X-axis is called the electrical axis and the Y-axis is called the mechanical axis. Quartz crystal can be cut in different ways. Crystal cut perpendicular to the X-axis is called *X-cut crystal* whereas that cut perpendicular to Y-axis is called *Y-cut crystal*. The piezoelectric properties of a crystal depend upon its cut. The type of the cut and physical dimensions of the crystal determines the frequency for which the crystal is used.

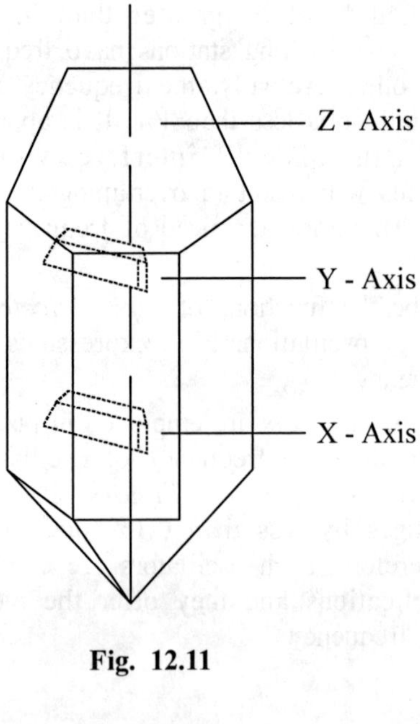

Fig. 12.11

X-cut and Y-cut crystal are not used now-a-days because of their large temperature co-efficient, *i.e.*, change in physical dimension with change in temperature. The cut which is most commonly used now is AT cut crystals whose temperature co-efficient is negligible small. AT cut crystals are suitable for generation of frequencies in the range of 50 KHz to about 10 MHz. AT cut crystal plate is cut from a plane rotated about X-axis, so that the angle Q made with Z-axis is approximately 35.5° as shown in Fig. 12.12.

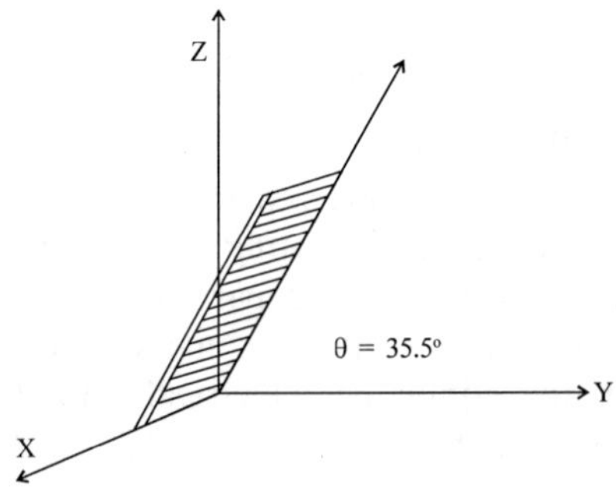

Fig. 12.12. At cut crystal

Frequency of crystal

Each crystal has a natural frequency like a pendulum. The natural frequency f of a crystal is given by

$$f = \frac{K}{t} \qquad ...(1)$$

where K is a constant that depends upon the cut and t is the thickness of the crystal. From equn. (1), it is obvious that the frequency is inversely proportional to crystal thickness. As t decreases, f increases. That means the thinner the crystal, the greater is its natural frequency and *vice versa*. However, extremely thin crystal may break due to vibrations. This puts a limit to the frequency obtainable. Practically, frequencies between 25 KHz to 5 MHz have been detained with crystals.

Working of quartz crystal

The quartz crystal is used in an electronic circuit by placing it between two metal plates. This arrangement forms a capacitor with crystal as the dielectric (Fig. 12.11). When an a.c. voltage is applied across the plates, the crystal will start vibrating at the frequency of the applied voltage. If the frequency of the applied voltage is made equal to the natural frequency of the crystal, then *resonance* occurs and the amplitude of vibrations becomes maximum.

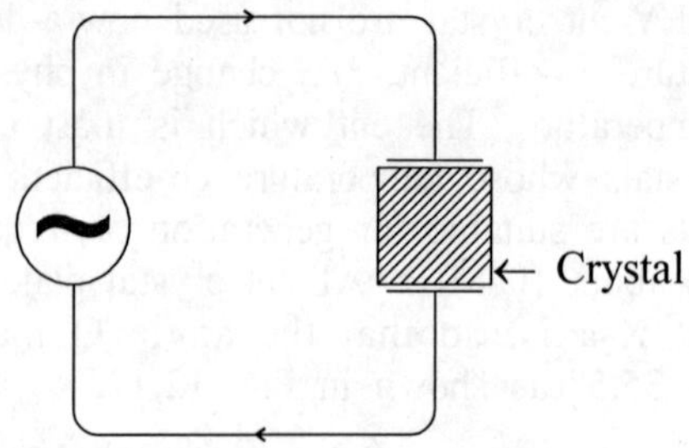

Fig. 12.13

Equivalent circuit of crystal

A crystal can be conveniently replaced by an electrical equivalent circuit. When the crystal is not vibrating, it is equivalent to capacitance C_m because it has two metal plates separated by a dielectric. However, when the crystal is vibrating, it is equivalent to series tank circuit R-L-C. Therefore, the equivalent electrical circuit of the crystal is given in Fig. 12.14(i) where

C_m = mounting capacitance

C_g = series capacitance introduced by air gap

R-L-C = electrical equivalent of vibrational characteristic of crystal.

The resonant frequency of R-L-C is the frequency of mechanical crystal resonance. However, due to presence of C_m, there will be parallel-resonant frequency also. But these two resonant frequencies are very close to each other as shown by crystal resonance curve in Fig. 12. 14(ii).

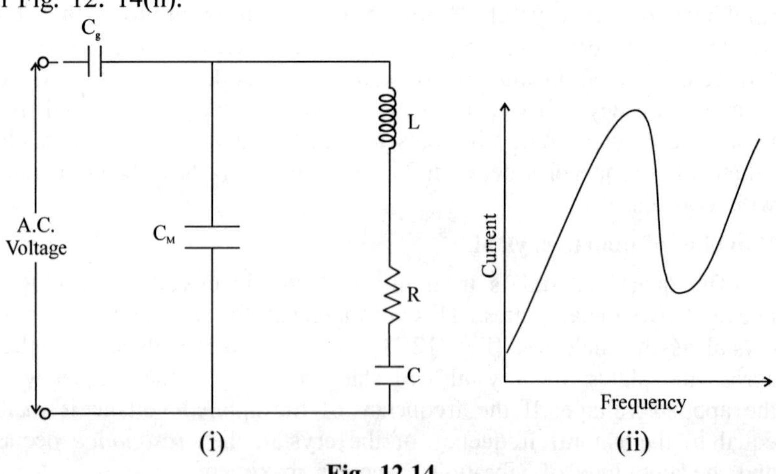

Fig. 12.14

Series Resonant Frequency

When the inductive reactance becomes equal to the capacitive reactance, then resonance occurs. The frequency of oscillation at this condition is known as the series resonant frequency and is represented by f_s.

So, $$X_L = X_C$$

or, $$\omega L = \frac{1}{\omega C}$$

or, $$\omega^2 = \frac{1}{LC}$$

or, $$\omega = \frac{1}{\sqrt{LC}}$$

or, $$2\pi f_s = \frac{1}{\sqrt{LC}}$$

$$f_s = \frac{1}{2\pi\sqrt{LC}} \text{ Hz} \qquad \text{...(2)}$$

where L is in henry and C is in farad.

Parallel Resonant Frequency

As the frequency is increased beyond f_s, the combined effective reactance of L and C becomes inductive. At some particular frequency, this inductive reactance becomes equal to the capacitive reactance of the capacitor Cm. At this condition, the crystal circuit acts as a parallel resonant circuit. The parallel resonant frequency is

$$f_p = \frac{1}{2\pi\sqrt{LC_T}} \text{ Hz}$$

where $$C_T = \frac{C \times C_m}{C + C_m}$$

So, $$f_p = \frac{1}{2\pi\sqrt{L}\sqrt{\frac{CC_m}{C+C_m}}} = \frac{1}{2\pi\sqrt{L}}\sqrt{\frac{C+C_m}{CC_m}}$$

$$= \frac{1}{2\pi\sqrt{L_C}} \frac{\sqrt{C+C_m}}{C_m}$$

$$= \frac{1}{2\pi\sqrt{LC}}\sqrt{1+C/C_m}$$

$$= f_s \sqrt{1 + \frac{C}{C_m}} \text{ Hz} \quad \text{...(3)}$$

Note

(a) Since C_T is less than C, f_p is always greater than f_s.

(b) When the crystal is used in an oscillator, the oscillation frequency will be between f_s and f_p.

(c) The frequencies f_p and f_s are very close to each other.

Q. 20. With a neat circuit diagram, explain the functions of a crystal oscillator.

Ans. *Circuit arrangement of crystal oscillator :* The circuit connections of a transistor crystal oscillator is shown in Fig. 12.13. A tank circuit $L_1 - C_1$ is placed in the collector and the crystal is connected in the base circuit. Feedback is obtained by coil L_2 inductively coupled to coil L_1. The crystal is connected in series with the feedback winding. The natural frequency of LC circuit is made nearly equal to the natural frequency of crystal.

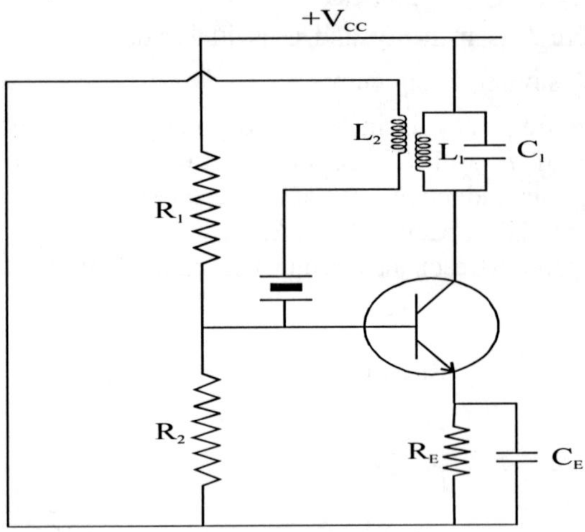

Fig. 12.15

Working

The capacitor C_1 is charged by applying the collector voltage. The discharging of this capacitor sets up oscillations. The voltage across L_1 is fed to coil L_2 due to mutual inductance. This positive feedback causes the oscillator to produce oscillations. The frequency of oscillations in the circuit is controlled by the crystal. Since the crystal is connected in the base circuit, its influence on the frequency of the circuit is much more than *LC* circuit. Consequently, the entire circuit vibrates at the natural frequency of the crystal. As the frequency of crystal is independent of temperature etc., therefore, the circuit produces a constant frequency.

Merits

(a) Crystal oscillators have a high order of frequency stability.

(b) The quality factor (Q) of the crystal is very high. The Q factor of the crystal may be as high as 10,000 compared to about 100 of *LC* tank.

Demerits

(a) A crystal is fragile. Hence, it can be used only in low power circuits.

(b) The frequency of the oscillator can not be varied.

EXERCISES

Numerical problems with solutions

1. A tuned collector oscillator circuit used in the local oscillator of a radio receiver makes use of an *LC* tuned circuit with L_1 = 64 μF and C_1 = 400 PF. Compute the frequency of oscillation.

Solution

$$L_1 = 64\ \mu\text{F} = 64 \times 10^{-6}\ \text{F}$$

$$C_1 = 400\ \text{PF} = 400 \times 10^{-12}\ \text{F}$$

Frequency of Oscillation f_r = ?

$$f_r = \frac{1}{2\pi\sqrt{L_1C_1}}$$

$$= \frac{1}{2\pi\sqrt{64\times10^{-6}\times400\times10^{-12}}}$$

$$= \frac{1}{2\pi\times8\times20\times10^{-9}}\ \text{Hz}$$

$$= \frac{10^9}{2\times3.142\times8\times20}\ \text{Hz}$$

$$= \frac{10^8}{32\times3.142} = 9.945\times10^5\ \text{Hz}$$

= 994.5 KHz

2. Calculate the operating frequency of a transistor Hartley oscillator if L_1 = 100 μH, L_2 = 1 mH, mutual inductance between the coils, M = 20 μH, and C = 20 PF.

Solution

$$L_1 = 100\ \mu H,\quad L_2 = 1\ mH = 1000\ \mu H$$

$$M = 20\ \mu H,\quad C = 20\ PF,\quad f_r = ?$$

$$\text{Total inductance, } L = L_1 + L_2 + 2\ M$$

$$= 100 + 1000 + 2 \times 20$$

$$= 1140\ \mu H = 1140 \times 10^{-6}\ H$$

$$C = 20 \times 10^{-12}\ F$$

∴ Frequency of oscillations,

$$f_r = \frac{1}{2\pi\sqrt{LC}}$$

$$= \frac{1}{2 \times 3.142\sqrt{1140 \times 10^{-6} \times 20 \times 10^{-12}}}$$

$$= \frac{1}{6.284 \times 10^{-9} \times 10\sqrt{228}}$$

$$= \frac{10^9}{62.84 \times 15}$$

$$= 1.060 \times 10^6 = 1060 \times 10^3 = 1060\ KHz$$

3. Find the operating frequency of a transistor Collpitt's oscillator if C_1 = 0.001 μF, C_2 = 0.01 μF and L = 15 μH.

Solution

$$C_T = \frac{C_1C_2}{C_1 + C_2} = \frac{0.001 \times 0.01}{0.001 + 0.01}$$

$$= 9.09 \times 10^{-4}\ \mu F = 909 \times 10^{-12}\ F$$

$$L = 15\ \mu H = 15 \times 19^{-6}\ H$$

$$f_r = ?$$

Frequency of oscillation, $f_r = \dfrac{1}{2\pi\sqrt{LC_T}}$

$$= \frac{1}{2\pi\sqrt{15 \times 10^{-6} \times 909 \times 10^{-12}}}$$

$$= \frac{1}{2\pi \times 10^{-9}\sqrt{15 \times 909}}$$

$$= \frac{10^9}{6.284 \times 116.769}$$

$$= 1.3628 \times 10^6 \text{ Hz}$$

$$= 1362.8 \times 10^3 \text{ Hz}$$

$$= 1362.8 \text{ KHz}$$

4. In a phase shift oscillator $R = 4.5$ KΩ, calculate the value of C needed for each section to achieve a frequency of oscillation of 3 KHz.

Solution

$R = 4.5 \times 10^3$ ohm, $f_r = 2$ KHz
$C = ?$

$$f_r = \frac{1}{2\pi RC\sqrt{6}} = \frac{1}{2 \times 3.142 \times 4.5 \times 10^3 \times C\sqrt{6}}$$

or, $$C = \frac{1}{6.284 \times 4.5 \times 10^3 \sqrt{6} \times f_r}$$

$$= \frac{10^{-3}}{28.278 \times 2.449 \times 2 \times 10^3}$$

$$= \frac{10^{-6}}{28.278 \times 4.898} \text{ F}$$

$$= \frac{10^{-6}}{138.5} \text{ F}$$

$$= 0.007 \times 10^{-6} \text{ F} = 0.007 \ \mu\text{F}$$

5. In a phase shift oscillator, $R_1 = R_2 = R_3 = 1$ MΩ and $C_1 = C_2 = C_3 = 64$ PF. At what frequency does the circuit oscillate?

Solution

$$R_1 = R_2 = R_3 = R = 1 \text{ M}\Omega = 10^6 \ \Omega$$

$$C_1 = C_2 = C_3 = C = 64 \text{ PF} = 64 \times 10^{-12} \text{ F}$$

$$f_r = ?$$

Frequency of oscillations,

$$f_r = \frac{1}{2\pi RC\sqrt{6}}$$

$$= \frac{1}{2\times 3.142\times 64\times 10^{-12}\times 10^{6}\cdot\sqrt{6}}$$

$$= \frac{1}{6.284\times 10^{-6}\times 2.449\times 64}$$

$$= \frac{10^{6}}{6.284\times 64\times 2.449}$$

$= 1.015 \times 10^3 = 1015$ Hz

6. A Wien bridge oscillator has $R_1 = R_2 = 220$ KΩ and $C_1 = C_2 =$ 250 PF. Find the frequency of oscillation.

Solution

$R_1 = R_2 = R = 220$ KΩ $= 220 \times 10^3$ ohm
$C_1 = C_2 = C = 250$ PF $= 250 \times 10^{-12}$ F

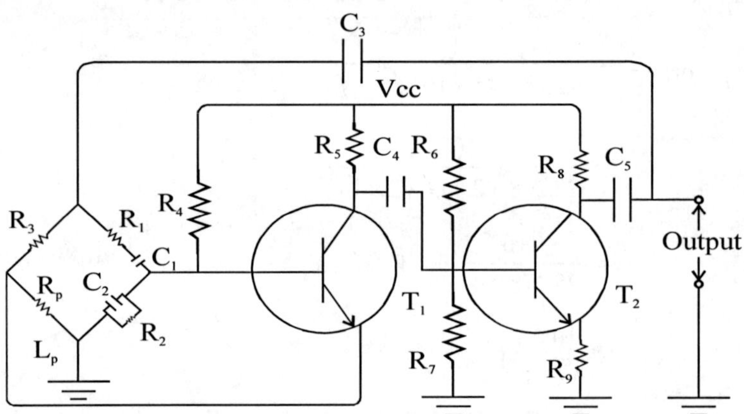

Fig. 12.16

Frequency of oscillations, $f_r = \dfrac{1}{2\pi RC}$

$$= \frac{1}{2\times 3.142\times 220\times 10^{3}\times 250\times 10^{-12}}$$

$$= \frac{10^{+9}}{6.284\times 22\times 25\times 10^{2}}$$

$$= \frac{10^7}{6.284 \times 550}$$

$$= \frac{10^6}{6.284 \times 55}$$

$$= 2.893 \times 10^3 \text{ Hz} = 2893 \text{ Hz}$$

7. A crystal has the parameter $L = 0.3$ H, $C = 0.6$ PF, $C_m = 2$ PF, $R = 5$ KΩ. Find (i) the series resonant frequency, (ii) the parallel resonant frequency and (iii) the Q of the crystal.

Solution

(i) $$f_s = \frac{1}{2\pi\sqrt{LC}} = \frac{1}{2\pi\sqrt{0.3 \times 0.6 \times 10^{-12}}}$$

$$= \frac{1}{2 \times 3.142 \times 10^{-6}\sqrt{1.8}} = \frac{10^{+6}}{6.284 \times 1.34}$$

$$= 0.1187 \text{ MHz}$$

(ii) $$f_P = f_S\sqrt{1 + C/C_m}$$

$$= 0.1187 \times 10^6\sqrt{1 + \frac{0.6 \times 10^{-12}}{2 \times 10^{-12}}}$$

$$= 0.1187 \times 10^6\sqrt{1 + 0.3}$$

$$= 0.1187 \times 10^6 \times 1.14$$

$$= 0.135 \times 10^6 \text{ Hz} = 0.135 \text{ MHz}$$

(iii) $$Q = \frac{\omega_r L}{R} = \frac{1}{\sqrt{LC}} \times \frac{L}{R} = \frac{1}{R}\sqrt{\frac{L}{C}}$$

$$= \frac{1}{5 \times 10^3}\sqrt{\frac{0.3}{0.6 \times 10^{-12}}}$$

$$= \frac{1}{5 \times 10^3} \times 10^6\sqrt{\frac{1}{2}}$$

$$= \frac{10^6}{5 \times 10^3 \times 1.414} = \frac{10^3}{7.070}$$

$$= 141$$

8. The a.c. equivalent circuit of a crystal has these values : L = 2H, C = 0.02 PF, R = 1000 Ω and C_m = 20 PF. Calculate f_s and f_p of the crystal.

Solution

L = 2H, C = 0.02, PF = 0.02 × 10^{-12} F

C_m = 20 PF = 20 × 10^{-12} F

f_s = ?, f_p = ?

$$f_s = \frac{1}{2\pi\sqrt{LC}} = \frac{1}{2\pi\sqrt{2\times 0.02\times 10^{-12}}}$$

$$= \frac{1}{2\times 3.142\times 10^{-6}\sqrt{0.04}}$$

$$= \frac{1}{6.284\times 10^{-6}\times 10^{-1}\times 2}$$

$$= \frac{10^7}{12.568} = 0.0795\times 10^7 \text{ Hz}$$

$$= 795 \times 10^3 \text{ Hz} = 795 \text{ KHz}$$

$$C_T = \frac{C\times C_m}{C+C_m} = \frac{0.02\times 10^{-12}\times 20\times 10^{-12}}{0.02\times 10^{-12}+20\times 10^{-12}}$$

$$= \frac{0.02\times 20\times 10^{-12}}{20.02} = \frac{0.4\times 10^{-12}}{20.02}$$

$$= 0.01998 \text{ F} = 19.98 \times 10^{-3} \text{ PF}$$

$$= 19.98 \times 10^{-15} \text{ F}$$

$$f_p = \frac{1}{2\pi\sqrt{LC_T}} = \frac{1}{2\times 3.142\sqrt{0.02\times 10^{-12}\times 19.98\times 10^{-15}}}$$

$$= \frac{1}{6.284\times 10^{-6}\sqrt{0.02\times 19.98\times 10^{-3}}} = \frac{10^7}{6.284\sqrt{0.02\times 1.998}}$$

$$= \frac{10^7}{6.284\times 0.19989} = 7961 \times 10^3 \text{ Hz} = 7961 \text{ KHz}$$

❑ ❑ ❑

QUESTIONS OF +2 Sc.

CHSE (ORISSA) QUESTION PAPERS

1999 (A)

ELECTRONICS – I

Full Marks : 75 Time : 3 Hours

Answer *all* questions from Group-A and Group-B and any *four* questions from Group-C.

Group-A

Answer very briefly

1. Answer *all* questions : (1 × 7)

   (a) Draw the circuit symbol of a variable resistor.

   (b) Give the colour code for 4.7 Ω ± 5%.

   (c) Mention the expression for time constant of RC circuit.

   (d) What is the meaning of cut-off bias for a triode?

   (e) Which transistor has more leakage current — made from silicon or from germanium?

   (f) What happens to the bandwidth of an amplifier when negative feedback is introduced?

   (g) What should be the minimum value of β of a transistor used in RC oscillator?

2. Choose the correct answer from those given in the bracket : (1 × 8)

   (a) Whose resistivity is more? (Silicon, Germanium)

   (b) Which of the following biasing provides greater stability? (Base resistor method, Feedback resistor method, Voltage divider method)

   (c) What is the peak voltage for 230 volt rms? (320, 325, 330 volt)

   (d) High voltage resistors are usually of which type? (Composition, Metal film, Wirewound)

   (e) What is the angle of conduction in a class-A amplifier? (360°, 180°, 90°)

(f) What happens to the gain of a common emitter RC coupled amplifier if the collector voltage is increased?

(Gain increases, Gain decreases, Gain is not changed)

(g) How does the frequency stability of a Hartley oscillator compare with that of a Collpitt oscillator?

(more, same, less)

(h) How does the PIV in a bridge rectifier compare with that of a full-wave rectifier using 2 diodes?

(more, less, the same)

Group-B

3. Answer any *ten* of the following briefly : (2 × 10)

(a) Calculate the minimum power rating of a 10 KΩ resistor to be used with a voltage source of 200 volt.

(b) What are the different processes of electron emission from metals? which of them is used in electronic valves?

(c) An a.c. source is connected in series with an inductor and a capacitor. Plot a graph to show the variation of the reactances with frequency and indicate the frequency at which both the reactances are equal.

(d) In a CE configuration the base current is 20 μA. Calculate the emitter current if $\beta = 100$.

(e) Explain, how a d.c. load line for a transistor amplifier is drawn.

(f) What is 'thermal runway' for a transistor? How is it reduced?

(g) What is meant by 'amplitude distortion' in an amplifier? Why is it caused?

(h) A negative feedback amplifier is provided with a voltage feedback factor 0.1. The cut-off frequencies of the amplifier without feedback are 60Hz and 8 KHz. Calculate these values when feedback is introduced. The voltage gain is 50.

(i) Explain the meaning of collector efficiency of an amplifier. In which class of amplifier, the collector efficiency is maximum?

(j) Draw the electrical equivalent circuit of a piezoelectric crystal and explain its features.

(k) Draw the VI characteristics of a PN junction diode and explain its features.

(l) In a class-C tuned amplifier, angle of conduction is less than 180° yet the output voltage is continuous. How is it possible?

(m) State the factors on which the co-efficient of self-inductance of a coil depends.

(n) Derive an expression for the average d.c. value of the current through a half-wave rectifier for one full cycle.

Group-C

Answer any *four* questions

4. Draw a parallel LCR circuit with an a.c. source. Discuss the variation of current in the circuit when frequency is changed.

   Derive the necessary expression. (10)

5. (a) Distinguish among conductor, semiconductor and insulator on the basis of band theory. (6)

   (b) How is the band diagram of a semiconductor change on doping? (4)

6. What are the causes of voltage fluctuation in the output of a rectifier? Explain, with a circuit diagram how the output voltage is regulated using Zener diode. (3 + 7)

7. Describe the method of base resistor biasing of a transistor amplifier. Explain its merits and demerits. (7 + 3)

8. Draw the circuit diagram of a single stage RC coupled amplifier. Draw its frequency response curve. Explain why the gain drops off at both ends. (4 + 2 + 4)

9. Mention the conditions necessary for maintenance of oscillation in an electronic oscillator.

   With a neat circuit diagram, explain the working of a Hartley oscillator. (2 + 8)

10. Write notes on any *two* : (5 × 2)

    (a) Richardson-Dushman equation

    (b) Characteristics of a tetrode.

    (c) Combination of capacitors.

1998 (A)

ELECTRONICS – I

Full Marks : 75 Time : 3 Hours

Answer *all* questions from Group-A and Group-B and any *four* questions from Group-C.

Group-A

Answer very briefly

1. Answer *all* questions : (1 × 7)
   - (a) Draw the circuit symbol of a Zener diode.
   - (b) Write the colour code for the resistor 2.2. Ω ± 20%.
   - (c) Write the expression for energy stored in a charged capacitor.
   - (d) Which type of vacuum tube can have negative resistance?
   - (e) Which of the three transistor configurations has minimum voltage gain?
   - (f) How is the upper cut-off frequency of a RC coupled amplifier affected if negative feedback is introduced?
   - (g) Write down the expression for frequency of oscillation in a Collpitt's oscillator.
2. Choose the *correct* answer out of those given inside the bracket : (1 × 8)
   - (a) Which type of transistor is suitable for high frequency operation?

     (n-p-n, p-n-p)
   - (b) Which amplifier output has minimum harmonic content?

     (Class A, Class B, Class C)
   - (c) In which LCR circuit, the impedance is minimum at the resonant frequency?

     (Series, parallel)
   - (d) You are given 8 capacitors of equal value. If you group them in parallel and then in series, what will be the ratio of the capacitances?

     (8, 32, 64)

(e) What is the angle of conduction in a class C amplifier?
(180°, 360°, less than 180°)

(f) What is the minimum limit for base-emitter bias in a silicon transistor?
(0.7 volt, 0.3 volt, 0.2 volt)

(g) The inductance in the tuned circuit in a Collpitt's oscillator is increased four-fold. What happens to the frequency of oscillation?
(halved, increased to four times, increased to two times)

(h) What is the ripple frequency in a half-wave rectifier used for 100 Hz a.c.?
(50, 100, 200)

Group-B

Answer briefly

3. Answer any *ten* questions : (2 × 10)

(a) Calculate the minimum power rating of a 25 KΩ resistor to be used with a voltage source of 250 volts.

(b) Mention the desirable characteristics of a metal for use as cathode in an electronic valve.

(c) What are the different types of capacitors used in electronic circuits? Which of them are suitable for use in RF and tuned circuits?

(d) A transistor is connected in CE configuration with a collector load of 2 KΩ. V_{CC} is 6 volt and the potential drop across the load is 0.8 volt. Determine the base current if $\alpha = 0.95$.

(e) Explain how there is a phase reversal in the output of a CE amplifier.

(f) What are the different types of distortion in amplifiers?

(g) Explain how a transformer is used for impedance matching between two stages of an amplifier.

(h) The total harmonic distortion of an amplifier is reduced from 10% to 5% when the amplifier is provided with 2% negative feedback in series with the input.
Find the voltage gain of the amplifier.

(i) Why does a power amplifier form the last stage in PA amplifier?

(j) A piezoelectric crystal has two resonant frequencies. Explain using the equivalent circuit.

(k) Draw the input characteristic curve of a transistor in CE configuration. Explain how input resistance is computed using it.

(l) How is positive feedback derived in a Hartley oscillator? Explain.

(m) What is the meaning of 'depletion layer' in a p-n junction?

(n) Derive an expression for efficiency of rectification in a half-wave rectifier.

Group-C

Answer any *four* questions

4. (a) What is the phase difference between the applied voltage and current in :

   (i) a resistor, (ii) an inductor and (iii) a capacitor placed separately in an a.c. circuit? Explain. (5)

   (b) Compare the characteristics of a series resonant circuit with that of a parallel resonant circuit. (5)

5. What is meant by 'doping' a semiconductor? Explain how p-type and n-type semiconductors are formed. Describe the mechanism of conduction of electricity through them. (10)

6. Draw the circuit diagram of a full-wave rectifier with π-filter and explain its operation.

   Show how a Zener diode can be placed in this circuit to regulate the voltage fluctuations. (3 + 6 + 1)

7. Describe the 'potential divider' method of biasing a transistor amplifier. Explain its merits and demerits. (8 + 2)

8. Draw the circuit diagram of a class B push-pull amplifier and explain its operation. Derive an expression for its efficiency. (3 + 5 + 2)

9. Draw the circuit of a RC phase-shift oscillator and explain its working. Is it suitable for use in RF range? (3 + 6 + 1)

10. Write notes on any *two* : (5 × 2)

    (a) Colour code for capacitor.

    (b) Space charge effect in a vacuum tube.

    (c) Relation between α, β and γ in a transistor.

1997 (A)

ELECTRONICS – I

Full Marks : 75 Time : 3 Hours

Answer *all* questions from Group-A and Group-B and any *four* questions from Group-C.

Group-A

Answer very briefly

1. Answer *all* questions : (1 × 7)

   (a) The bands on a resistor have colour in the order Blue, Violet, Gold and Silver.
   What is its resistance?

   (b) A 100 watt — 220 V bulb is connected to a source of 110 V. What is the power consumed by it?

   (c) How does the cut-off bias for a triode change when plate potential is increased?

   (d) Which is more in a transistor amplifier in common collector configuration — Input impedance or Output impedance?

   (e) Name a tuned amplifier stage and an untuned amplifier stage in a radio receiver.

   (f) The output from two-stage R-C coupled amplifier is coupled to the input of the first stage. What is the type of feedback?

   (g) If the capacitance is doubled while the inductance is halved in a parallel tuned circuit, what happens to the resonant frequency?

2. Fill in the blanks choosing correct answer from the bracket :

   (a) Direction of the induced electro-motive force is given by :
   (Lenz's law, Faraday's law, Ohm's law)

   (b) A series resonant circuit is ____________________ beyond resonance.
   (capacitive, inductive, resistive)

   (c) As r_p increases due to change in triode voltages, g_m ________________. (increases, decreases, remains constant)

(d) The value of α for a bipolar transistor ______________.
(is always more than one, is always less than one, may have any value)

(e) The stability factor is maximum for ____________ amplifier.
(CC, CE, CB)

(f) A tuned amplifier is suitable for use ______________.
(as class–C in RF, as class–B in AF, as class–C in AF)

(g) Zener is used in ______________ biased condition.
(Forward, reverse, either forward or reverse)

(h) The ripple frequency for a full wave rectifier is ______ Hz for the line frequency of 50 Hz.
(50 Hz, 100 Hz, 200 Hz)

Group-B

3. Answer briefly any *ten* questions of the following : (2 × 10)

(a) Why copper is not used for cathode in a thermionic tube?

(b) Mention the factors on which the inductance of a coil depend?

(c) Discuss, how the bandwidth of a tuned circuit is affected by the Q-of the coil?

(d) The cathode temperature of a diode is increased from T_1 degree to T_2 degree Kelvin whereby the saturation current is trebled. What is the ratio $T_2 : T_1$?

(e) Draw the energy-band diagram for an N-type germanium.

(f) Derive the relation :

$$\beta = \frac{\alpha}{1 - \alpha}$$

(g) Mobility of electron and holes in an extrinsic semi-conductor are not same. Why?

(h) What is thermal runway? How is it checked?

(i) Draw the frequency response curve of R-C coupled amplifier and explain the reason for reduction of gain for high frequencies.

(j) The emitter by-pass capacitor in R-C coupled CE amplifier is removed. How does it affect the gain and bandwidth of the amplifier?

(k) What is 'Zener Knee'? Explain.

(l) In class-C tuned amplifier, the angle of conduction is less than 180°, yet the output voltage across the tuned load is sinusoidal. Explain.

(m) Explain, efficiency of a full-wave rectifier.

(n) State the reasons for frequency drift in an oscillator.

Group-C

Answer any *four* questions

4. Draw the vector diagram for voltage drop across each section of a series RLC circuit containing a.c. course.

   Discuss the variation of current with frequency in this circuit. Derive necessary expressions.

5. (a) What is meant by depletion layer in P-N junction diode? Explain its formation and discuss its dependence on doping level.

   (b) What is Dynatron effect? Explain.

6. Draw the circuit diagram of a Bridge rectifier and explain its working. Show, how the ripple content in it is eliminated by using a filter (any type) in the output.

   What have the merits of this rectifier over a full-wave rectifier using 2-diodes.

7. What is load line?

   Explain its use in choosing the operating point in an amplifier circuit.

8. What are the different methods of biasing a transistor? Describe the base-resistor bias circuit with d.c. stabilization in an amplifier.

   Explain, how stabilization is achieved.

9. What is meant by negative feedback in amplifiers? Show, how the gain of feedback amplifier is affected by the feedback factor.

   The voltage gain of an amplifier reduces from 100 to 20 with introduction of negative feedback. Calculate the feedback factor.

10. State the conditions of sustained oscillation in an electronic oscillator. With a circuit diagram, explain the functioning of a tuned collector oscillator.

    Discuss its merits and demerits.

1997 (S)

ELECTRONICS – I

Full Marks : 75 Time : 2 Hours

Answer *all* questions from Group-A and Group-B and any *four* questions from Group-C.

Group-A

Answer very briefly

1. Answer *all* questions : (1 × 7)
   - (a) Draw the circuit symbol of p-n-p transistor.
   - (b) Define Q-factor.
   - (c) What is the value of carbon resistor having colour bands of grey, blue, gold and gold?
   - (d) Indicate the passive component(s) out of the following :
     - (i) Transformer (ii) Loudspeaker
     - (iii) Transistor (iv) Tunnel Diode
   - (e) Name the three possible transistor connections.
   - (f) In which circuit positive feedback is normally used?
   - (g) What is the maximum efficiency of a full-wave rectifier?
2. Fill up the blanks choosing correct answer from the brackets : (1 × 8)
   - (a) Maximum power will be transferred from a source of 10 Ω resistance to a load of ______________. (100 Ω, 10 Ω, 1 Ω)
   - (b) Unit of reactance of a circuit is ______________. (Ampere, Ohm, Farad)
   - (c) The screen grid potential is somewhat ____________ than plate potential. (lower, higher)
   - (d) In an n-type semiconductor, current conduction takes place by ________________. (holes, free electrons)
   - (e) The base of a transistor is ________________ doped. (lightly, heavily, moderately)

(f) If the operating point changes __________________ results.
(Thermal runway, unfaithful amplification)

(g) Negative feedback reduces ________________ in amplifiers
(distortion, stability)

(h) Output transformer is always a _____________ transformer.
(step-down, step-up)

3. Briefly answer any *ten* questions of the following : (2 × 10)

(a) State the formula for energy stored in a capacitor.

(b) How much is the resonance frequency for a series circuit containing L, R and C ?

(c) What are the limitations of a triode ?

(d) If a triode has mutual conductance of 1.5 mA/V and anode slope resistance of 12 KΩ, calculate the amplification factor.

(e) A half-wave rectifier uses a transformer of turn ratio 4 : 1. If the primary voltage is 240 V (rms), find the d.c. output voltage.

(f) Find the value of β, if $\alpha = 0.99$.

(g) Name different methods of transistor biasing.

(h) Name different types of transistor amplifiers according to coupling methods used.

(i) If the absolute voltage gain of an amplifier is 73, find its decibel gain.

(j) We are to match a 8 Ω speaker load to an amplifier through a transformer. What should be the turn ratio of the transformer?

(k) A class-A power amplifier has a maximum a.c. power output of 30 W. Find the power rating of the transistor.

(l) An amplifier has a gain of 10,000 without feedback. With negative feedback the gain reduces to 50. Find the feedback fraction.

Group-C

Answer any *four* questions

4. Three coils are connected in series across a 100 V, 50 Hz supply. Values of their resistances and inductances are as follows :

$R_1 = 15\ \Omega,\ L_1 = 0.01$ H, $R_2 = 10\ \Omega,\ L_2 = 0.03$ H,

$R_3 = 3\ \Omega,\ L_3 = 0.06$ H.

Find the current in the circuit. (10)

5. Draw a full-wave bridge rectifier circuit with filter and explain the operation. Calculate the efficiency of the rectifier. (10)

6. What do you understand by feedback? Why is negative feedback applied in high gain amplifiers? Discuss the principles of negative feedback in amplifiers with neat diagram. (10)
7. What is a transistor? Why is it so called? Explain the operation of transistor as an amplifier. (10)
8. Write short notes on any *two* : (5 × 2)
   (a) Knee voltage
   (b) PIV
   (c) Complementary-symmetry Amplifier.

1996 (A)

ELECTRONICS – I

Full Marks : 75 Time : 3 Hours

Answer *all* questions of Group-A and Group-B and any *four* questions from Group-C.

Group-A

Answer very briefly

1. Answer *all* questions : (1 × 7)
   (a) What should be the maximum voltage applied across a 100 Ω, 1 watt resistor?
   (b) You have sufficient stock of 8 μf, 10 V capacitors. Your requirement in the circuit is a 8 μf, 20 V capacitor. Show, how can you arrange the capacitors available with you to get the required capacitance.
   (c) Show graphically, the variation of reactance of an inductor with frequency.
   (d) Draw the circuit symbol of an iron-core inductor.
   (e) Which electrode in a triode is closer to the grid-plate or cathode?
   (f) What is the minimum forward bias voltage for a silicon p-n junction diode to conduct?
   (g) Which of the following type of transistor suitable for high frequency operation — PNP or NPN?
2. Answer *all* questions : (1 × 8)
   (a) Which is higher in a CE amplifier — Input impedance or Output impedance?
   (b) Which type of biasing offers better stabilization in amplifiers?
   (c) What is the ripple frequency of a half-wave rectifier for a line frequency of 50 Hz?
   (d) What is the phase-shift of output relative to the input in a CB amplifier?
   (e) In which frequency range, R-C coupled amplifiers are usually used?

(f) How is the lower cut-off frequency of an amplifier influenced with negative feedback?

(g) What is the most important characteristic of a crystal oscillator?

(h) What is the time constant of a series RL circuit?

Group-B

3. Answer any *ten* questions : (2 × 10)

(a) Mention the factors governing the inductance of an inductor.

(b) What is meant by a leaky capacitor? How is the characteristics of a capacitor affected by leakage?

(c) Define mutual inductance. How is the co-efficient of mutual inductance related to the self-inductance of the coils?

(d) What are the different methods of electron emission from metal. Give one example from each.

(e) Explain, Dynatron effect.

(f) Draw the plate characteristics of a triode and explain how the plate resistance is calculated using it.

(g) Explain the energy bands in a p-type semiconductor.

(h) Define "Zener impedance". Discuss its significance.

(i) Define β for a bipolar transistor. How is it related to α?

(j) What is thermal runway? How is it controlled?

(k) State the advantages of a bridge rectifier over a full-wave rectifier using 2 diodes.

(l) Explain the significance of gain-band width product.

(m) Following are the data for a CE amplifier. Calculate the voltage and power gain.

R_L = 10 KΩ, R_{in} = 200 Ω, β = 50

(n) The feedback factor in a negative feedback amplifier is 0.05. What is the percentage reduction in distortion?

Given, voltage gain of the amplifier without feedback is 100.

Group-C

Answer any *four* of the following questions (10 × 4)

4. Discuss the characteristics of a parallel resonant circuit containing L, C and R. How is the dynamic resistance of this circuit related to the Q of the circuit?

5. (a) Distinguish among conductor, insulator and semiconductor on the basis of band theory.

(b) Discuss, with a two dimensional lattice picture formation of a N-type semiconductor.

6. What is a Zener diode? How does its VI characteristics curve differ from a common type of PN junction diode?

   With a neat circuit diagram, explain how a Zener diode is used to regulate the output derived from a rectifier.

7. Draw a circuit for study of transistor characteristics in common base configuration. Explain, the nature of output and input characteristic curves drawn using it. Show, how the input and output impedances are calculated from them.

8. What is meant by negative feedback in amplifier? What are the standard techniques for achieving it? Explain, with a neat circuit diagram, the principle of any one method for it.

9. State the conditions of sustained oscillation in an electronic oscillator. With a circuit diagram, explain the operation of a Collpitt oscillator.

   Compare its performance characteristics with that of a Hartley oscillator.

10. Write short notes on any *two* : (5 × 2)

    (a) Distortion in amplifier.

    (b) VI characeristics of a thermionic diode.

    (c) Colour code for resistor.

1996 (S)

ELECTRONICS – I

Full Marks : 75 Time : 3 Hours

Answer *all* questions from Group-A and Group-B and any *four* questions from Group-C.

Group-A

1. Answer *all* questions : (1 × 7)
   (a) Sketch the circuit symbol of a Zener diode.
   (b) How much is the phase difference between input and output signals of transistor amplifier in CE mode?
   (c) Define β of a transistor.
   (d) Write the colour code of a carbon resistor of 100°K with 5% tolerance.
   (e) Which is the minority carrier in a p-type semiconductor?
   (f) Write Richardson's Thermionic equation.
   (g) How does the width of depletion layer in a p-n junction change with reverse bias?

2. Fill up the blanks choosing correct answers from the brackets. (1 × 8)
   (a) Unit of inductive reactance of a coil is ______________.
   (Ampere, Ohm, Henry)
   (b) A capacitor has ____________ reactance at high frequency.
   (high, low)
   (c) Bandwidth of an amplifier ________ with negative feedback.
   (increases, decreases)
   (d) Pentavalent impurity atoms are added in pure silicon to make ___________________ semiconductor.
   (n-type, p-type)
   (e) Lenz's law indicates ____________ of the induced current.
   (magnitude, direction)
   (f) Primary current in a step-up-transformer is ________ than secondary current.
   (more, less)

(g) Electron-volt is a unit of ______________.

(voltage, current, energy)

(h) ______________ mode is the most commonly used transistor amplifier.

(CB, CE, CC)

Group-B

3. Briefly answer any *ten* questions of the following : (2 × 10)
   (a) Why does the conductivity of a semiconductor increase with increase of temperature?
   (b) For a transistor $\alpha = 0.95$, $I_e = 2$ mA. Calculate I_b.
   (c) There is 25 V voltage drop across a 200 Ω resistor. How much power is consumed in it?
   (d) Define 'g_m' of a triode.
   (e) Establish the relation between α and β of a transistor.
   (f) Explain the function of suppressor grid in a pentode.
   (g) Sketch the circuit diagram of a bridge-rectifier.
   (h) What do you mean by 'bandwidth' of an amplifier?
   (i) In a transistor circuit, $V_{CC} = 9$ V, $R_c = 2$ KΩ and $V_{CE} = 6$ V. Calculate I_c.
   (j) What is 'thermal runway' in a transistor?
   (k) What is Q of a coil?
   (l) Explain, class-A operation of an amplifier.

Group-C

Answer any *four* of the following questions

(10 × 4)

4. Explain working of a p-n junction under forward and reverse bias conditions with reference to the band structure and VI characteristics.
5. In a series L-C-R circuit, how is Z estimated? Explain the phenomenon of series resonance.
6. On the basis of band theory, explain the electrical conduction process in insulators, semiconductors and good conductors.
7. With a circuit diagram, explain working of a push-pull power amplifier and discuss its merits.
8. Derive Barkshausen criterion for oscillation in a circuit. Explain the working of Collpitt oscillator.

9. Draw the circuit diagram of transistor amplifier in the three standard configuration.

   Explain the working of an amplifier in CE configuration. Derive expressions for its voltage and power gain.

10. Write notes on any *two* :
    (a) Feedback amplifier.
    (b) Crystal oscillator.
    (c) Filter circuits.

1995 (A)

ELECTRONICS – I

Full Marks : 75 Time : 3 Hours

Answer *all* questions from Group-A and Group-B and any *four* questions from Group-C.

Group-A

1. Answer *all* questions : (1 × 8)
   (a) Express 1 pf in terms of Farad.
   (b) Give the colour code for a resistor of value 1 MΩ ± 10%.
   (c) Give one example of a variable resistor.
   (d) What is the unit of work function.
   (e) Define α in a transistor.
   (f) Name any two impurities to make n-type silicon.
   (g) Which type of feedback is used in an oscillator?
   (h) Name the electrodes of a pentode.
2. Answer *all* questions : (1 × 7)
   (a) Define ripple factor.
   (b) What are the meanings of the symbols i_e and i_E for a transistor?
   (c) Name the charges present in the depletion zone of a p–n junction.
   (d) Why resistance of a semiconductor decreases with increase of temperature?
   (e) Which mode of transistor amplifier connection gives the highest power gain?
   (f) Give the circuit for non-transistor with proper bias arrangement.
   (g) Define the resonance frequency for a series resonant circuit.

Group-B

3. Answer any *ten* questions : (2 × 10)
   (a) Discuss the characteristics of a carbon resistor.
   (b) What do you mean by wattless component of current?
   (c) Calculate the ripple factor for a FW rectifier of peak ac voltage is 200 V and dc voltage is 100 V.
   (d) Prove $\alpha = \beta/(1 + \beta)$ for a transistor.
   (e) Define BW for RC coupled amplifier.
   (f) What is the significance of PIV in a rectifier?
   (g) State and explain Faraday's law.
   (h) What is a self-bias? Explain.
   (i) Give the principle of Zener diode.
   (j) Why amplification factor is more in pentode than that in triode?
   (k) What are the advantages of CE amplifier over CB and CC circuits?
   (l) Distinguish between n and p-type semiconductors?
   (m) Why gain in a RC amplifier falls at high frequency.

Group-C

Answer any *four* questions

(10 × 4)

4. Discuss the characteristics of series resonance circuit containing L, C and R. What will happen to resonance curve at high values of R?
5. Discuss the behaviour of p-n junction under forward and reverse bias conditions. Discuss its characteristics. Show the difference in band structure under both these conditions.
6. Explain the operation of a p-n-p transistor. Define α and β. Why base is made thin for transistors?
7. Describe a FW rectifier with a neat circuit diagram. Show how filters are useful for a rectifier.
8. Discuss the operation of a push-pull amplifier. Show how distortion is reduced in it.
9. Draw a circuit for a RC coupled amplifier and explain the function of each component. Discuss its frequency response curve.
10. Write short notes on any *two* :
    (a) Crystal oscillator,
    (b) Band theory,
    (c) Mutual inductance.

1994 (A)

ELECTRONICS – I

Full Marks : 75 Time : 3 Hours

Answer *all* questions from Group-A and Group-B and any *four* questions from Group-C.

Group-A

1. Answer *all* questions : (1 × 10)
   (a) Give the colour code for resistor
   6.8 KΩ ± 5%
   (b) Define one Farad.
   (c) Give the correct relation between Q, C and V.
   (d) What is meant by wattless components?
   (e) What is the unit of work function?
   (f) How much will be the maximum efficiency of a full-wave rectifier?
   (g) In which class the voltage amplifiers are operated?
   (h) Name the current carrier in n-p-n transistor.
   (i) Express the power gain of 100 in decibel.
   (j) Which type of feedback is employed by an oscillator?
2. Fill up the blanks with correct word from bracket : (1 × 5)
   (a) A parallel resonant circuit is known as __________________.
   (rejector/acceptor)
   (b) A Zener diode uses ________ portion of diode characteristics.
   (reverse/forward)
   (c) Efficiency of power amplifier is maximum for class ______________ operation. (B, C)
   (d) In transistor $\alpha = 0.9$ and $\beta =$ ____________. (9, 0, 9)
   (e) A semiconductor material is formed by ___________ bonds.
   (covalent/electrovalent)

Group-B

3. Answer any *ten* questions : (2 × 10)
   (a) State and explain Lenz's law.
   (b) Write down the Richardson-Dushman equation on thermionic emission and narrate the symbols.
   (c) Explain the V-I characteristics of a p-n junction.
   (d) Why thermionic emission is not possible at room temperature?
   (e) Explain the function of a filter in a rectifier circuit.
   (f) Establish a relation between μ, g_m and r_p for a triode.
   (g) What are the advantages of a pentode over a triode?
   (h) Explain what do you mean by a Q-factor of resonant circuit.
   (i) The gain of an amplifier is 50. Calculate its feedback gain for negative feedback factor of 5%.
   (j) Describe briefly how a crystal oscillator works.
   (k) Why the gain in a RC amplifier falls at high frequency?
   (l) Give by circuit diagrams the normal biasing for both n-p-n and p-n-p transistors.
   (m) Define ripple factor by an equation. Calculate ripple factor for full-wave rectifier.
   (n) Draw output characteristics of a CE amplifier and indicate its different regions.

Group-C

Answer any *four* questions

(10 × 4)

4. Derive an expression for impedance of L, C and R series circuit when an a.c. is applied to it. Obtain the condition for minimum impedance. How a parallel resonance circuit differs from a series resonance circuit?
5. Describe an experiment to obtain different characteristics of a triode valve. How can these characteristics be used to define the triode constants.
6. How does a Zener diode function? Explain its use as voltage stabiliser with a diagram.
7. What do you mean by average and r.m.s. values of an a.c. potential? Derive an expression for average and r.m.s. values over a half-wave. Give the operation of a half-wave rectifier.
8. Discuss the operation and advantages of a push-pull amplifier.

9. Give a circuit for Hartley oscillator and explain its operation. Why this is not suitable as audio-oscillator?

10. Write short notes on any *two* :

   (a) Filter.

   (b) RC time constant.

   (c) Gain-Band-width product.

According to 1993 Syllabus

Group-A

1. Answer *all* questions : (1 × 8)
   (a) The colour code of a resistor is Orange, Red and Yellow. What is its value?
   (b) What is Henry? Define it.
   (c) Give a relation between C, Q and V.
   (d) Name two substances which are used as thermionic emitter.
   (e) What is the current rating of a 0.25 W 2.5 KΩ resistor?
   (f) Why base of transistor is made thin?
   (g) Express transistor β in terms of α.
   (h) Give the symbol of FET.
2. Answer *all* questions : (1 × 7)
   (a) Give the expression for ripple factor.
   (b) Define Q-factor of a LCR series circuit.
   (c) Give an application of Zener diode.
   (d) The peak inverse voltage of diode rectifier is 200 V. What is the rms value of this?
   (e) What type of feedback is used in an oscillator?
   (f) What type of biasing given to emitter junction of transistor?
   (g) What is the band gap energy of silicon?

Group-B

3. Answer any *ten* questions :
   (a) Name the active and passive elements in electronic circuit (two each).
   (b) A 100 V, 1 KHz a.c. supply is connected to a 100 mH inductor. Calculate the inductive reactance.
   (c) Distinguish between apparent power and true power.
   (d) Define the constants of a triode valve.
   (e) Draw the characteristics of a CE amplifier.
   (f) What are Barkhausen conditions of oscillations
   (g) Explain the function of control grid in a triode.

(h) Derive the expression for resonance frequency of a LCR parallel circuit.

(i) Why CE mode is preferred for a transistor amplifier?

(j) Explain Zener breakdown.

(k) Why JFET is called an unipolar transistor?

(l) Write Richardson's equation for thermionic emission.

(m) Why gain in RC amplifier falls at low frequency?

Group-C

Answer any *four* questions

(10 × 4)

4. Draw a full-wave rectifier circuit and explain its operation. Why a filter is necessary for a FW rectifier circuit? Calculate its ripple factor.
5. Explain the working of a triode as an amplifier.
6. Show how distortion is reduced in push-pull amplifier. Explain its working with a diagram.
7. Discuss operation of Hartley oscillator.
8. Draw the characteristics of a JFET. Explain its operation. How a JFET differs from BJT?
9. Explain what do you mean by p-type and n-type semiconductors. How a p-n junction is formed?
10. Write shot notes on any *two* :

    (a) Class A amplifier

    (b) Tunnel diode

    (c) Series resonant circuit.

1993 (A)

APPLIED ELECTRONICS

PAPER – I

Full Marks : 75 Time : 3 Hours

The figures in the right-hand margin indicate marks.

Answer *all* question from Group-A and Group-B and any *four* questions from Group-C.

Group-A

1. Answer briefly any *seven* of the following : (1 × 7)
   - (i) Give the unit of frequency.
   - (ii) What is farad?
   - (iii) What is work function of a metal?
   - (iv) What is plate resistance of a triode?
   - (v) What is PIV of a diode?
   - (vi) What is a class A amplifier?
   - (vii) How would you connect two 0.1 μF capacitor to get a capacitance of 0.2 μF?
   - (viii) Name of majority and minority carriers in an N-type semiconductor.
   - (ix) What is electromagnetic induction?

2. Fill in the blanks with correct words from the brackets. Answer any *eight* : (1 × 8)
   - (i) The work function of thoriated tungsten is ______________ than that of pure tungsten. (lower, higher)
   - (ii) If a diode is heavily doped, the depletion layer width will be ______________. (more, less)
   - (iii) More a.c. component in the rectifier output means ______________ ripple factor. (smaller, greater)
   - (iv) A crystal diode can withstand reverse voltage much________ than vacuum diodes. (higher, lesser)

(v) In full-wave rectification, if the input frequency is 50 Hz, then the output has a frequency of ________ Hz. (50, 100, 150)

(vi) A triode is normally operated with the grid at ________ potential. (positive, negative)

(vii) A Hartley oscillator uses ________ feedback. (positive, negative)

(viii) If the value of α of a transistor is 0.9, then its β is ________. (90, 9, 0.9)

(ix) In a transistor the base is ________ doped. (heavily, lightly)

(x) Negative feedback ________ the bandwidth of an amplifier. (increases, decreases)

Group-B

3. Answer any *ten* in brief : (2 × 10)

(i) Why are thermionic emitters heated in vacuum?

(ii) Why is Zener diode always reverse connected across the supply?

(iii) Why is bridge type circuit commonly used in rectifier?

(iv) Why in a vacuum tube amplifier, the plate is placed at a positive potential?

(v) Draw the characteristic curve of a tunnel diode.

(vi) Write down Richardson's law for thermionic emission. What do the different symbols stand for ?

(vii) A 2mH inductor resonates with a capacitor at a signal frequency of 10 KHz. What is the value of the capacitance?

(viii) Show the circuit arrangement for drawing the static characteristics of a transistor in the CB mode.

(ix) State and explain the conditions of oscillation.

(x) What are the advantages of negative feedback in amplifiers?

(xi) Why are transistors mostly connected in CE mode?

(xii) In a transistor $I_E = 10$ mA and $I_B = 50$ μA. Find its α.

Group-C

Answer *four* questions : (10 × 4)

4. What is a Zener diode? Explain its principle of operation. How is it used as a voltage regulator?

5. Discuss the principle of classification of solids into conductors, semiconductors and insulators. Explain, how free electrons and holes can be created in a semiconductor by adding suitable impurities.
6. Draw the diagram of a full-wave rectifier and explain its working. Discuss how the use of a capacitor filter reduces ripple.
7. Give the circuit diagrams of transistor amplifiers in three configurations and explain the operation in any one configuration.
8. With a neat circuit diagram, explain the operation of a Hartley oscillator.
9. Draw the diagram of a circuit containing a resistance and a capacitance in series with an alternating source.

   Derive the relation between voltage and current. Draw the voltage and current wave forms to show their phase relationship.
10. Write notes on any *two* :
    (i) Tunnel diode.
    (ii) Thermionic emission.
    (iii) MOSFET
    (iv) Crystal controlled oscillator.

ANSWERS TO OBJECTIVE TYPE QUESTIONS

CHSE, ORISSA

1999 (A)

Group-A

1. (a)

(Circuit symbol of a variable resistor)

(b) Yellow, violet, black, gold

(c) $\lambda = RC$

where, λ is the time constant of RC circuit.

(d) If the grid is made very much negative then electrons emitted from cathode are repelled back, consequently making the plate current zero, even if the plate is positive with respect to cathode. In this way, the plate current is controlled by grid potential. When the plate current is zero, the tube is said to be cut off and the negative grid voltage responsible for this is called *cut off grid bias*.

(e) A transistor made from Germanium has more leakage current.

(f) The bandwidth of an amplifier increases when negative feedback is introduced.

(g) The minimum value of β of a transistor used in RC oscillator is 29 to achieve oscillation.

2. (a) Resistivity of silicon is more.

(b) Voltage-divider method of biasing provides greater stability.

(c) $V_{peak} = V_{rms} \times \sqrt{2}$

$$= 230 \text{ volt} \times 1.414$$

$$= 325 \text{ volt}$$

So, the peak voltage for 230 volt rms is 325 volt.

(d) High wattage resistors are usually wirewound type.

(e) The angle of conduction in a class-A amplifier is 360°.

(f) If the voltage is increased, the gain of a common emitter RC coupled amplifier increases.

(g) The frequency stability of a Hartley oscillator compared with that of a Collpitt oscillator is less.

(h) The PIV in a bridge rectifier compared with that of a full-wave rectifier using 2 diodes is less.

Group-B

3. (a) $R = 10\ \text{K}\Omega = 10 \times 10^3\ \Omega$

$V = 200$ volt

$$P = \frac{V^2}{R} = \frac{200 \times 200}{10 \times 10^3} = \frac{4 \times 10^4}{10^4} = 4 \text{ watts}$$

(b) There are four different methods of electron emission from metals. These are :

(i) Thermionic emission

(ii) Field emission

(iii) Photo-electric emission

(iv) Secondary emission

Thermionic emission is used in electronic valves.

(c) $X_L = 2\pi f L$, Also $\vec{X}_L = j\omega L$

$X_c = 1/2\pi fc$ $\vec{X}_C = j/\omega c$

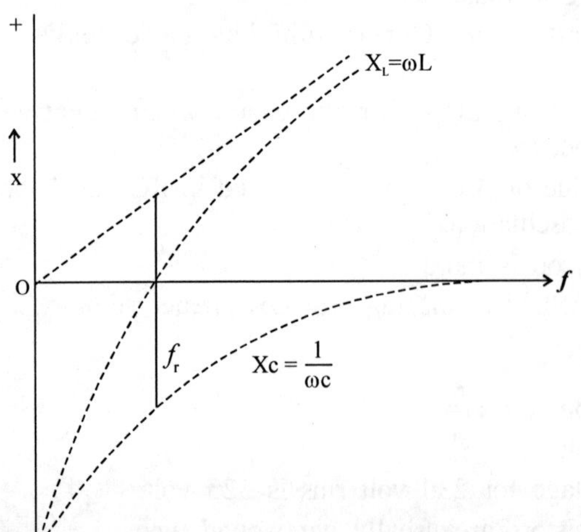

Capacitive reactance is negative in nature if inductive reactance is considered positive. At resonant frequency 'f_r', the inductive reactance becomes equal to capacitive reactance.

Fig. 1

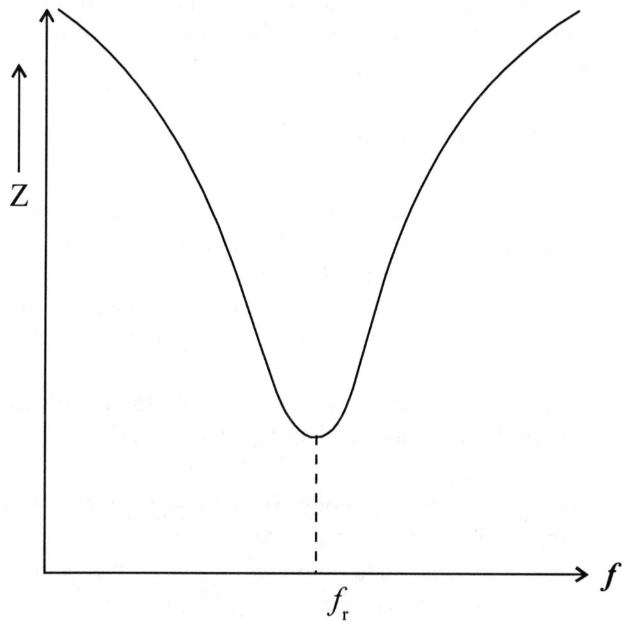

Fig. 2

At f_r, $\omega L = \dfrac{1}{\omega C}$

i.e., $X_L = X_C$

Thus, at resonant frequency, the inductive reactance is equal to the capacitive reactance.

(d) $I_B = 20\ \mu A = 20 \times 10^{-6}$ A

$\beta = 100$, $I_E = ?$

$$\begin{aligned} I_E &= I_B + I_C = I_B + \beta I_B \\ &= I_B (1 + \beta) \\ &= 20 \times 10^{-6} (1 + 100) \\ &= 20 \times 10^{-6} \times 101 \\ &= 202 \times 10^{-5} \text{ A} \\ &= 2.02 \times 10^{-3} \text{ A} \\ &= 2.02 \text{ mA} \end{aligned}$$

(e) D.C. load line for a transistor amplifier

Let us consider a common npn transistor circuit shown in Fig. 3, where no signal is applied. Therefore, d.c. conditions prevail in the circuit. The output characteristics of this circuit are shown in Fig. 4.

The value of collector-emitter voltage V_{CE} at any time is given by

$$V_{CE} = V_{CC} - I_C R_C, \quad \text{...(1)}$$

As V_{CC} and R_C are fixed values, therefore, equn. (1) can be represented by a straight-line on the output characteristics. This is known as *d.c. load line* and determines the locus of $V_{CE} - I_C$ points for any given value of R_C. To construct load line, we require two end points of the straight-line. These two points can be located as under :

(i) When the collector current $I_C = 0$, then collector-emitter voltage is maximum and is equal to V_{CC}, *i.e.*

$(V_{CE})_{max} = V_{CC} \qquad (\because \; I_C = 0)$

This gives the first point B (OB = V_{CC}) on the collector-emitter voltage axis as shown in Fig. 4.

(ii) When collector-emitter voltage $V_{CE} = 0$, the collector-current is maximum and is equal to V_{CC}/R_c, *i.e.*,

$$V_{CE} = V_{CC} - I_C R_C$$

$$O = V_{CC} - I_C R_C$$

$$(I_C)_{max} = \frac{V_{CC}}{R_C}$$

This gives the second point A (OA = V_{CC}/R_C) on the collector-current axis as shown in Fig. 4. By joining these two points, *d.c. load line* AB is constructed.

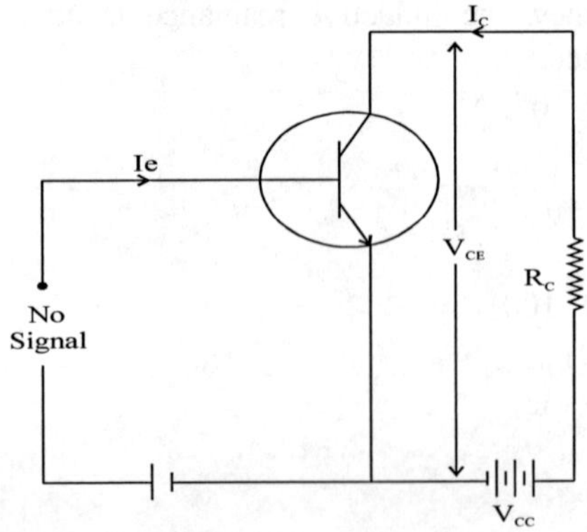

Fig. 3

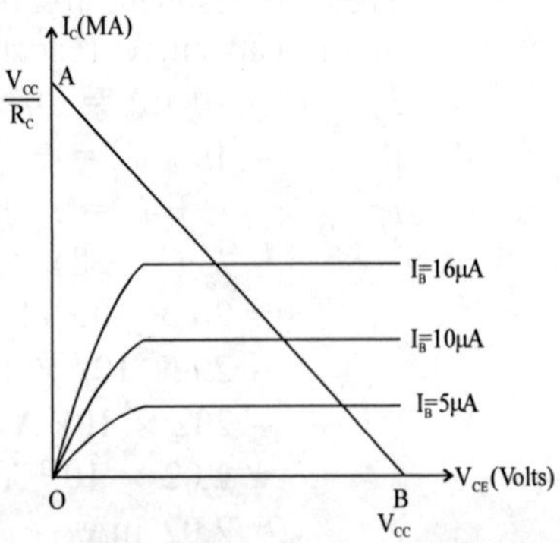

Fig. 4

(f) Thermal runway

The Self-destruction of an unstabilised transistor is known as thermal runway.

The collector-current for a CE configuration is given by

$$I_C = \beta I_B + (\beta + 1)\, I_{CBO}, \qquad \text{...(1)}$$

The collector leakage current I_{CBO} is strongly dependent on temperature. The flow of collector-current produces heat within the transistor. This raises the transistor temperature and if no stabilisation is done, the collector leakage current I_{CBO} also increases. It is clear from the equn. (1) that if I_{CBO} increases, the collector-current I_C increases by $(\beta + 1)\, I_{CBO}$. The increased I_C will raise the temperature of the transistor, which in turn will cause I_{CBO} to increase. This effect is cumulative and in a short time, the collector current may become very large, causing the transistor to burn out. This is called thermal runway.

Remedy

In order to eliminate thermal runway, it is very essential that operating point is stabilised, *i.e.*, I_C is kept constant. Practically, this is achieved by causing I_B to decrease automatically with the increase of temperature by circuit modification. Then decrease in βI_B will compensate for the increase in $(\beta + 1)\, I_{CBO}$, keeping I_C nearly constant. Actually, this is taken into consideration while building and designing a biasing circuit.

(g) Amplitude distortion in an amplifier

The change of output wave shape from the input wave shape of an amplifier is known as *distortion*.

A transistor like other electronic devices is essentially a non-linear device. Therefore, whenever a signal is applied to the input of the transistor, the output signal is not exactly like the input signal, *i.e.*, distortion occurs. For small signals, distortion is not a problem, since transistor is a linear device for small variations about the operating. point. However, a power amplifier handles large signals and, therefore, the problem of distortion immediately arises.

Remedy

A large signal stage has non-linear distortion because its voltage gain changes at various points in the cycle. The negative feedback reduces the non-linear distortion in large signal amplifiers. It can be proved mathematically that,

$$D_{fb} = \frac{D}{1 + Am}, \qquad \text{...(1)}$$

Where,

D is the distortion in amplifier without feedback,

D_{fb} is the distortion in amplifier with feedback.

It is seen that distortion reduced by a factor $(1 + Am)$ when negative feedback is applied to an amplifier.

(h) $f_1 = 60$ Hz, $f_2 = 8$ KHz $= 8 \times 10^3$ Hz, $\alpha = 0.1$

$A = 50$

$$f_1' = \frac{f_1}{1+\alpha A} = \frac{60}{1+0.1\times 50} = \frac{60}{1+5}$$

$$= \frac{60}{6} = 10 \text{ Hz}$$

$$f_2' = f_2\ (1+\alpha A)$$

$$= 8 \times 10^3\ (1 + 1\ 0.1 \times 50)$$

$$= 8 \times 10^3\ (1 + 5) = 8 \times 6 \times 10^3 \text{ Hz}$$

$$= 48 \text{ KHz}$$

Thus, the lower cut-off frequency becomes 10Hz while the upper cut-off frequency becomes 48 KHz.

(i) Collector efficiency of an amplifier

The ratio of a.c. output power to the zero signal power (i.e., d.c. power) supplied by the battery of a power amplifier is known as collector efficiency.

Hence, the ability of a power amplifier to convert d.c. power from supply into a.c. output power is a measure of its effectiveness. This is called collector efficiency. It is thus clear that for power amplifiers, maximum collector efficiency is the desired goal.

In class C amplifier, the collector efficiency is maximum.

(j) Electrical equivalent circuit of a piezoelectric crystal

The electrical equivalent circuit of a piezoelectric crystal is shown in Fig. 5 below. When the crystal is not vibrating, it is equivalent to capacitance C_m because it has two metal plates separated by a dielectric. On the other hand, when the crystal is vibrating, it is equivalent to series tank circuit R-L-C. In the equivalent electrical circuit of the crystal, L is the electrical equivalent of crystal mass, C is the electrical equivalent of elasticity and R is electrical equivalent of mechanical friction. In this circuit,

C_m = mounting capacitance
C_g = series capacitance introduced by air gap
R-L-C = electrical equivalent of vibrational characteristic of crystal

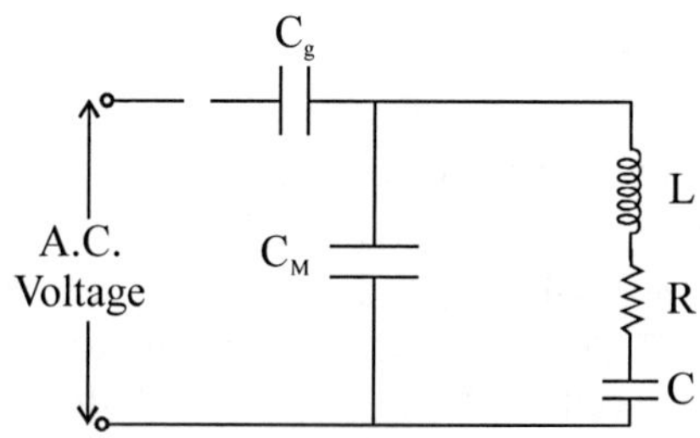

Fig. 5

The resonant frequency of R-L-C is the frequency of mechanical crystal resonance. However, due to presence of C_m, there will be parallel-resonant frequency also. But these two resonant frequencies are very close to each other as shown by crystal resonance curve of Fig. 6.

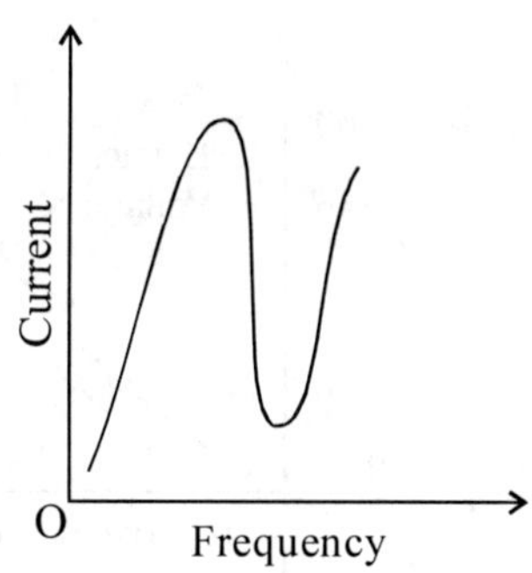

Fig. 6

Series and parallel resonance

A crystal has two resonant frequencies, *viz.* series and parallel resonant frequency.

The series resonant frequency is given by

$$f_s = \frac{1}{2\pi\sqrt{LC}} \text{ H}_z$$

and the parallel resonant frequency is given by

$$f_p = \frac{1}{2\pi\sqrt{LC_T}} \text{ H}_z$$

where, $$C_T = \frac{C \times C_m}{C + C_m}$$

(k) VI characteristics of PN junction diode

The nature of variation of current through the PN junction placed in the circuit with applied voltage is called the VI characteristic of a PN junction. Fig. 7 shows typical characteristic curve for a PN junction. It can be seen from this curve that the current increases slowly for small applied voltage and then very rapidly with the increase of applied forward bias voltage. The upward surge occurs at about 0.6 – 0.8 volts for silicon and 0.2 – 0.3 volts for germanium. But, the value of current for the reverse bias voltage is very small and increase is small enough to regard the current very nearly constant. This constant reverse bias current is known as reverse *saturation current*. As the reverse bias voltage increases to sufficiently high value, breakdown occurs as shown in the figure.

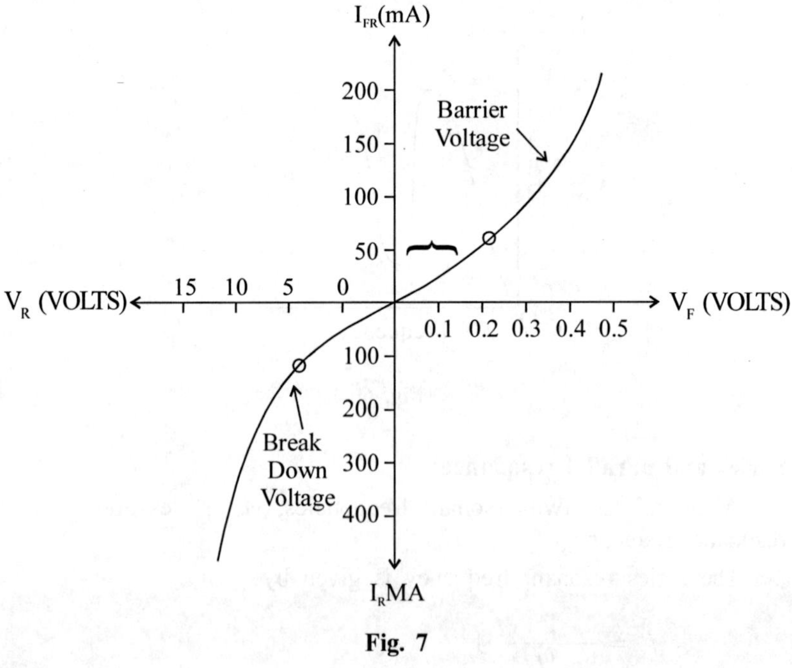

Fig. 7

(l) Class-C tuned amplifier

Class-C operation means that collector current flows for less than 180°. In a practical tuned class C amplifier, the collector current flows for much less than 180°; the current looks like narrow pulses as shown in Fig. 8. Here, the collector current wave form is a train of narrow pulses. When these narrow current pulses drive a high-Q resonant (*i.e.* LC) circuit, the output voltage is a perfect sine wave. This can be easily explained.

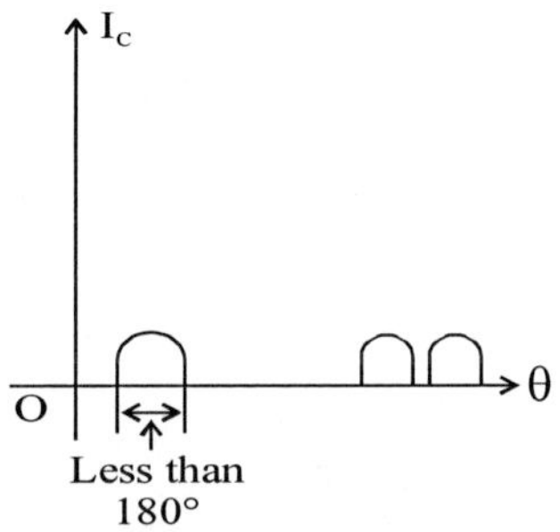

Fig. 8

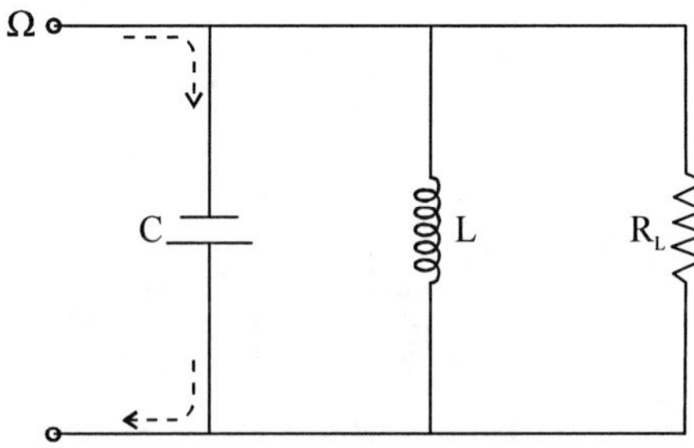

Fig. 9

Since the pulse is narrow, inductor looks like high impedance and the capacitor like a low impedance. Consequently, most of the current charges the capacitor as shown in Fig. 9. When the capacitor is fully charged, it will discharge through the coil and the load resistor, setting up oscillation just as one oscillatory circuit does. Consequently, sine wave output is obtained.

(m) The value of self-inductance of a coil

It depends upon :

(i) Number of turns in the coil.
(ii) Gap between the two adjacent turns of the coil.
(iii) Geometrical shape of the coil.
(iv) Permeability of the material used in the core of the coil.
(v) Diameter of the wire used in the coil.

Hence, the expression for L of a coil is

$$L = \frac{\mu N^2 A}{l}, \quad ...(1)$$

where, μ is the permeability of the core on which the coil is wound,

l is the length of the coil,

A is the area of each turn of the coil,

N is the no. of turns of the coil (inductor).

(n) Expression for the average d.c. value of current through a half-wave rectifier

In a half-wave rectifier, the diode conducts during positive half-cycles of a.c. supply while no current conduction takes place during negative half-cycles. So, the average d.c. value of the current through a half-wave rectifier for one full cycle is given by

$$I_{av} = I_{dc} = \frac{1}{2\pi}\int_0^{\pi} i d\theta = \frac{1}{2\pi}\int_0^{\pi} \frac{V_m \sin\theta}{r_f + R_L} d\theta$$

$$= \frac{V_m}{2\pi(r_f + R_L)} \int_0^{\pi} \sin\theta d\theta$$

$$= \frac{V_m}{2\pi(r_f + R_L)} [-\cos\theta]_0^{\pi}$$

$$= \frac{V_m}{2\pi(r_f + R_L)} \times 2 = \frac{V_m}{\pi(r_f + R_L)}$$

$$= \frac{I_m}{\pi} \quad \text{...(1)}$$

where, $I_m = \frac{V_m}{(r_f + R_L)}$

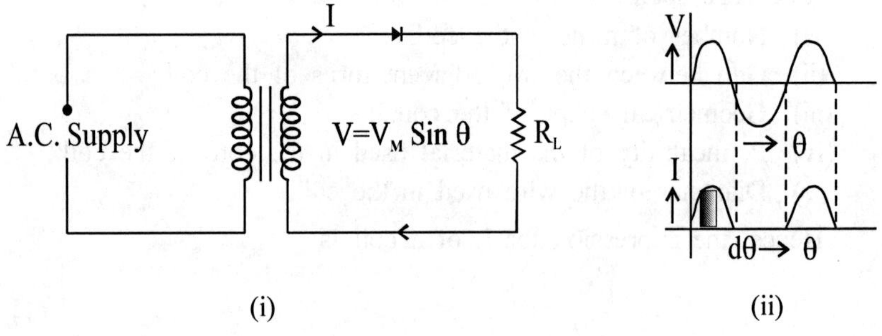

Fig. 10. (i) Half-wave rectifier, (ii) Output current

1998 (A)

Group-A

1. (a) 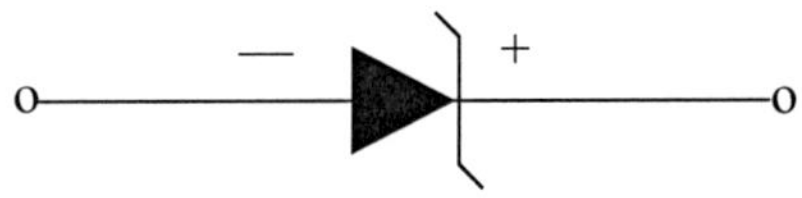

(b) 2.2 ohm ± 20% = 22 × 10^{-1} ohm ± 20%. The colours are Red, Red, Gold.

(c) The energy stored in a charged capacitor is given by

$$W = \frac{1}{2} CV^2 = \frac{1}{2} QV = \frac{1}{2}\frac{Q^2}{C}$$

Where, $C = \frac{Q}{V}$ = Capacitance of the charged capacitor.

Q = Charge on the positive plate of the capacitor.

V = P.d. between the plates of the capacitor.

(d) Vacuum Tetrode has negative resistance.

(e) Common collector (cc) configuration of a transistor amplifier has minimum voltage gain (< 1).

(f) Upper cut off, frequency increases from f_2 to f'_2 due to negative feedback. Because

$f'_2 = f_2 (1 + mA)$

where, A = Gain without feedback

m = feedback factor

(g) $$f = \frac{1}{2\pi\sqrt{L\left(\frac{C_1C_2}{C_1 + C_2}\right)}}$$

where, L = inductance

C_1 and C_2 are the capacitances.

2. (a) npn transistor is suitable for high frequency operation.

(b) Class A amplifier output has minimum harmonic content.

(c) In series LCR circuit the impedance is minimum at the resonant frequency.

(d) $C_S = \frac{C}{8}, C_P = 8C$

where, C_s is the combined capacitance of series grouping and C_p is that of parallel grouping, C is the capacity of each capacitor

So, $\frac{C_P}{C_S} = 64$

(e) Angle of conduction in class C amplifier is less than 180°.

(f) 0.7 Volt.

(g) halved.

(h) 100 Hz.

Group-B

3. (a) $P = \frac{V^2}{R} = \frac{250 \times 250}{25 \times 1000} = 2.5$ Watt.

(b) A Cathode is made of a metal of
   (i) Low work function.
   (ii) High melting point.
   (iii) High emission efficiency.
   (iv) High mechanical strength

(c) Parallel plate type capacitors used in electronic circuits are of the types :
   (i) Air Capacitors
   (ii) Paper Capacitors
   (iii) Mica Capacitors
   (iv) Ceramic Capacitors and
   (v) Electrolytic Capacitors

   Air and mica capacitors are suitable for use in RF circuits.

(d) $R_C = 2\ K\Omega = 2000\Omega$

$V_{CC} = 6$ Volt

P.d. across the load = $I_C R_C = 0.8$ Volt.

From a CE amplifier circuit, it is followed that

$V_{CE} = V_{CC} - I_C R_C$

$= 6 - 0.8 = 5.2$ Volts.

$I_C R_C = 0.8$

$\therefore\ I_C = \frac{0.8}{R_C} = \frac{0.8}{2000} = 4 \times 10^{-4}$ amp.

or, IC = 0.4 m

Again, $\alpha = \dfrac{I_C}{I_E}$

$$\therefore I_E = \frac{I}{\alpha} = \frac{0.4\,mA}{0.95} = 0.421\,mA$$

Now, $I_B = I_E - I_C = 0.421 - 0.4 = 0.021$ mA

(e) We have,

$$I_b = \frac{V_{BB}}{R_B} \text{ (approx.)} \qquad ...(1)$$

$V_{BB} \approx V_{BE} = V_{in}$

$$I_C = \beta I_B \qquad ...(2)$$

and $$V_{CE} = V_{CC} - I_C R_C \qquad ...(3)$$

(here $V_{CE} = V_{out}$)

When V_{BB} is increased, I_b increases and hence I_C increases. As I_C increases, $I_C R_C$ increases and hence V_{CE} decreases.

That means if V_{in} increases, V_{out} decreases. Hence, a positive going input signal becomes a negative going output signal leading to a phase reversal (180° phase difference) of the output with respect to input.

(f) **Distortions in amplifiers**

There are two types of distortions in Amplifiers, namely :

(i) Non-linear distortion and

(ii) Linear distortion.

Types of non-linear distortion

(i) *Amplitude distortion* : In this case, the output waveform is not the exact replica of the input.

(ii) *Harmonic distortion* : Here the output contains more than one frequency although the input is of single frequency (sine wave type).

(iii) *Inter modulation distortion* : In this case, the input has more than one frequency and of large amplitude.

Types of linear distortion

(i) *Frequency distortion* : It arises due to unequal amplification of different frequencies present in the input signal.

(ii) *Phase distortion* : It is caused due to unequal shifting of phase of the various components of the input signal.

(g) transformers are used for impedance matching between two stages of an amplifier.

The impedance of the primary of the transformer is made equal to the output impedance of the previous stage. The impedance of the secondary of the transformer is equal to the input impedance of the following stages.

(h) We know

$$D_f = \frac{D}{1+Am}$$

where, D = Distortion without feedback = 10%

D_f = Distortion with feedback = 5%

m = feedback factor = 2%

A = gain of the amplifier without feedback.

So, we have

$$\frac{5}{100} = \frac{10/100}{1+A\left(\frac{2}{100}\right)}$$

or, $$0.05 = \frac{0.10}{1+0.02A}$$

or, 0.05+0.001 A = 0.10

or, $$A = \frac{0.10-0.05}{0.001} = \frac{0.05}{0.001} = 50$$

(i) In public address system, the human voice is to be spread over a particular region. Normally the power level of the human voice is insufficient to activate the loudspeaker. Hence, the last stage of a PA system used before the loudspeaker is a power amplifier stage which will increase the power level of the human voice and help in actuating the loudspeaker.

(j)

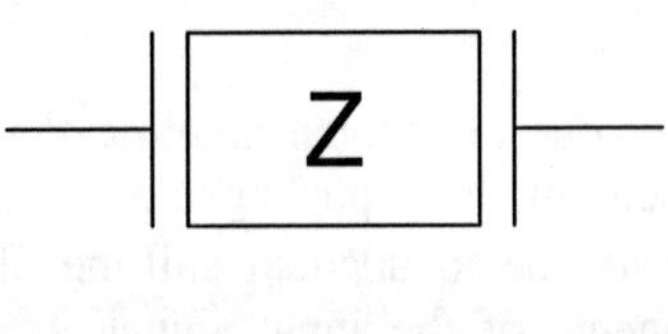

Fig. A

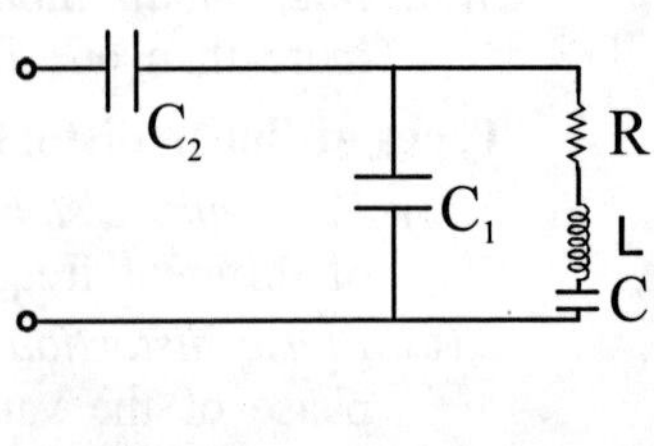

Fig. B

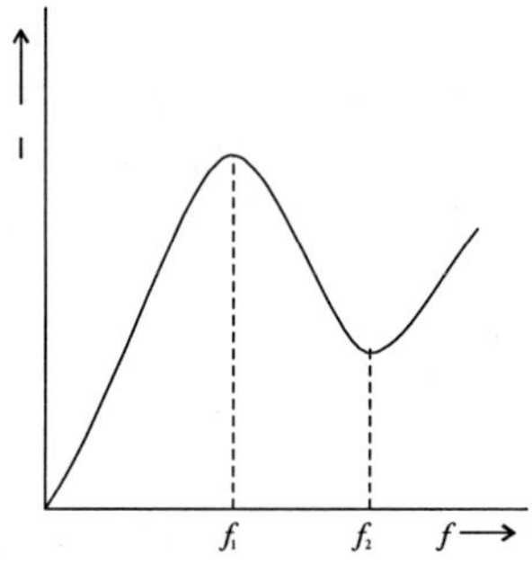

Fig. C

The piezo-electric crystal, its equivalent circuit and its frequency response curve are shown in Fig. A, Fig. B and Fig. C respectively. When the crystal is not vibrating, it is equivalent to capacitor C_1 (with metal plates P_1 and P_2 and crystal as dielectric).

When the crystal vibrates, it is equivalent to series R-L-C circuit in resonance. Capacity C_2 is due to the air gap in between crystal and metal plates.

The frequency response curve Fig. C indicates that it is a series resonant circuit (R-L-C) with resonant frequency f_1 and a parallel resonant circuit (R-L-C_1) with frequency f_2. Hence, the vibrating piezo-electric crystal is equivalent to a series-cum-parallel resonant circuit with two resonant frequencies f_1 and f_2.

(k)

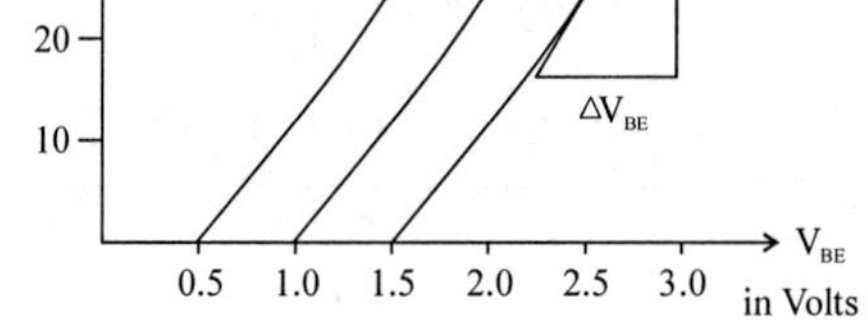

Fig. D

The input resistance is computed as

$$r_i = \frac{\Delta V_{BE}}{\Delta I_B}$$

where, V_{CE} = Constant

ΔV_{BE} = Change in Base Emitter Voltage

ΔI_B = Change in Base Current

ΔV_{CE} = Collector-Emitter Voltage

(l) Positive feedback is achieved in Hartley oscillator circuit as follows :

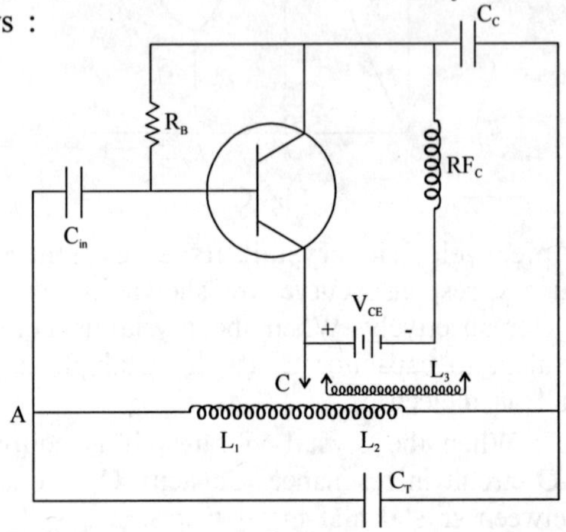

Fig. E

Transistor operates in the CE mode and hence there is a change of phase of 180° between output and input.

A part of the output voltage developed across L_1 is fed-back to input. The voltage across the inductor is 180° out of phase in accordance with Lenz's Law.

Hence, the net phase shift of 0° or 360° is developed and hence the feedback is positive.

(m) Depletion layer (D) :

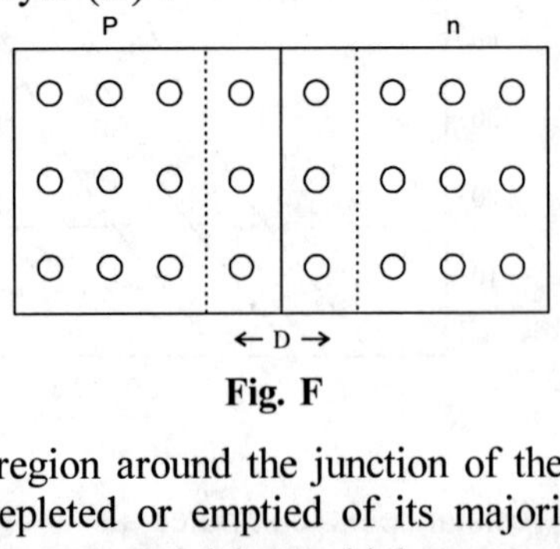

Fig. F

It is a region around the junction of the PN junction diode which is depleted or emptied of its majority carriers.

The p-type material has a high concentration of holes and the n-type material has a high concentration of electrons. At the junction, there is a tendency of holes to diffuse from p to n-side and electrons to diffuse from n to p-side. Therefore, near the junction net negative charge is established on the p-side and net positive charge on the n-side. The depletion layer is also called Barrier layer because the electric field developed prevents further diffusion of the majority carriers.

(n) Efficiency of a rectifier is given by

$$\eta = \frac{d.c.\text{ output power}}{a.c.\text{ input power}}$$

$$= \frac{(P_{dc})_{out}}{(P_{ac})_{in}} \qquad ...(1)$$

Now, $P_{dc} = I^2_{dc}\ R_L$ ...(2)

$$I_{dc} = \frac{1}{2\pi}\int_0^{\pi} I\, d\,\theta$$ for a half-wave rectifier.

$$I_{dc} = \frac{1}{2\pi}\left[\frac{E_0}{r + R_L}\right]\int_0^{\pi} Sin\theta\ d\theta$$

$$= \frac{E_o}{2\pi\,(r + R_L)}\left[-\,Cos\ \theta\right]_0^{\pi}$$

$$= \frac{E_o}{2\pi(r + R_L)} \times 2$$

$$= \frac{E_o}{\pi(r + R_L)} = \frac{I_o}{\pi} \qquad ...(3)$$

Where, r is the resistance of diode.

R_L is the load resistance.

Using (3) in (1), we have

$$P_{dc} = I^2_{dc}\ R_L = \left(\frac{I_0}{\pi}\right)^2 .R_L \qquad ...(4)$$

Now, $P_{ac} = I^2_{rms}\ (r + R_L)$, ...(5)

and $$I^2_{rms} = \frac{1}{2\pi}\int_0^{\pi} I_0^2\ Sin^2\ \theta\, d\,\theta$$

for half-wave rectifier.

So, $I^2_{rms} = \dfrac{I_o^2}{4}$ ...(6)

Using (6) in (5), we have

$$P_{ac} = \frac{I_o^2 \,(r + R_L)}{4} \qquad ...(7)$$

Using (4) and (7) in (1), we obtain

$$\eta = \frac{(I_{o/\pi})^2 R_L}{\frac{I_o^2}{4}(r + R_L)}$$

$$= \left(\frac{4}{\pi^2}\right) \frac{R_L}{r + R_L}$$

$$= \frac{0.406\, R_L}{r + R_L}$$

$$= \frac{0.406}{1 + (r / R_L)} \qquad ...(8)$$

1997 (A)

Group-A

1. (a) Blue, Violet, Gold, Silver —

   Resistance is $67 \times 10^{-1} \pm 10\% = 6.7$ ohm $\pm$ 10%

   (b) P = 100 Watt, V = 220 Volt

   $$R = \frac{V^2}{P} = \frac{220 \times 220}{P} = 484\ ohm$$

   $$\text{Power consumed} = \frac{V^2}{R}$$

   $$= \frac{110 \times 110}{484}$$

   = 25 Watt.

   (c) When plate potential increases, the cut off bias becomes more negative.

   (d) In common collector configuration of a transistor amplifier the input impedance is more than the output impedance.

   (e) In a Radio Receiver, the name of tuned amplifier stage is the R.F. amplifier and untuned amplifier stage is the A.F. amplifier, which may be 2 to 3 stages depending on signal strength.

   (f) The type of feedback is negative feedback.

   (g) The resonant frequency is given by

   $$f = \frac{1}{2\pi\sqrt{LC}}$$

   If C is doubled and L is halved, then the value of resonant frequency remains unaltered.

2. (a) Direction of the induced electromotive force is given by *Lenz's law.*

   (b) A series resonant circuit is *inductive* beyond resonance.

   (c) As r_p increases due to change in voltages, g_m *decreases.*

   (d) The value of α for a bipolar transistor is *always less than one.*

(e) The stability factor is maximum in *C.E.* amplifier.
(f) A tuned amplifier is suitable for use as *class C in RF stage.*
(g) Zener diode is used in *reverse biased condition.*
(h) The ripple frequency for a full wave rectifier is 100 Hz for the line frequency of 50 Hz.

Group-B

3. (a) The work function of copper is very high (about 6.5 eV). So, it is not used as thermionic emitter in vacuum.
(b) Refer to the Answer of Q. 3 (a) of 1996 (A).
(c) The bandwidth of a tuned circuit is given by

$$BW = \frac{\text{Resonant frequency}}{Q}$$

So, bandwidth varies inversely with Q. As Q increases the bandwidth decreases.

(d) From the Richardson – Dushman equation, we have

$J_S = AT^2 e^{-b/T}$

Here we have $$\frac{(J_s)_2}{(J_s)_1} = \frac{AT_2^2\, e^{-b/T_2}}{AT_1^2\, e^{-b/T_1}}$$

$$= \left(\frac{T_2}{T_1}\right)^2 \left[\frac{e^{-b/T_2}}{e^{-b/T_1}}\right]$$

or, $$\left(\frac{T_2}{T_1}\right)^2 = \frac{(J_s)_2}{(J_s)_1}\left[\frac{e^{-b/T_1}}{e^{-b/T_2}}\right]$$

or, $$\left(\frac{T_2}{T_1}\right)^2 = 3\left[e^{-b\left(\frac{1}{T_1}-\frac{1}{T_2}\right)}\right]$$

Unless the value of 'b' is known, $\left(\frac{T_2}{T_1}\right)$ cannot be evaluated.

Roughly, $\frac{T_2}{T_1} < \sqrt{3}$

(e) The concentration of the conduction electrons in the conduction band is more than the concentration of holes in the valence band.

E_F = Energy of Fermi Level.

E_D = Energy of Donor Level.

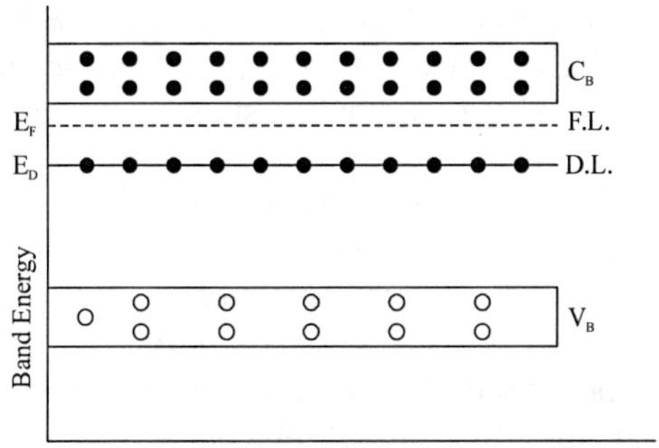

Fig. G

(f) Refer to the Answer of Q. 3 (i) of 1996 (A).

(g) We have

$$\frac{\text{Mobility of electrons}}{\text{Mobility of holes}} = \frac{\mu_e}{\mu_h}$$

$$= \frac{[(V_d)_e / E]}{[(V_d)_h / E]} = \frac{(V_d)_e}{(V_d)_h}$$

and $$\frac{(V_d)_e}{(V_d)_h} = \frac{\text{Drift velocity of electrons}}{\text{Drift velocity of holes}} > 1$$

Hence, mobility of electrons is more than the mobility of holes in an extrinsic semiconductor.

Here, E is the electric field intensity of the external field.

(h) Refer to the Answer of Q. 3 (j) of 1996 (A).

(i) Frequency response Curve of RC Coupled Amplifier :

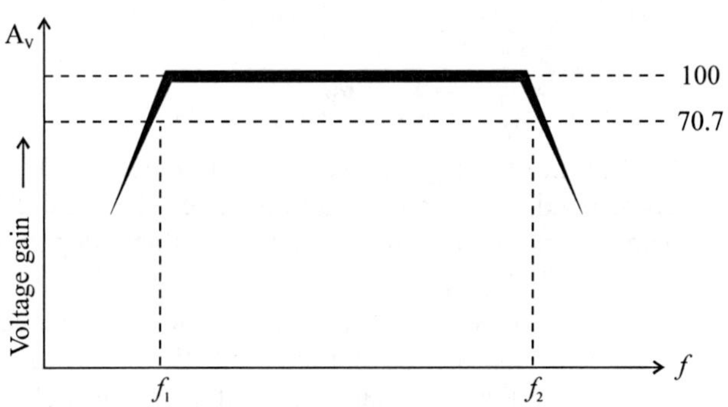

Fig. H

The gain at high frequencies (>20 KHz) reduces due to the following reasons :

(a) The reactance of the coupling capacitor is very small and provides a short circuit path. This increases the loading effect of the following stages and hence reduces the voltage gain.

(b) At high frequency, the reactance of the emitter bypass capacitor is low, which increases the base current. As a result, the current amplification factor β decreases and hence the voltage gain is reduced.

(j) When the emitter bypass capacitor C_E is removed the R–C coupled amplifier circuit in common emitter configuration reduces to a negative current feedback (or emitter follower) circuit. As a result of negative feedback the voltage gain reduces and hence bandwidth increases.

(k) Zener Knee :

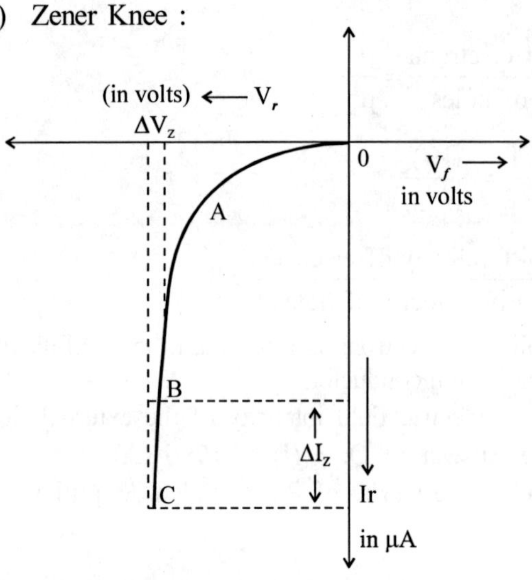

Fig. I

In the above figure, the Zener Knee is at A and the corresponding reverse voltage is called the breakdown voltage. The Zener breakdown or Knee voltage depends upon the amount of doping.

(l) In class C tuned amplifiers, the period of conduction is less than 180° as the output current flows for a part of the positive half cycle. But, the load is a parallel resonant LC circuit whose impedance is very high. The frequency response curve of the amplifier is of the same shape as the resonance curve of the tuned circuit. Only narrow band of frequencies around the resonant frequency f_o will be amplified whereas other frequencies are not amplified.

Non-linear distortion is eliminated due to high selectivity of the load impedance. Hence, the output is nearly sinusoidal.

(m) The efficiency of a rectifier is given by

$$\eta = \frac{P_{dc}}{P_{ac}} \times 100$$

Now, $P_{dc} = I^2_{dc} \times R_L = \left(\frac{2I_o}{\pi}\right)^2 R_L$

$P_{ac} = I^2_{rms} (r + R_L)$

Where, r is the forward resistance of junction diode or plate resistance of vacuum diode.

$I_{rms} = I_o/V_2$

So, $P_{ac} = \left(\frac{I_o}{\sqrt{2}}\right)^2 (r + R_L)$

$$\eta = \frac{P_{dc}}{P_{ac}} \times 100$$

$$= \frac{\left(\frac{2I_o}{\pi}\right)^2 R_L}{\left(I_o / \sqrt{2}\right)^2 + R_L} \times 100$$

$$= \frac{8}{\pi^2}\left(\frac{R_2}{r + R_L}\right) \times 100$$

$$= \frac{0.812}{\left(1 + \frac{r}{R_L}\right)} \times 100 = 81.2\%$$

$$\left(\because \frac{r}{R_L} << 1\right)$$

Thus, $\eta_{max} = 81.2\%$

(n) The frequency drift in an oscillator is due to the reason that the values of resistors, capacitors and inductors change with temperature. A change is R and C or L and C causes a change in the frequency of oscillation of the oscillator circuit.

1997 (S)

Group-A

1. (a)

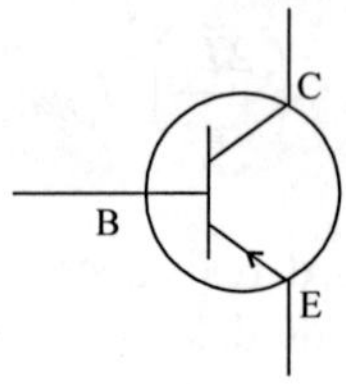

Fig. J : Circuit symbol of pnp transistor

(b) Q-factor or quality factor of a coil is defined as the ratio of the reactance (X_L) of the coil at resonance to the resistance (R) of the coil.

$$Q = \frac{\omega_o L}{R} = \frac{2\pi f_o L}{R}$$

(c) Grey, Blue, Gold, Gold :
The resistance is $86 \times 10^{-1} \pm 5\%$
$= 8.6\ \Omega \pm 5\%$

(d) The passive component is the transformer.

(e) Three possible transistor connection are :
(i) Common Base (CB)
(ii) Common Emitter (CE)
and (iii) Common Collector (CC)

(f) Positive feedback is employed in oscillator circuit.

(g) The maximum efficiency of a full-wave rectifier is 81.2%.

2. (a) Maximum power will be transferred from a source of 10 ohm resistance to a load of *10 ohm.*

(b) Unit of reactance of a circuit is *ohm.*

(c) The screen grid potential is somewhat *lower* than plate potential.

(d) In an n-type semiconductor, current conduction takes place by *free electrons.*

(e) The base of a transistor is *lightly* doped.

(f) If the operating point is changed, *unfaithful amplification* results.

(g) Negative feedback reduces *distortion* in amplifiers.

(h) Output transformer is always a *step down* transformer.

Group-B

3. (a) The energy stored in a charged capacitor is given by

$$W = \frac{1}{2} QV = \frac{1}{2} CV^2 = \frac{1}{2} \frac{Q^2}{C}$$

where, $C = \frac{Q}{V}$ = Capacitance of the Charged capacitor.

Q = Charge on the positive plate of the capacitor.

V = p.d. between the plates of the capacitor.

(b) The resonant frequency in a series resonant circuit is given by

$$f_r = \frac{1}{2\pi\sqrt{LC}}$$

Where, L is the inductance and C is the capacitance.

(c) A triode valve has following limitations :

(i) Interelectrode capacitance :

The reactance of the plate to grid capacitance

$$X_c = \frac{1}{2\pi f\, C_{gp}}$$

reduces at high frequencies and provide an easy feedback path for the output energy to be feedback to the input and the feedback is negative. Thus, the gain is reduced.

(ii) Insufficient amplification :

The amplification factor of a triode is usually small and less than 100. So, the amplification is not high.

(d) We have $g_m = 1.5 \text{ mA/V} = 1.5 \times 10^{-3} \text{ A/V}$

$= 1.5 \times 10^{-3} \text{ ohm}^{-1}$

$r_p = 12 \text{ K}\Omega = 12 \times 10^3 \text{ ohm}$

So, $\mu = r_p \times g_m = 18$

(e) We know $\frac{N_P}{N_S} = \frac{V_P}{V_S}$

Here, $\frac{4}{1} = \frac{240}{V_S}$

$$\therefore V_S = \frac{240}{4} = 60 \text{ Volt (rms)}$$

Hence, $(V_S)_{peak} = 60 \times V_2 = 84.84$ Volts.

(f) $\beta = \frac{\alpha}{1-\alpha} = \frac{0.99}{1-0.99} = \frac{0.99}{0.01} = 99$

(g) Various methods of biasing a transistor are :

(i) Base Resistor method

(ii) Feedback method

(iii) Voltage divider method

(iv) Method of self-biasing

(h) The different transistor amplifiers are :

(i) Direct coupled amplifier

(ii) R-C coupled amplifier

(iii) L-C coupled amplifier

(iv) Transformer coupled amplifier.

(i) Voltage gain = 73

We know

Voltage gain = 20 $\log_{10}$ 73 dB = 37.27 dB

(j) We know

$$\left(\frac{N_P}{N_S}\right)^2 = \frac{Z_P}{Z_L} = \frac{0.9 \times 10^3\ \Omega}{9} = 100$$

So, $\frac{N_P}{N_S}$ = Turns ratio = $\sqrt{100}$ = 10.

(k) Maximum a.c. power output = 30 watt.

The power rating of the transistor = ?

η = 50% = 0.5

So, $0.5 = \frac{P_{out}}{P_{in}} = \frac{30 \text{ Watt}}{P_{in}}$

or, P_{in} = 60 Watts.

(l) Gain without feedback = A = 10^4

Gain with feedback = A_f = 50

If feedback fraction is β, then

$$A_f = \frac{A}{1 + A\beta}$$

Since, the gain is reduced with feedback. We have negative feedback here.

$$\text{So,} \quad 50 = \frac{10^4}{1 + 10^4 \beta}$$

or, $50 + (50 \times 10^4)\ \beta = 10^4$

or, $5 \times 10^5\ \beta = 10000 - 50 = 9950$

$$\therefore \beta = \frac{9950}{500000} = \frac{0.0995}{5}$$

$$= 0.0199 \approx 0.02.$$

1996 (A)

Group-A

1. (a) R = 100 Ω, P = 1 watt, V = ?

$$V = \sqrt{P \times R} = \sqrt{1 \times 100} = 10 \text{ Volt.}$$

(b)

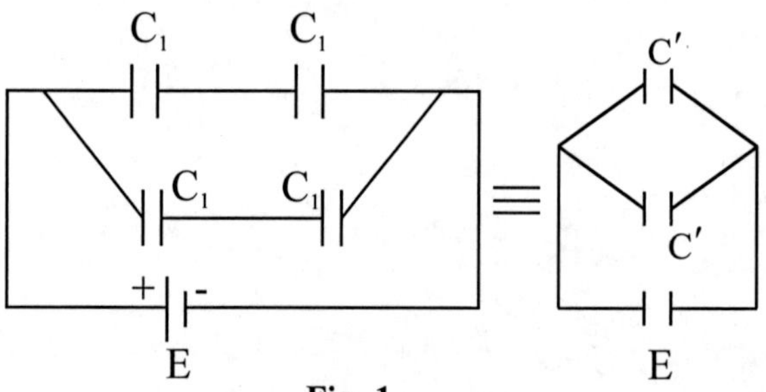

Fig. 1

$$C_1 = 8\ \mu f,\quad E_1 = 10\ \text{V}$$
$$C = 8\ \mu f,\quad E = 20\ \text{V}$$

when two capacitors of 8 μf capacitance each are joined in series, we get the combined capacitance of

$$\frac{1}{C'} = \frac{1}{C_1} + \frac{1}{C_1} = \frac{1}{8} + \frac{1}{8} = \frac{2}{8} = \frac{1}{4}$$

or $$C' = 4\ \mu f$$

Again, two 4 μf capacitors are joined in parallel, then we have the combined capacitance

$$C = C' + C'$$
$$= 4\ \mu f + 4\ \mu f = 8\ \mu f$$

Regarding the voltage, we have $E' = 10\ \text{V} + 10\ \text{V} = 20\ \text{V}$

for two capacitors joined in series. When two emfs 20 V each joined in parallel, then we get

$$E = E' = 20\ \text{V}$$

(c) $X_L = \omega L = 2\pi f L \Rightarrow X_L \alpha\ f$

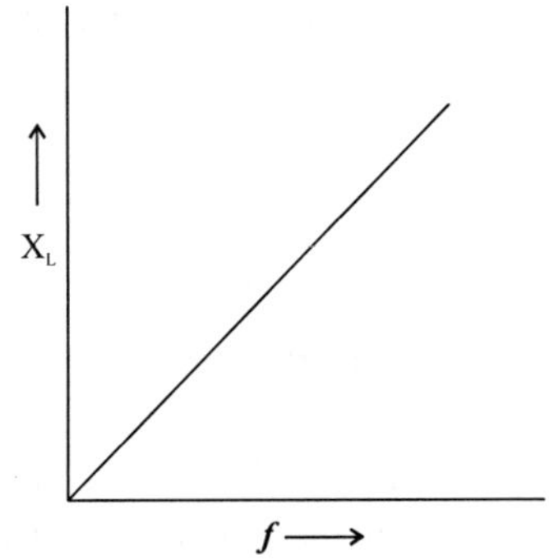

Fig. 2. Variation of reactance of an induction with frequency

(d) Circuit symbol of an iron-core inductor :

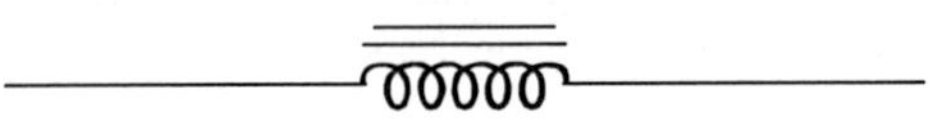

(e) Cathode is closer to the grid in a triode.

(f) The minimum forward bias voltage for a silicon p-n junction diode to conduct is 0.7 V.

(g) NPN type of transistor is suitable for high frequency operation as the majority charge carriers are electrons.

2. (a) In a CE amplifier, Output impedance is higher than input impedance.

(b) Voltage divider method of biasing offers better stabilization in amplifiers.

(c) Ripple frequency = Line frequency × Ripple factor
= (50 × 1.21) Hz
= 60.50 Hz = 60.5 Hz

(d) The phase-shift of output relative to the input in a CB amplifier is zero.

(e) In audio frequency range (20 Hz to 20 KHz), R-C coupled amplifiers are usually used.

(f) Lower cut-off frequency of an amplifier with negative feedback decreases by a factor (1 + αA), *i.e.*,

$$f = \frac{f_1}{1 + \alpha A}$$

where, f_1 is the lower cut-off frequency without negative feedback, f is the lower cut-off frequency with negative feedback, α is the feedback factor and A is the gain without feedback.

(g) The most important characteristic of a crystal oscillator is its high order of frequency stability.

(h) The time constant of a series RL circuit is given by :

$$\lambda = L/R$$

Group-B

3. (a) The factors governing the inductance of an inductor are :
   (i) Number of turns in the coil,
   (ii) Geometrical shape of the coil, and
   (iii) Permeability of the material used in the core of the coil.

(b) A leaky capacitor

A capacitor having paper dielectric has considerable leakage and hysteresis loss. Such a capacitor is known as a leaky capacitor. Capacitance of this capacitor decreases by leakage.

(c) Mutual Inductance

Mutual inductance M between two closely packed coils is defined as the emf induced in the secondary coil by unit rate of change of current in the primary. Mathematically,

$$M = \frac{E_2}{\frac{dI_1}{dt}} \qquad ...(1)$$

where, E_2 is the induced emf in the secondary due to rate of change of current (dI_1/dt) in the primary.

When $dI_1/dt = 1$, then $M = E_2$ ...(2)

i.e., mutual inductance M is defined as the emf induced in the secondary by unit rate of change of current in the primary.

Similarly, we have the induced emf. E_1 in the primary is due to the rate of change of current in the secondary. Hence,

$$M = \frac{E_1}{dI_2/dt} \qquad ...(3)$$

where, E_1 is the induced emf in the primary due to the rate of change of current dI_2/dt in the secondary.

If $dI_2/dt = 1$, then $M = E_1$ ...(4)

Thus, the definition of M is followed.

Unit : The MKSa unit of mutual inductance is Henry (H).

The mutual inductance between two coils or circuits is 1 henry if a rate of change of current of 1 amp/sec in the primary produces an induced emf of 1 volt in the secondary.

Relation between mutual inductance and self-inductance of the coils :

The mutual inductance is given by,

$$M = K\sqrt{L_p L_s}$$

where, K is the co-efficient of coupling

L_p is the self-inductance of the primary

L_s is the self-inductance of the secondary

(d) There are four methods of electron emission from metal, namely

(i) Thermionic emission $\rightarrow$ Diode valve

(ii) Photoelectric emission $\rightarrow$ Photo cell

(iii) Field emission $\rightarrow$ Discharge tubes

(iv) Secondary emission $\rightarrow$ Dynatron effect

(e) Dynatron effect

The phenomenon of reduction of plate current in a tetrode due to secondary emission is known as *dynatron effect*. Under this condition, the tetrode is said to be having *negative resistance* because I_p decreases with increase of V_p. This property is utilized in designing negative resistance or *dynatron oscillator*.

(f) Plate characteristics of a triode

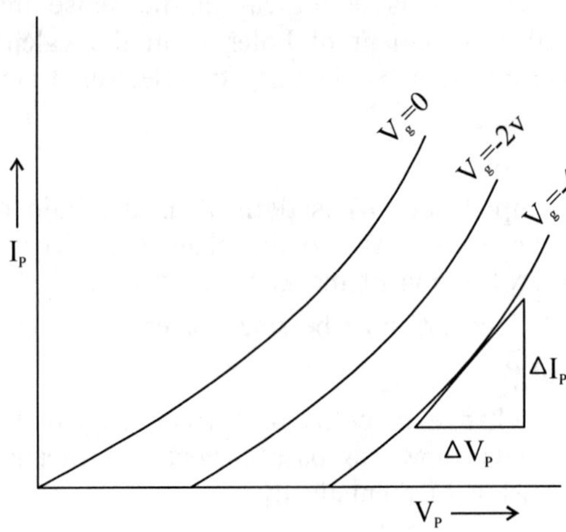

Fig. 3

$$R_p = \left(\frac{\Delta V_p}{\Delta I_p}\right)_{V_g} = \text{plate resistance}$$

(g)

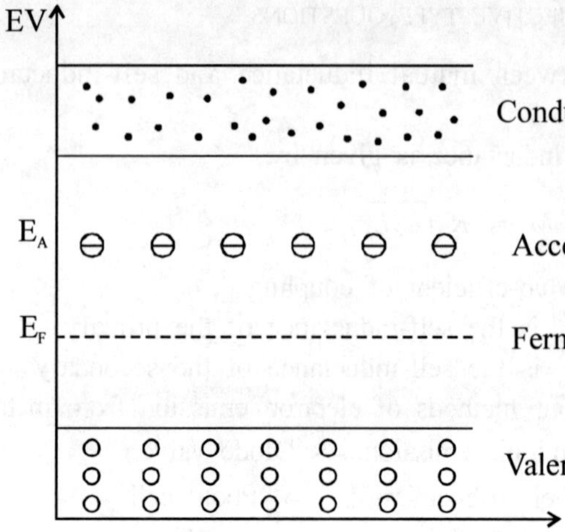

Fig. 4. P-type semiconductor

Explanation

In p-type semiconductor, acceptor level is above the fermi level by the addition of a small amount of acceptor trivalent impurity to the pure semiconductor. Hence, by the process of doping, extra energy level appears in the forbidden band above the valence band. The new level is indicated by (—) sign in the Fig. 4. The new level is nearer to the valence band. This is logical in the sense that in P-type semiconductor, the movement of holes is in the valence band and addition of impurity atom would raise the electron from the valence band.

(h) Zener Impedance

The Zener impedance (Z) is defined as the ratio of change in voltage across the Zener (ΔV) to the change in Zener current (ΔI) in the Zener control region of the Zener diode.

For an ideal Zener, Z must be small or equal to zero.

(i) Definition of β

The ratio of change in collector current (ΔI_c) to the change in base current (ΔI_B) is known as base current amplification factor, 'β' of a bipolar transistor. Mathematically,

$$\beta = \frac{\Delta I_C}{\Delta I_B} \qquad ...(1)$$

If d.c. values are considered,

$$\beta = \frac{I_C}{I_B} \qquad ...(2)$$

Relation between β and α :

$$\beta = \frac{\Delta I_C}{\Delta I_B}, \quad \alpha = \frac{\Delta I_C}{\Delta I_E}$$

Now, $I_E = I_B + I_C$

$\Delta I_E = \Delta I_B + \Delta I_C$

$\Delta I_B = \Delta I_E - \Delta I_C$

So, $$\beta = \frac{\Delta I_C}{\Delta I_E - \Delta I_C} = \frac{\Delta I_C / \Delta I_E}{\frac{\Delta I_E}{\Delta I_E} - \frac{\Delta I_C}{\Delta I_E}}$$

$$= \frac{\alpha}{1-\alpha} \qquad ...(3)$$

∴ $$\beta = \frac{\alpha}{1-\alpha}$$

where, α is the current amplification factor of CB transistor configuration.

(j) Thermal runway

The self-destruction of an unstabilised transistor is known as thermal runway.

Causes

The collector current for a CE configuration is given by

$$I_C = \beta I_B + (\beta + 1)\, I_{CBO} \qquad ...(1)$$

The collector leakage current I_{CBO} is strongly dependent on temperature. The flow of collector current produces heat within the transistor. This raises the transistor temperature and if no stabilisation is done, the collector leakage current I_{CBO} also increases. It is clear from equn. (1) that if I_{CBO} increases, the collector current I_C increases by (β +1) I_{CBO}. The increased I_C will raise the temperature of the transistor, which in turn will cause I_{CBO} to increase. This effect is cumulative and in a very short time, the collector current may rise to a high value, causing the transistor to burn out.

Remedy

Thermal runway is controlled by proper stabilization and transistor biasing. For this, I_C is kept constant. In practice, this is done by causing I_B to decrease automatically with temperature increase by circuit modification. Consequently, the decrease in βI_B will compensate for the increase in $(\beta + 1)\ I_{CBO}$, maintaining I_C nearly constant.

(k) Advantages of a bridge rectifier over a full-wave rectifier using 2 diodes

(i) In a bridge rectifier, the output voltage is twice that of 2-diode rectifier.

(ii) The PIV of a bridge rectifier is half that of 2-diode rectifier.

(l) The gain-bandwidth product for a given transistor is a constant.

(m) $R_L = 10\ \text{K}\Omega = 10 \times 10^3\ \Omega$

$R_{in} = 200\ \Omega,\ \ \beta = 50$

$A_V = ?,\ \ A_p = ?$

$$\text{Voltage gain}\quad A_v = \frac{\beta R_L}{R_{in}}$$

$$= \frac{50 \times 10 \times 10^3}{2 \times 10^2} = 2500$$

$$\text{Power gain}\quad A_p = \frac{\beta^2 R_L}{R_{in}}$$

$$= \frac{(50)^2 \times 10 \times 10^3}{2 \times 10^2}$$

$$= (50)^2 \times 50$$

$$= (50)^3 = 125000$$

(n) $m = 0.05,\ \ A = 100$

Reduction in distortion is given by

$$\frac{D}{D_{fb}} = (1 + mA)\%$$

$$= (1 + 0.05 \times 100)\%$$

$$= (1 + 5)\% = 6\%$$

1996 (S)

Group-A

1. (a) Circuit symbol of Zener diode :

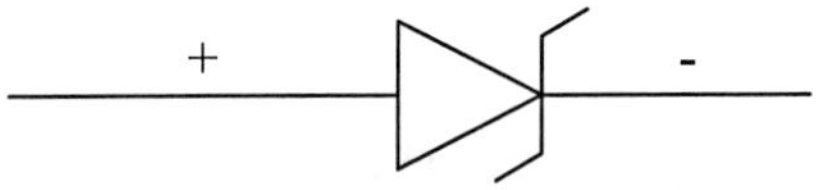

(b) The phase difference between input and output signals of transistor amplifier in CE more is 180° or π radian.

(c) The base current amplification factor 'β' :

The base current amplification factor 'β' for a CE transistor is defined as the ratio of change in collector current (ΔI_C) to the change in base current (ΔI_B). Mathematically,

$$\beta = \frac{\Delta I_C}{\Delta I_B} \qquad ...(1)$$

where, ΔI_C is the change in collector current and ΔI_B is the change in base current.

If d.c. values are considered, then

$$\beta = \frac{I_C}{I_B} \qquad ...(2)$$

(d) The colour code of a carbon resistor of 100 KΩ with 5% tolerance is Brown, black, yellow, gold

(e) Electron is the minority carrier in a p-type semiconductor.

(f) Richardson's thermionic equation :

The emission current density is given by Richardson-Dushman equation given below;

$$J_s = AT^2\, e^{-b/T} \text{ amp/m}^2 \qquad ...(1)$$

where, J_s = emission current density *i.e.*, current per square metre of the emitting surface.

T = absolute temperature of emitter in °K

A = constant which depends upon the type of emitter and is measured in amp/m²/°K²

b = a constant for the emitter
e = natural logarithmic base

The value of b is constant for a metal and is given by;

$$b = \frac{\Phi e}{K}$$

where, ϕ = Work function of emitter
e = electron charge = 1.602×10^{-19} coulomb
K = Boltzmann's constant = 1.38×10^{-23} J/°K

So, $$b = \frac{\phi \times 1.602 \times 10^{-19}}{1.38 \times 10^{-23}} = 11600\,\phi \text{ °K}$$

(g) The width of depletion layer in a p-n junction increases with reverse bias.

2. (a) Unit of inductive reactance of a coil is *ohm.*
   (b) A capacitor has *low* reactance at high frequency.
   (c) Bandwidth of an amplifier *increases* with negative feedback.
   (d) Pentavalent impurity atoms are added in pure silicon to make *n-type* semiconductor.
   (e) Lenz's law indicates *direction* of the induced current.
   (f) Primary current in a step-up transformer is more than secondary current.
   (g) Electron-volt is a unit of *energy*.
   (h) *CE* mode is the most commonly used transistor amplifier.

Group-B

3. (a) The conductivity of a semiconductor :
   As the semiconductor has negative temperature co-efficient of resistance, its conductivity increases with increase of temperature.

   (b) $\alpha = 0.95$, $I_e = 2$ mA, $I_B = ?$

$$\frac{I_c}{I_e} = \alpha = 0.95$$

$I_c = \alpha\, I_e = 0.95 \times 2 \text{ mA} = 1.90 \text{ mA}$
$I_B = I_e - I_C = 2 \text{ mA} - 1.90 \text{ mA} = 0.1 \text{ mA}$

   (c) $V = 25$ V, $R = 200$ ohm
   $P = ?$

$$P = \frac{V^2}{R} = \frac{(25)^2}{200} = \frac{625}{200} = \frac{312.5}{100} = 3.125 \text{ watts}$$

(d) **Definition of g_m of a triode**

The transconductance (g_m) of a triode valve is defined as the ratio of change in plate current to the change in grid voltage for a constant plate voltage. Mathematically, it is given by,

$$g_m = \left(\frac{\Delta I_p}{\Delta V_g}\right)_{V_p}$$

where, g_m is the mutual conductance or transconductance of the triode,

ΔI_p is the change in plate current and ΔV_g is the change in grid voltage.

Unit. The MKSa unit of g_m is *mho.*

(e) **Relation between α and β of a transistor**

A simple relation exists between α and β of a transistor. This can be derived as follows :

$$\beta = \frac{\Delta I_C}{\Delta I_B} \quad \text{...(1)}$$

$$a = \frac{\Delta I_C}{\Delta I_E} \quad \text{...(2)}$$

Now, $I_E = I_B + I_C$

or, $\Delta I_E = \Delta I_B + \Delta I_C$

or, $\Delta I_B = \Delta I_E - \Delta I_C$

Substituting the value of ΔI_B in equn. (1), we get

$$\beta = \frac{\Delta I_C}{\Delta I_E - \Delta I_C} \quad \text{...(3)}$$

or, $$\beta = \frac{\Delta I_C / \Delta I_E}{\frac{\Delta I_E}{\Delta I_E} - \frac{\Delta I_C}{\Delta I_E}}$$

$$= \frac{\alpha}{1-\alpha} \qquad \left[\because \alpha = \frac{\Delta I_C}{\Delta I_E}\right]$$

$$\therefore \beta = \frac{\alpha}{1-\alpha} \quad \text{...(4)}$$

where, α is the current amplification factor for CB transistor connection and β is the current amplification factor for CE transistor connection.

(f) **Functions of suppresser grid in a pentode**

The suppressor grid in a pentode performs the following functions :

(i) Suppresses secondary emission effects

As the suppressor grid is connected to the cathode, therefore, it is at substantial negative potential with respect to plate. The electrons from cathode still produce secondary emission as they hit the plate with high speeds. But, now the secondary electrons are no longer attracted to the screen grid. As soon as the secondary electrons are emitted by the plate, they are repelled back by the negative suppressor grid. Thus, the effects of secondary emission are suppressed and the 'dip' in the plate characteristic of tetrode is eliminated.

(ii) Increases amplification factor

The presence of suppressor grid provides further shielding between plate and control grid. This puts the control grid in a more commanding position and hence amplification factor is increased.

(iii) Reduces C_{gp}

The suppressor grid also reduces grid to plate capacitance. This completely eliminates the feedback action.

(g) **Circuit diagram of a bridge rectifier**

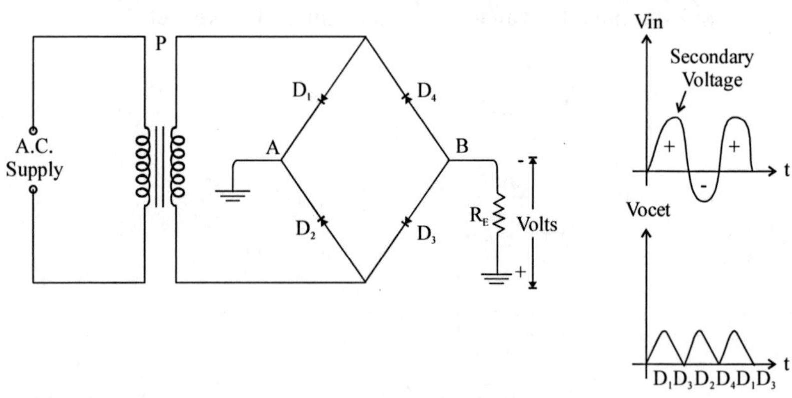

Fig. 1. Circuit diagram of a Full-wave Bridge Rectifier

(h) **Bandwidth of an amplifier**

The range of frequency over which the gain is equal to or greater than 70.7% of the maximum gain is known as bandwidth.

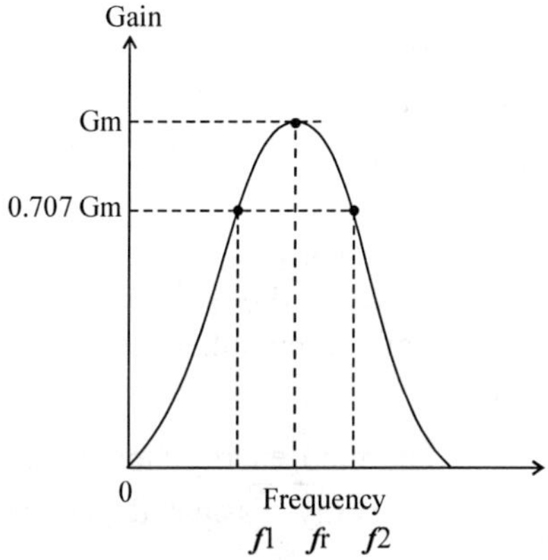

Fig. 2

The voltage gain of an amplifier changes with frequency. Referring to the frequency response in Fig. 2 above, it is clear that for any frequency lying between f_1 and f_2, the gain is equal to or greater than 70.7% of the maximum gain. Therefore, f_1–f_2 is the *bandwidth*. It may be seen that f_1 and f_2 are the limiting frequencies. The former (f_1) is called *lower cut-off frequency* and the latter (f_2) is known as *upper cut-off frequency*. For distortionless amplification, it is important that signal frequency range must be within the bandwidth of the amplifier.

The bandwidth of an amplifier can also be defined in terms of *decibel* (*db*). Suppose the maximum voltage gain of an amplifier is 100. Then 70.7% of it is 70.7.

Hence, fall in voltage gain from maximum gain

$$= 20 \log_{10} \frac{100}{70.7} \text{ db}$$

$$= 20 \log_{10} 1.4142 \text{ db}$$

$$= 3 \text{ db}$$

Thus, *bandwidth* of an amplifier is the range of frequency at the limits of which its voltage gain falls by 3 db from the maximum gain.

(i) $V_{CC} = 9$ V, $R_C = 2$ KΩ, $V_{CE} = 6$ V, $I_C = ?$

$$V_{CE} = V_{CC} - I_C R_C$$

or, $$I_C = \frac{V_{CC} - V_{CE}}{R_C}$$

$$= \frac{9-6}{2\times10^3} = \frac{3}{2}\times10^3 \text{ A}$$

$$= 1.5 \text{ mA}$$

(j) Please refer to the answer of Q. No. 3 (j) of 1996 (A).

(k) **Q of a coil**

The quality factor 'Q' determines the bandwidth of a selective circuit. In other words, the sharpness of resonance is governed by the Q of the circuit. From energy consideration, Q is defined as,

$$Q = 2\pi \frac{\text{Maximum energy stored per cycle}}{\text{Energy dissipated per cycle}}$$

For a coil or inductance, $Q_L = \frac{X_L}{R} = \frac{\omega L}{R}$ ...(1)

where, L is the inductance and R is the resistance of the coil. X_L is called *inductive reactance* which is equal to ωL.

Thus, a coil with higher resistance has smaller Q. So, when a resistance is used in series, it produces lower Q. When a resistance is used in parallel, the effective Q of the circuit will vary directly with the value of the resistance. A high shunting resistance means a high Q; a low parallel resistance results in low Q.

(l) **Class-A Operation of an amplifier**

If the collector current flows at all times during the full cycle of the signal the power amplifier is known as class-A power amplifier.

Obviously, for this to happen, the power amplifier must be biased in such a way that no part of the signal is cut off. To achieve this, a transformer is used in the collector circuit as load. This transformer permits impedance matching, resulting in the transference of maximum power to the load, *i.e.*, loudspeaker. Fig. 3 shows the circuit of class-A power amplifier.

In order to explain class-A operation of an amplifier, let us consider a CE amplifier using a NPN transistor. The base-emitter bias voltage is such that the operating point

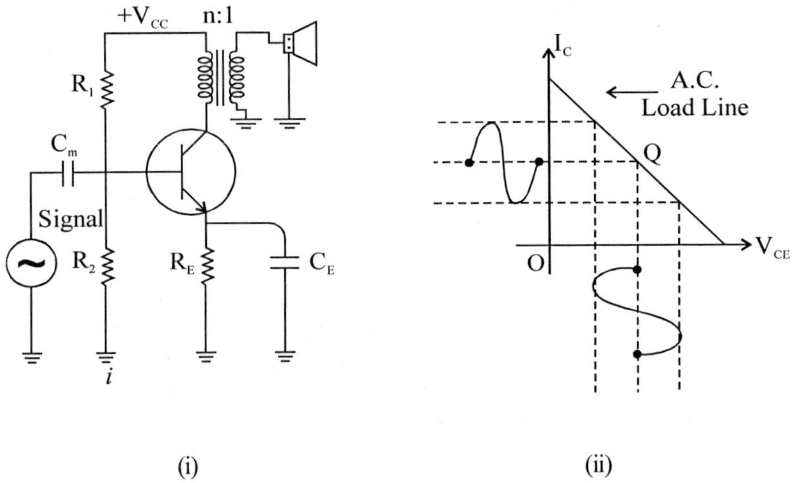

Fig. 3. Circuit of class-A power amplifier

corresponds to a base current of 15 μA (Fig. 4). During the positive half-cycle of the applied signal, the base-emitter bias becomes more positive resulting in increase of base current. Suppose the base current increases to 20 μA at the positive peak of the input signal. Similarly, during negative half cycle, the base-emitter voltage becomes less positive and the base

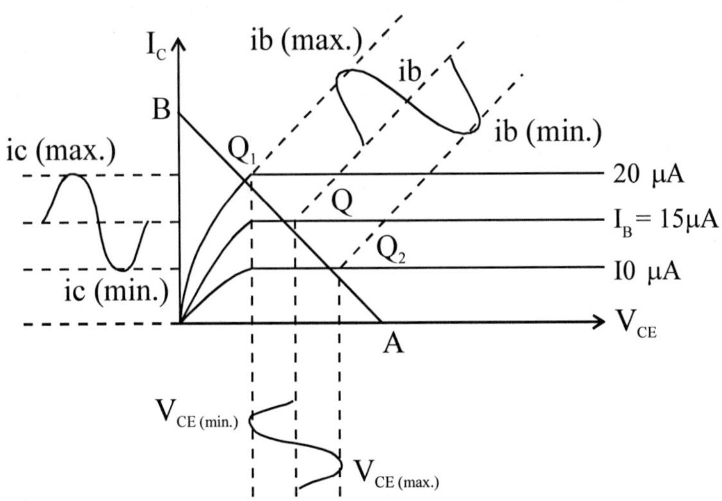

Fig. 4. Operating condition for class-A amplifier

current reduces, say, to 10 μA at the negative peak. So, the base current and hence the collector current flows for the full cycle of the input signal.

This type of amplifier where the input bias voltage is such that the collector current flows for the full cycle of the input signal is called a class-A amplifier.

There is an alternative way for this statement. If a complete cycle is regarded equivalent to 2π or 360°, the fraction of the cycle for which collector current flows or conduction takes place through the transistor can also be expressed in terms of an equivalent angle. This angle 'θ' is called the angle of conduction. In a class-A amplifier, conduction takes place for the full cycle of the signal. Hence, the angle of conduction $\theta = 360°$.

1995 (A)

Group-A

1. (a) 1 pf = 10^{-12} farad

   (b) 1 MΩ = 16^6 Ω = 10 × 10^5 Ω ± 10%

   Colour code = Brown, black, green, silver

   (c) Rheostat is the example of a variable resistor.

   (d) The unit of work function is electron volt (ev).

   (e) The current amplification factor (α) in a CB transistor configuration is defined as the ratio of change in collector current to the change in emitter current at constant collector-base voltage V_{CB}. Mathematically,

$$\alpha = \frac{\Delta I_C}{\Delta I_E} \text{ at constant } V_{CB}$$

   where, ΔI_C is the change in collector current and ΔI_E is the change in emitter current.

   The current amplification factor is hence less than unity. Practical values of α in commercial transistors range from 0.9 to 0.99.

   (f) For making n-type silicon, pentavalent impurities are required. Such impurities are (i) arsenic and (ii) phosphorous.

   (g) In an oscillator, positive feedback is used.

   (h) Electrodes of a pentode are cathode, control grid, screen grid, suppressor grid and plate.

2. (a) *Ripple factor :*

   The ratio of rms value of a.c. component to the d.c. component in the rectifier output is known as ripple factor. That is

$$\text{Ripple factor} = \frac{\text{rms value of a.c. component}}{\text{Value of d.c. component}}$$

$$= \frac{I_{ac}}{I_{dc}}$$

   Mathematically, it is given by

$$\text{Ripple factor} = \sqrt{\left(\frac{I_{rms}}{I_{dc}}\right)^2 - 1}$$

(b) $i_e \rightarrow$ Emitter current due to signal alone

$i_E \rightarrow$ Resultant emitter current due to signal and the d.c. bias.

(c) No charges are present in the depletion zone of a p-n junction. Only +ve and –ve ions are present.

(d) As the temperature is raised, some of the valence electrons acquire sufficient energy to enter into the conduction band and thus become free electrons. Under the influence of electric field, these free electrons will constitute electric current. So, the resistance of a semiconductor decreases with the rise in temperature, *i.e.* the semiconductor has negative temperature co-efficient of resistance.

(e) CE transistor configuration gives the highest power gain.

(f) *Circuit for npn transistor :*

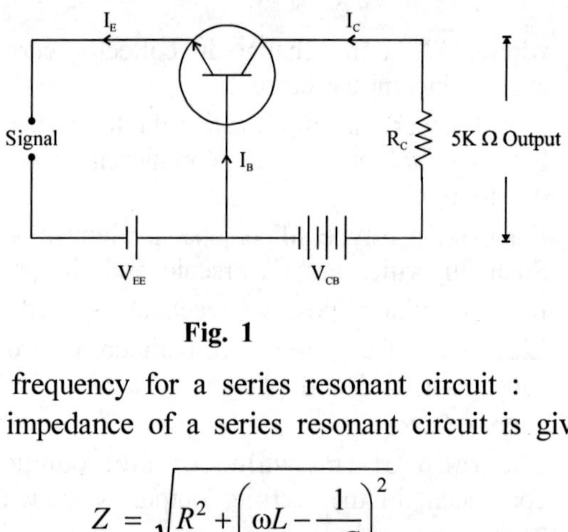

Fig. 1

(g) Resonant frequency for a series resonant circuit :

The impedance of a series resonant circuit is given by

$$Z = \sqrt{R^2 + \left(\omega L - \frac{1}{\omega C}\right)^2}$$

Resonance occurs when inductive reactance becomes equal to the capacitive reactance, *i.e.*

$$X_L = X_C$$

and hence, the impedance (Z) becomes minimum.

or, $$\omega L = \frac{1}{\omega C}$$

$$\text{or,} \qquad \omega^2 = \frac{1}{LC}$$

$$\text{or,} \qquad \omega = \frac{1}{\sqrt{LC}}$$

$$\text{or,} \qquad 2\pi f = \frac{1}{\sqrt{LC}}$$

$$\text{or,} \qquad f = \frac{1}{2\pi\sqrt{LC}} = f_r$$

where, f_r is the resonant frequency.

Group-B

3. (a) *Carbon Resistor :*

It is made from powdered carbon mixed with a clay binding material and baked into a small, hard cylinder with a wire attached to each end (Fig. 2).

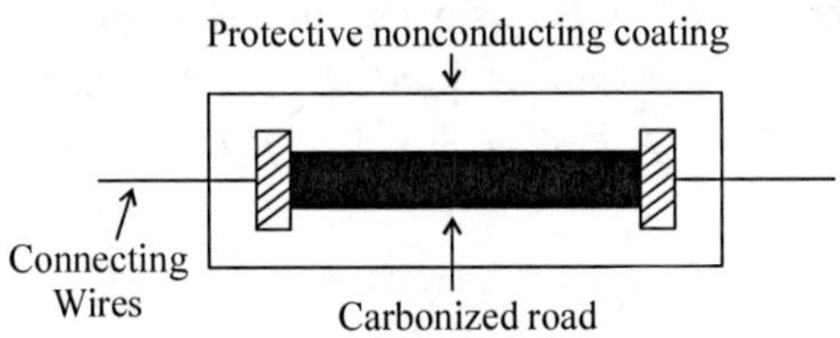

Fig. 2. Features of a fixed type carbon resistor.

The value of resistance depends on the proportion of carbon in the mixture used. Carbon resistors range in value from a fraction of an ohm to several megaohms. Size has no bearing on resistance value, it only affects the power rating of the resistor. Carbon resistors have a positive temperature co-efficient, *i.e.* the resistance increases with increase of temperature.

Common power ratings of carbon resistors are $\frac{1}{8}, \frac{1}{4}, \frac{1}{2}$, 1, 2, 3 watts.

(b) In a pure reactive a.c. circuit, current either leads or lags the emf applied through an angle ϕ (say).

In an inductive a.c. circuit, the current lags behind the emf by an angle ϕ. So,

$$E = E_o \sin \omega t \qquad ...(1)$$

$$I = I_o \sin (\omega t - \phi) \qquad ...(2)$$

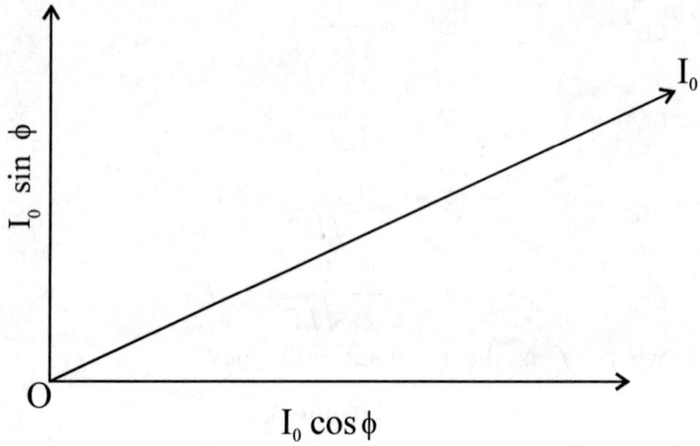

Fig. 3

$$\text{and the power} = EI = E_o I_o \sin \omega t \sin (\omega t - \phi)$$
$$= E_o I_o \sin \omega t [\sin \omega t. \cos \phi - \cos \omega t \sin \phi]$$
$$= E_o I_o \sin^2 \omega t \cos \phi - E_o I_o \sin \omega t. \cos \omega t \sin \phi$$
$$= E_o I_o \sin^2 \omega t \cos \phi — \frac{1}{2} EI_o \sin \phi \sin^2 \omega t$$

The mean value for a complete cycle of $\sin^2 \omega t$ is 1/2 and that of $\sin^2 \omega t = 0$. Hence, the average power is

$$\overline{P} \frac{1}{2} E_0 I_0 \cos \phi \qquad ...(3)$$

Where, E_o represents the maximum emf and I_o the maximum current. The angle of lag is ϕ. Fig. 3 above shows the directions of E_o and I_o. Resolving I_o along and $\perp$*er* to E_o, we get the component $I_o \cos \phi$ called *wattfull* or *active* component of current which is in phase with the emf giving the rate of working $E_o I_o \cos \phi$.

The component $I_o \sin \phi$ is called *wattless* or *idle* or *reactive* component of current, since it does not contribute to the mean power of the circuit.

(c) $E_o = 200$ V, $E_{dc} = 100$ V, Ripple factor = ?

$$\text{Ripple factor} = \frac{\text{rms value of a.c. component}}{\text{d.c. component}}$$

$$= \frac{E_0 / \sqrt{2}}{100}$$

$$= \frac{200}{\sqrt{2}\ 100} = \frac{2}{\sqrt{2}} = 1.414$$

(d) **Proof of** $\alpha = \frac{\beta}{1+\beta}$ **for a transistor**

For a transistor $\alpha = \frac{\Delta I_C}{\Delta I_E}$ ...(1)

$$\beta = \frac{\Delta I_C}{\Delta I_B} \quad ...(2)$$

and $I_E = I_B + I_C$

or, $\Delta I_E = \Delta I_B + \Delta I_C$

Dividing throughout by ΔI_E, we get

$$1 = \frac{\Delta I_B}{\Delta I_E} + \frac{\Delta I_C}{\Delta I_E}$$

$$= \frac{\Delta I_B}{\Delta I_C} \times \frac{\Delta I_C}{\Delta I_E} + \frac{\Delta I_C}{\Delta I_E}$$

$$= \frac{1}{\beta} \times \alpha + \alpha$$

$$= \alpha \left(1 + \frac{1}{\beta}\right) = \alpha \left(\frac{\beta + 1}{\beta}\right)$$

or $\alpha = \frac{\beta}{1+\beta}$ (proved)

(e) Bandwidth of *RC* coupled amplifier is given by

$$BW = f_2 - f_1$$

where, f_2 is the upper cut-off frequency and f_1 is the lower cut-off frequency.

(f) **Significance of PIV in a rectifier**

Peak inverse voltage is maximum reverse voltage that a diode can withstand without destroying the junction.

If the reverse voltage across a diode exceeds this value, the reverse current increases sharply and breaks down the junction due to excessive heat. Peak inverse voltage is significant when diode is used as a rectifier.

In case of a rectifier, it has to be ensured that reverse voltage across the diode does not exceed its PIV during the negative half-cycle of input a.c. voltage. Consequently, in diode rectifier circuits, PIV consideration is generally the deciding factor.

(g) **Faraday's laws of electromagnetic Induction**

Faraday, formulated two laws of electromagnetic induction.

Law 1 : Whenever there is a change in the magnetic flux linked through the coil, an induced current flows through it. Hence, increase in magnetic flux produces an inverse (opposite sense) current and decrease in magnetic flux produces a direct (same sense) current.

Law 2 : The magnitude of the induced emf is directly proportional to the number of turns of the coil and the rate of change of flux linked through the coil.

Mathematically,

$$E \propto N$$

and $$E \propto \frac{d\phi}{dt}$$

$\therefore$ $$E \propto N\frac{d\phi}{dt}$$

or, $$E = KN\frac{d\phi}{dt} = N\frac{d\phi}{dt} \qquad ...(1)$$

where, the constant of proportionality K = 1, using suitable units.

'$d\phi$' is the change in flux linked through the coil in time interval 'dt'. In MKS system, '$d\phi$' is expressed in Weber, 'dt' in second and 'E' in volt.

The emf induced in a single turn of coil is 1 volt when the rate of change of flux linked through it is 1 weber per sec.

(h) **Self-bias**

The feedback resistor method of biasing is called self-bias. This method of biasing is shown in Fig. 4 below.

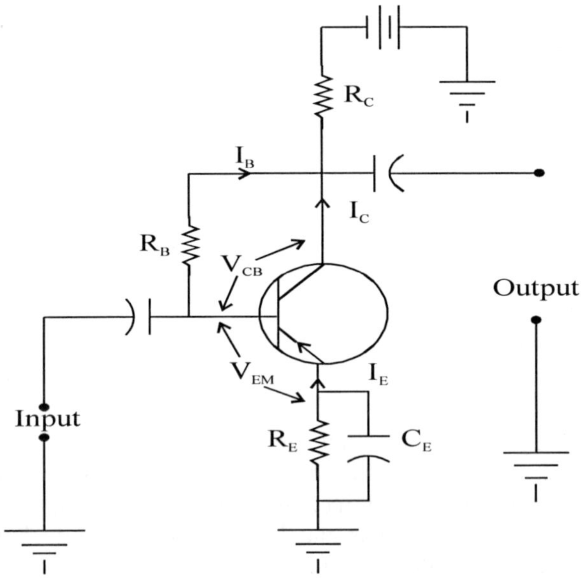

Fig. 4. Feedback resistor biasing method

Here the resistor R_B connected between the collector and the base provides the bias. It is known that the zero signal base current is determined by the forward biasing voltage V_{BE} available between the emitter and the base. But, V_{BE} here is affected by the voltage across R_B, because I_B flows from base to collector through R_B. Increase in the voltage drop V_{CB} across R_B will result in decrease of forward bias voltage V_{BE} available at the base-emitter junction. Similarly, decrease in V_{CB} results in increase of V_{BE}. Consequently, base-collector current (output side) ultimately controls the base-emitter current (input side) through R_B. This type of action is called feedback. Hence, the zero-signal base-collector current is not governed by V_{CC} but by the feedback voltage V_{CB} developed across R_B. So, R_B is called the feedback resistor. Value of R_B can be varied to determine the operating condition.

Value of R_B

The required value of R_B needed to give the zero signal current I_C can be determined as follows. Referring to the circuit diagram, we have

$$V_{CC} = (I_B + I_C)\, R_C + I_B\, R_B + V_{BE}$$
$$= I_C\, R_C + I_B\, R_B + V_{BE}$$

Since, $I_B << I_C$ and for reasonable approximation we can neglect I_B.

$$\therefore \qquad R_B = \frac{V_{CC} - V_{BE} - I_C R_C}{I_B}$$

$$= \frac{V_{CC} - V_{BE} - \beta I_B R_C}{I_B} \qquad (\because I_C = \beta I_B)$$

For this method of biasing, the stability factor $S < (\beta + 1)$ and hence this method provides better thermal stability than the fixed bias.

(i) **Principle of Zener diode**

A properly doped crystal diode which has a sharp breakdown voltage is known as a Zener diode.

When the reverse bias on a crystal diode is increased, a critical voltage, called *breakdown voltage* is reached where the reverse current increases sharply to a high value. The breakdown region is the knee of the reverse characteristic as shown in Fig. 5 below.

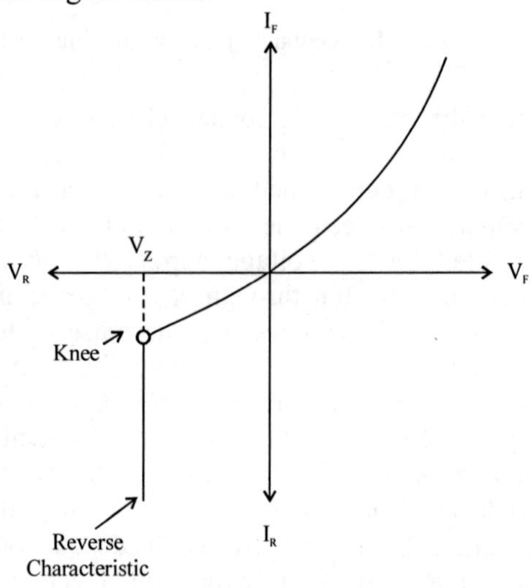

Fig. 5. V~I Characteristics of Zener diode

American Scientist C. Zener first gave the satisfactory explanation of this breakdown of the junction. So, the breakdown voltage is also called as Zener voltage and the sudden rise in current is known as Zener current.

The breakdown or Zener voltage depends upon the amount of doping. If the diode is heavily doped, depletion layer will be thin and as a result of which the breakdown of the junction will occur at a lower reverse voltage. On the other hand, a lightly doped diode has a higher breakdown voltage. When an ordinary crystal diode is properly doped, so that it has a sharp breakdown voltage, it becomes a Zener diode.

(j) Amplification factor is given by

$$\mu = \frac{\text{change in plate voltage}}{\text{change in grid voltage}}$$

$$= \left(\frac{\Delta V_p}{\Delta V_g}\right)$$

The tetrode has control grid and screen grid, whereas the pentode has control grid, screen grid and suppressor grid. The presence of suppressor grid provides further shielding between plate and control grid. So, the plate potential in a pentode has even less effect on the plate current than in a tetrode. This enables the pentode to possess a higher amplification factor and plate resistance than a tetrode.

In comparison to triode, the pentode has higher amplification factor 'μ' due to its two extra grids, the screen grid and the suppressor grid. A small change in grid voltage, there is a greater change in plate voltage. So, amplification factor μ is more than that of triode.

(k) **Advantages of CE amplifier over CB and CC circuits**

(i) The current amplification factor β of CE amplifier is more than that of CB and CC amplifier.

(ii) *High current gain :* As 'β' is very large in CE amplifier, the output current I_c is much more than the input current I_B, because

$$I_C = \beta I_B + I_{CEO}$$

Thus, the current gain in CE configuration is very high. It ranges from 20 to 500.

(iii) *High voltage and power gain :* Due to high current gain, the common emitter circuit has the highest voltage and power gain of the three transistor connections. The voltage gain of CE transistor is about 500.

(iv) *Moderate output to input impedance ratio :* In a common emitter circuit, the ratio of output impedance to input impedance is small (about 50). This makes this circuit arrangement an ideal one for coupling between various transistor stages. However, in other configurations, the ratio of output impedance to input impedance is very large and hence coupling becomes highly inefficient due to gross mismatching.

(l) **Distinction between n-type and p-type semi-conductor**

n-type	p-type
(i) n-type semiconductor is formed by adding pentavalent impurity to a pure semiconductor.	(i) p-type semiconductor is formed by adding trivalent impurity to a pure semi-conductor.
(ii) Electrons are the majority current carriers in n-type semiconductor.	(ii) Holes are the majority current carriers in p-type semiconductor.
(iii) The holes are the minority carriers in n-type semi-conductor	(iii) The electrons are the minority charge carriers in p-type semiconductor.

(m) **Fall of gain of RC amplifier at high frequency**

At high frequencies (720 KHz), the reactance of C_C is very small and it behaves as a short circuit. This increases the loading effect of next stage and serves to reduce the voltage gain. Moreover, at high frequency, capacitive reactance of base-emitter junction is low which increases the base current. This reduces the current amplification factor β. So, voltage gain falls at high frequency.

1994 (A)

Group-A

1. (a) The colour code for a resistor of resistance 6.8 KΩ ± 5% is Blue, Grey, Red, Gold.

(b) **Farad**

One farad is the capacitance of a capacitor when 1 coulomb of charge stored in the plates at a potential difference of 1 volt.

$$1f = \frac{1 \text{ coulomb}}{1 \text{ volt}}$$

(c) $C = \frac{Q}{V}$

(d) Please refer to the answer of Q. No. 3(b) of 1995 (A).

(e) The unit of work function is electron-volt (ev).

(f) The maximum efficiency of a fullwave rectifier is 81.2%.

(g) The voltage amplifiers are operated in Class *A*.

(h) In n-p-n transistor, electrons are the current carriers.

(i) $10 \log_{10}^{100} = 10 \log_{10} 10^2 = 20$ db.

(j) Positive feedback is employed by an oscillator.

2. (a) A parallel resonant circuit is known as *rejector*.

(b) A Zenner diode uses *reverse* portion of diode characteristics.

(c) Efficiency of power amplifier is maximum for Class *B* operation.

(d) If transistor $\alpha = 0.9$, then β is *9*.

(e) A semiconductor material is formed by *covalent* bonds.

Group-B

3. **(a) Lenz's Law**

Statement. The direction of induced emf is in such a direction that it opposes the very cause to which it is due.

Explanation. If the magnetic flux linked through a coil increases, the induced current developed in the coil due to this shall create a flux in opposite direction to the original magnetic flux. On the other hand, if the magnetic flux linked through the coil decreases, the induced current developed creates a flux in the direction of original flux. So, using Lenz's law, we can write the induced emf as

$$E = -N\frac{d\phi}{dt} \qquad \text{...(1)}$$

where, E is the induced emf,

N is the number of turns of the coil

and $\frac{d\phi}{dt}$ is the time rate of change of flux.

Lenz's law follows from the principle of conservation of energy and can be understood by Faraday's experiments shown in Figs. 1, 2 and 3 below.

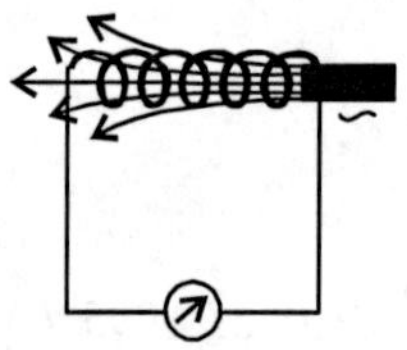

Fig. 1
Magnet moving towards the coil

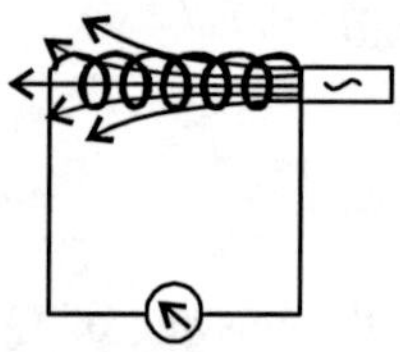

Fig. 2
Magnet moving away from the coil

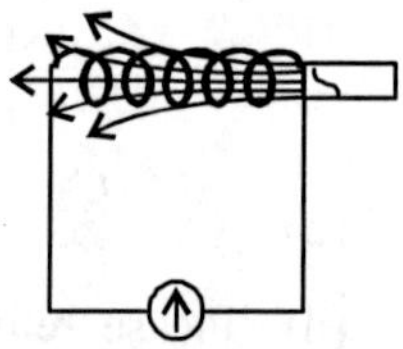

Fig. 3
Magnet is stationary

Faraday performed a series of experiments to show the induction effects. He took a coil, a magnet and a galvanometer. In the first case, when the magnet is moved towards the coil, the galvanometer is deflected in one direction showing the direction of flow of current in the coil. This current is the induced current. The emf responsible for this current is called induced emf. The process generating the induced emf is called electromagnetic induction.

In the second case, when the magnet is moved away from the coil, the galvanometer deflects in opposite direction showing that the induced current flows in a direction opposite to that of the first case.

In the third case, when the magnet is stationary, there is no deflection in the galvanometer indicating the absence of induced current in the coil.

The above effects are also observed when the magnet is held stationary and the coil is moved towards or away from the magnet. Hence, the rate of change of flux linked with the coil due to the relative motion of the coil and the magnet induces a current in the coil.

An induced current can produce heat to do chemical or mechanical work. The energy must come from the work done in inducing the current. When induction is due to the motion of a magnet or a coil, work is done; therefore the motion must be resisted by a force. This opposing force comes from the action of the magnetic field of the induced current. This is in accordance with the principle of conservation of energy.

(b) **Richardson-Dushman equation**

Richardson-Dushman equation gives a relation between the thermionic emission current density and the emitter temperature. The amount of thermionic emission increases rapidly as the emitter temperature is raised. This equation is given as :

$$J_s = AT^2 e^{-\frac{b}{T}} \text{ amp/m}^2 \quad \text{...(1)}$$

where, J_s = emission current density *i.e.* current per square metre of the emitting surface.

T = absolute temperature of emitter in °K.

A = constant, depending upon the type of emitter and is measured in amp/m^2/°K^2

b = a constant for the emitter

e = natural logarithmic base

The value of b is constant for a metal and is given by :

$$b = \frac{\phi e}{K} \quad \text{...(2)}$$

where, ϕ = work function of emitter

e = electron charge = 1.602×10^{-19} coulomb

K = Bottzmann's constant

= 1.38×10^{-23} J/°K

$$b = \frac{\phi \times 1.602 \times 10^{-19}}{1.38 \times 10^{-23}} = 11600\,\phi \text{ °K}$$

Putting the value of b in exp. (1), we get,

$$J_s = AT^2 e^{-\frac{11600\phi}{T}} \quad \text{...(3)}$$

Equation (3) reveals that :

(i) The emission is markedly affected by temperature changes. Doubling the temperature of an emitter may increase electron emission by more than 10^7 times. For instance, emission from pure tungsten metal is about 10^{-6} ampere/cm^2 at 1300°C but rises to enormous value of about 100 amps when temperature is raised to 2900°C.

(ii) Small changes in the work function of the emitter can produce enormous effects on emission. If the work function is halved, it produces the same effect as doubling the temperature.

(c) **V-I characteristics of p-n junction**

Please refer to the answer of Q. 3(k) of 1999 (A).

(d) Thermionic emission is not possible at room temperature because at ordinary temperature, the energy possessed by free electrons in the metal is inadequate to cause them to escape from the surface. At room temperature, the electrons cannot overcome the potential barrier and hence cannot leave the metal surface.

(e) **Function of a filter in a rectifier circuit**

A filter circuit is a device which removes the a.c. component of rectifier output but allows the d.c. component to reach the load.

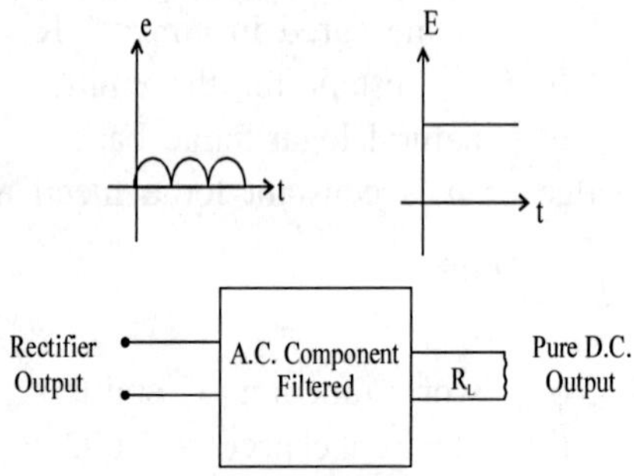

Fig. 4. Filter Circuit

Obviously, a filter circuit should be installed between the rectifier and the load as shown in Fig. 4. A filter circuit is generally a combination of inductors (L) and capacitors (C). The filtering action of L and C depends upon the basic electrical principles. A capacitor passes a.c. readily but does not pass d.c. at all. On the other hand, an inductor opposes a.c. but allows d.c. to pass through it. It is hence clear that suitable network of L and C can effectively remove the a.c. component, allowing the d.c. component to reach the load.

(f) **Relation between μ, r_p and g_m in a triode**

In a triode valve, amplification factor 'm' is given by

$$\mu = \left(\frac{\Delta V_p}{\Delta V_g}\right)_{I_P} \quad ...(1)$$

$$= \frac{\text{change in plate voltage}}{\text{change in plate current}}$$

Plate resistance 'r_p' is defined as

$$r_p = \left(\frac{\Delta V_P}{\Delta I_P}\right)_{V_g} \quad ...(2)$$

$$= \frac{\text{change in plate voltage}}{\text{change in plate current}}$$

Transconductance 'g_m' is written as

$$g_m = \left(\frac{\Delta I_P}{\Delta V_g}\right)_{V_P} \quad ...(3)$$

$$= \frac{\text{change in plate current}}{\text{change in grid voltage}}$$

So, $$\mu = \frac{\Delta V_P}{\Delta V_g} = \frac{\Delta V_P}{\Delta I_P} \times \frac{\Delta I_P}{\Delta V_g}$$

or, $$\mu = r_p \times g_m \quad ...(4)$$

This is the relation between r_p and g_m.

(g) **Advantages of a pentode over a triode**

(i) Amplification factor of a pentode is more than a triode.
(ii) Plate resistance of a pentode is higher than triode.
(iii) Transconductance of a pentode is more than a triode.

Table of constants

Valve constants	Triode	Pentode
Amplification factor (μ)	10 to 100	1000 to 5000
Plate resistance (r_p)	3000 Ω to 1000 KΩ	0.5 to 2 MΩ
Transconductance (g_m)	About 2500 μmho	1000 to 9000 μmho

(h) **Q - factor of a resonant circuit**

The quality factor Q determines the bandwidth of a selective circuit. In other words, the sharpness of resonance is governed by the Q of the circuit. From energy consideration,

$$Q = 2\pi \frac{\text{Max}^{\text{m}} \text{ energy stored per cycle}}{\text{Energy dissipated per cycle}}$$

The Q of a series resonant circuit is given by :

$$Q_s = \frac{\omega_r L}{R_s} = \frac{1}{R_S}\sqrt{\frac{L}{C}} \quad ...(1)$$

where, R_s is the total series resistance of the circuit, L is the inductance and C is the capacitance.

The Q of a parallel resonant circuit is given by,

$$Q_P = \frac{R}{\sqrt{L/C}} \quad ...(2)$$

Thus, it is seen that the quality factor of a series resonant circuit becomes more if L/C becomes higher, whereas the Q of a parallel resonant circuit increases if L/C decreases.

While designing tuned circuits, Quality factor is taken into consideration. The bandwidth of a tuned circuit is also related to its Q by the relation,

$$Q = \frac{f_r}{\Delta f} \quad ...(3)$$

where, f_r is the resonant frequency and Δf is the bandwidth. Hence, high Q circuit results in narrow bandwidth.

(i) $A = 50, \; m = .05, \; A_f = ?$

$$A_f = \frac{A}{1+mA}$$

$$= \frac{50}{1+(0.05\times 50)} = \frac{50}{1.25}$$

$$= \frac{5000}{125} = 40$$

(j) **Working of a crystal oscillator**

The circuit of a transistor crystal oscillator is shown in Fig. 5 below. A tank circuit $L_1 - C_1$ is placed in the collector and the crystal is connected in the base circuit. Feedback is obtained by coil L_2 inductively coupled to coil L_1. The crystal is connected in series with the feedback winding. The natural frequency of LC circuit is made nearly equal to the natural frequency of crystal.

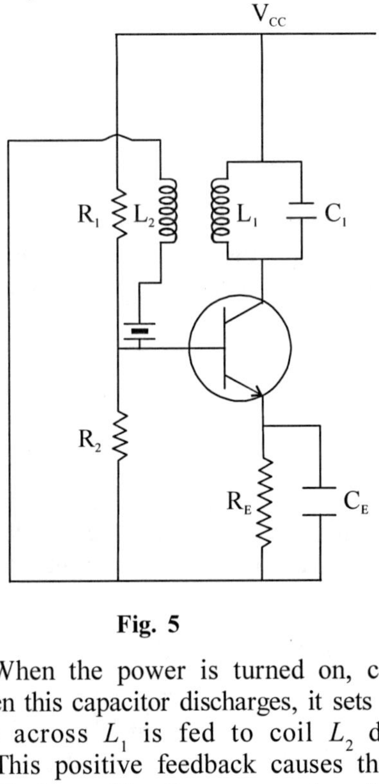

Fig. 5

Functions. When the power is turned on, capacitor C_1 is charged. When this capacitor discharges, it sets up oscillations. The voltage across L_1 is fed to coil L_2 due to mutual inductance. This positive feedback causes the oscillator to produce oscillations. The frequency of oscillations in the circuit is controlled by the crystal. It is because the crystal is connected in the base circuit and hence its influence on the frequency of the circuit is much more than LC circuit. As a result of which the entire circuit vibrates as the natural frequency of the crystal. Since the frequency of the crystal is independent of temperature etc., the circuit generates a constant frequency.

(k) **Fall of gain in a RC coupled amplifier at high frequency**

Please refer to the answer of Q. 3(m) of 1995 (A).

(l)

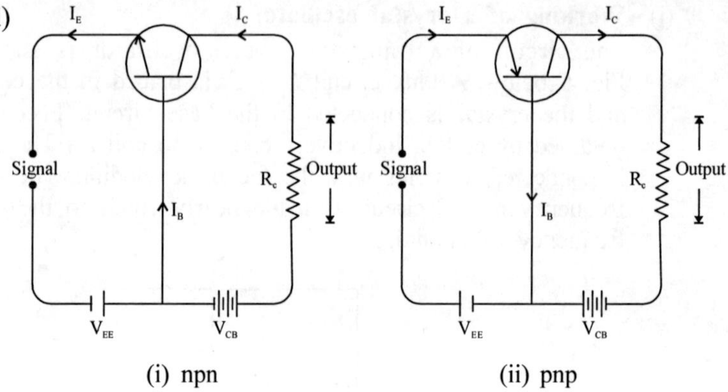

Fig. 6. (i) npn transistor connections (ii) pnp transistor connections

(m) **Ripple factor**

The ratio of rms value of a.c. component to the d.c. component in the rectifier output is known as ripple factor, *i.e.*,

$$\text{Ripple factor} = \frac{\text{rms value of a.c. component}}{\text{value of d.c. component}}$$

$$= \frac{I_{ac}}{I_{dc}} \quad \text{...(1)}$$

Therefore, ripple factor is very important in deciding the effectiveness of a rectifier. The smaller the ripple factor, the lesser the effective a.c. component and hence more effective is the rectifier.

By definition, $I_{rms} = \sqrt{I_{dc}^2 + I_{ac}^2}$

So, $I_{ac} = \sqrt{I_{rms}^2 - I_{dc}^2}$

Dividing throughout by I_{dc}, we get

$$\frac{I_{ac}}{I_{dc}} = \frac{\sqrt{I_{rms}^2 - I_{dc}^2}}{I_{dc}}$$

or, $$\text{Ripple factor} = \sqrt{\left(\frac{I_{rms}^2 - I_{dc}^2}{I_{dc}}\right)}$$

$$= \sqrt{\left(\frac{I_{rms}}{I_{dc}}\right)^2 - 1}$$

For full-wave rectification,

$$I_{rms} = \frac{I_m}{\sqrt{2}}; \qquad I_{dc} = \frac{2I_m}{\pi}$$

$$\therefore \quad \text{Ripple factor} = \sqrt{\left(\frac{I_m/\sqrt{2}}{2I_m/\pi}\right)^2 - 1} = 0.48$$

Thus, the d.c. component in the output of a full-wave rectifier is more than the a.c. component.

(n) **Output characteristics of a CE amplifier**

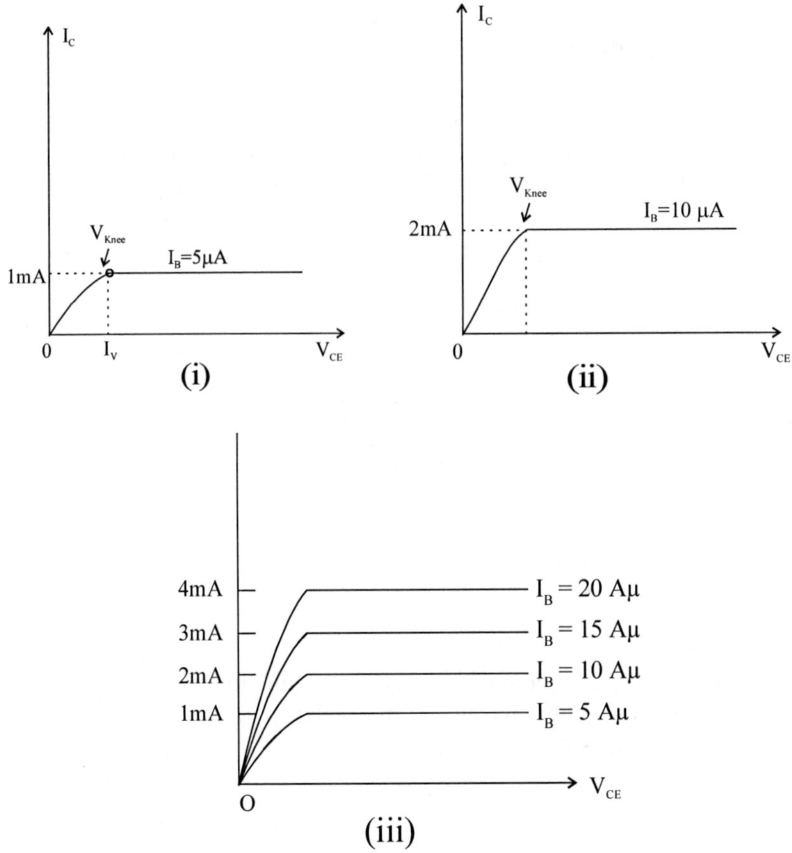

Fig. 7

According to 1993 Syllabus

Group-A

1. (a) The colour code of a resistor is Orange, Red and Yellow. Its value is 32×10^4 ohms or 320 kilo-ohms.

   (b) *Henry is the unit of inductance. Definition :* The inductance of an inductor or a circuit is 1 henry if a rate of change of current 1 amp/sec. produces an induced emf of 1 volt in that circuit.

   (c) $C = Q/V$

   where, C is the capacitance, Q is the charge and V is the potential difference between two plates of a capacitor.

   (d) Thermionic emitters : Tungsten, Thoriated tungsten.

   (e) $$P = 0.25 \text{ W}, \quad R = 2.5 \text{ K}\Omega = 2.5 \times 10^3 \ \Omega$$

   $$P = \text{VI} = (RI)\ I = RI^2$$

   So, $$I^2 = \frac{P}{R} \Rightarrow I = \sqrt{\frac{P}{R}} = \sqrt{\frac{0.25}{2.5 \times 10^3}}$$

   $$= \sqrt{\frac{25 \times 10^{-2}}{25 \times 10^2}}$$

$$= \sqrt{10^{-4}} = 10^{-2} \text{ amp}$$

$$= 0.01 \text{ amp}$$

   (f) The base of a transistor is made thin in order to pass most of the emitter injected charge carriers to the collector.

   (g) Transistor β is given by

   $$\beta = \frac{\alpha}{1-\alpha}$$

   where, β is the base current amplification factor of CE configuration and α is the current amplification factor of CB connection.

(h) Symbol of *FET*

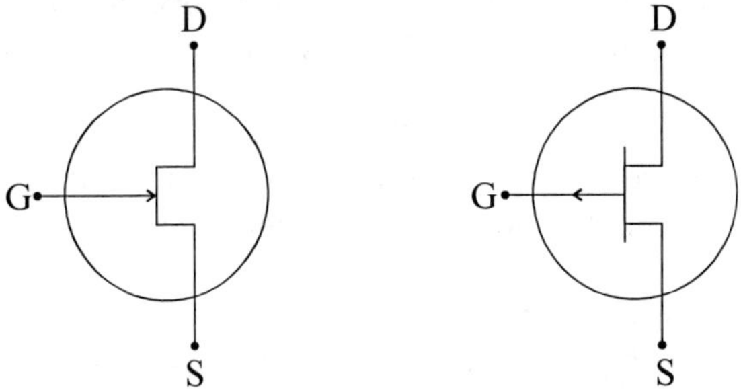

n-channel FET (i)   p-channel FET (ii)

Fig. 1

2. (a) Ripple factor is given by $\frac{I_{ac}}{I_{dc}}$

$$\text{Ripple factor} = \frac{I_{ac}}{I_{dc}} = \frac{1}{I_{dc}}\sqrt{I_{rms}^2 - I_{dc}^2}$$

$$= \sqrt{\left(\frac{I_{rms}}{I_{dc}}\right)^2 - 1}$$

where, I_{rms} is the rms value of total current and I_{dc} is the value of d.c. component.

(b) Q-factor of a LCR series circuit is given by

$$Q_S = \frac{\omega_r L}{R_S} = \frac{\left(1/\sqrt{LC}\right)L}{R_S} = \frac{1}{R_S}\sqrt{\frac{L}{C}}$$

where, R_s is the total series resistance of the circuit, L is the inductance and C is the capacitance. ω_r is the angular frequency at resonance, which is equal to $2\pi f_r$, f_r being the resonant frequency.

(c) *Application of Zener diode :*
Zener diode can be used as a voltage stabiliser.

(d) $PIV = 200$ V, $V_{rms} = ?$

$$V_{rms} = \frac{PIV}{\sqrt{2}} = \frac{V_m}{\sqrt{2}} = \frac{200}{\sqrt{2}} = \frac{200\sqrt{2}}{2} \text{ volt}$$

$$= 100\sqrt{2} \text{ volts}$$

(e) Positive feedback is used in an oscillator.
(f) Forward biasing is given to the emitter junction of transistor.
(g) The band gap energy of silicon is 1.1 ev.

Group-B

3. (a) Active elements : voltage source or current source.
Passive elements : Resistors, Inductors and Capacitors.

(b) $V = 100$ V, $f = 1$ KHz $= 10^3$ Hz
$L = 100$ mH $= 100 \times 10^{-3}$ H, $X_L = ?$
$X_L = \omega L = 2\pi f L = 2 \times 3.142 \times 10^3 \times 10^{-1}$ ohms
$= 2 \times 3.142 \times 10^2$ ohms
$= 2 \times 314.2$ ohms
$= 628.4$ ohms

(c) *Apparent power :* In an a.c. circuit, the product of virtual emf and virtual current, *i.e.*,

$$P_a = E_v I_v = \frac{E_0}{\sqrt{2}} \times \frac{I_0}{\sqrt{2}} \quad ...(1)$$

where, P_a is the apparent power,
E_v is the effective or virtual or rms value of a.c. emf,
I_v is the effective or virtual or rms value of a.c. current.

True power : In an a.c. circuit, the true power is the product of E_v, I_v and cos ϕ, *i.e.*

$$P_t = E_v I_v \cos\phi$$

$$= \frac{E_0}{\sqrt{2}} \times \frac{I_0}{\sqrt{2}} \times \cos\phi \quad ...(2)$$

where, P_t is the true power and Cos ϕ is called the power factor. The true power is less than the apparent power because of the presence of reactive elements. Hence, true power is given by

$$P_t = I^2 R \quad ...(3)$$

and apparent power is given by

$$P_a = I^2 Z \quad ...(4)$$

where, Z is the impedance of the a.c. circuit.

(d) *Constants of a triode valve :* There are three constants of a triode valve, viz. plate resistance, transconductance and amplification factor.

Plate resistance : The ratio of small change in plate voltage to the small change in plate current at constant grid voltage is known as a.c. plate resistance, *i.e.*

$$r_p = \left(\frac{\Delta V_p}{\Delta I_p}\right)_{V_g} \quad ...(1)$$

where, 'r_p' is the plate resistance.

Transconductance or mutual conductance : The ratio of small change in plate current to the small change in grid voltage at constant plate voltage is known as transconductance or mutual conductance, *i.e.*

$$g_m = \left(\frac{\Delta I_p}{\Delta V_g}\right)_{V_p} \quad ...(2)$$

where, 'g_m' is the transconductance.

Amplification factor : The ratio of small change in plate voltage to a small change in grid voltage of a triode at constant plate current is known as amplification factor, *i.e.*

$$\mu = \left(\frac{\Delta V_p}{\Delta V_g}\right)_{I_p} \quad ...(3)$$

where, μ is the amplification factor.

(e) *Characteristics of a CE amplifier :* The input and output characteristics of a CE amplifier are shown in Fig. 2 and Fig. 3 respectively.

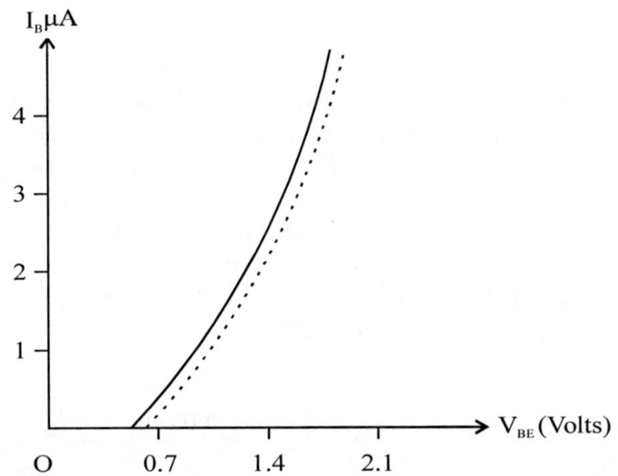

Fig. 2. Output characteristics

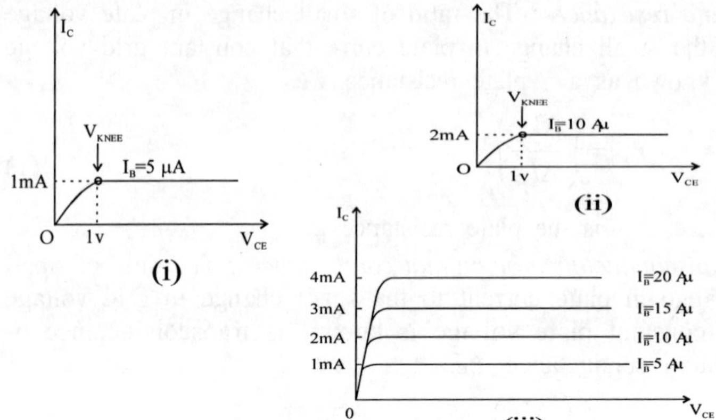

Fig. 3

(f) *Barkhausen conditions of oscillations* : The condition for the maintenance of oscillations is

$$\beta A = 1 \qquad ...(1)$$

where, β is the fraction of the output introduced to the input by the feedback network, A is the voltage gain of the feedback amplifier without feedback.

Since, $$A = -\frac{\mu Z_L}{r_p + Z_L}$$

$$\beta = -\frac{r_p + Z_L}{\mu Z_L}$$

$$= -\left(\frac{1}{\mu} + \frac{1}{g_m Z_L}\right) \qquad ...(2)$$

This equation is known as '*Barkhausen Criterion*' for self-sustained oscillations. The negative sign here indicates that the feedback network must also introduce a phase shift of 180° in the signal (in addition to a phase shift of 180° introduced by the valve), so that overall phase shift suffered by the signal is 360° or zero. Condition (2) can also be put in another useful form by defining a quantity Z_T, the transfer impedance of the circuit.

$$Z_T = \frac{\text{Input voltage}}{\text{Output current}} = -\frac{e_g}{I_b}$$

where, $$I_b = g_m e_g + \frac{e_0}{r_p} \qquad ...(3)$$

So, we have

$$Z_T \left(g_m + \frac{A}{r_P} \right) = -1$$

or, $$Z_T \, g_m \left(1 + \frac{A}{\mu} \right) = -1 \qquad \text{...(4)}$$

This is another form of Barkhausen condition for sustained oscillations.

(g) *Function of control grid in a triode :* The function of the control grid in a triode valve is to control the flow of electrons between the cathode and plate. When the grid is given a *negative* potential, the electrons moving from cathode to plate experience a retarding force. Consequently, the space charge increases and the plate current decreases. Plate current decreases continuously as the grid bias is made more and more negative. For some value of negative grid bias, the plate current can be reduced completely to zero. The minimum negative grid voltage which reduces the plate current to zero is called the *cut off grid bias.*

If the grid is made *positive* with respect to cathode, it aids the plate voltage in attracting more electrons from the space charge and hence the plate current increases. At the same time, a small grid current (current from grid to cathode) also flows, which dissipates power in the grid circuit. It is an undesirable feature because it heats up the grid structure which may damage the wire mesh. It also loads the source supplying grid potential. To avoid it, the grid is always kept at a negative potential with respect to cathode.

(h) *Resonant frequency of a LCR parallel circuit :* A circuit consisting of a coil associated with a resistance R in parallel with a capacitance is shown in Fig. 4. Resistance of the capacitance is assumed to be negligible.

The total impedance 'Z' of the circuit is given by

$$Z = \frac{(R + X_L) X_C}{R + X_L + X_C}$$

$$= \frac{(R + j\omega L)\left(-j/\omega C\right)}{R + j\left(\omega L - 1/\omega C\right)}$$

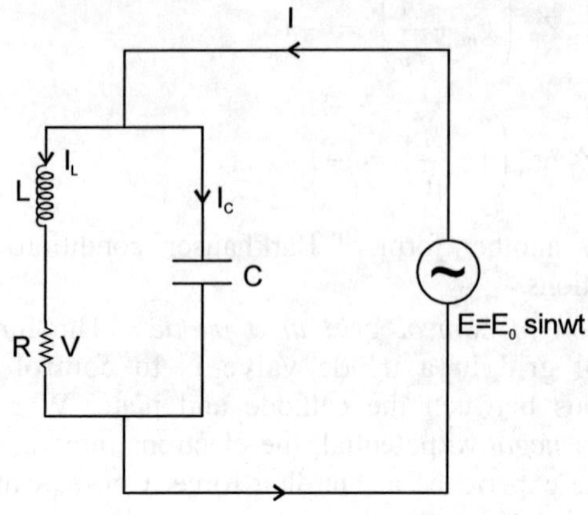

Fig. 4. Parallel resonant circuit

$$= \frac{(R + j\omega L)\left(-j/_{\omega C}\right)\left[R - j\left(\omega L - 1/_{\omega C}\right)\right]}{R^2 + \left(\omega L - 1/_{\omega C}\right)^2}$$

$$= \frac{(R + j\omega L)\left[\frac{jR}{\omega C} - \frac{1}{\omega C}\left(\omega L - \frac{1}{\omega C}\right)\right]}{R^2 + \left(\omega L - 1/_{\omega C}\right)^2}$$

$$= \frac{\frac{-jR^2}{\omega C} + \frac{LR}{C} - \frac{R}{\omega C}\left(\omega L - \frac{1}{\omega C}\right) - \frac{jL}{C}\left(\omega L - \frac{1}{\omega C}\right)}{R^2 + \left(\omega L - \frac{1}{\omega C}\right)^2}$$

$$= \frac{\frac{LR}{C} - \frac{LR}{C} + \frac{R}{\omega^2 C^2} + \frac{j\omega}{\omega^2 C^2}\left[-R^2 C - \omega^2 L^2 C + L\right]}{R^2 + \left(\omega L - \frac{1}{\omega C}\right)^2}$$

$$= \frac{\frac{R}{\omega^2 C^2} + \frac{j\omega}{\omega^2 C^2}\left[L - C\left(\omega^2 L^2 + R^2\right)\right]}{R^2 + \left(\omega L - \frac{1}{\omega C}\right)^2}$$

$$= \frac{R + j\omega\left[L - C\left(\omega^2 L^2 + R^2\right)\right]}{\omega^2 C^2 R^2 + \left(\omega^2 LC - 1\right)^2}$$

Separating the real and imaginary parts, we get,

effective resistance $R_0 = \dfrac{R}{\omega^2 C^2 R^2 + \left(\omega^2 LC - 1\right)^2}$

and effective reactance $X_0 = \dfrac{\omega\left[L - C\left(\omega^2 L^2 + R^2\right)\right]}{\omega^2 C^2 R^2 + \left(\omega^2 LC - 1\right)^2}$

The magnitude of the total impedance Z can be written as

$$Z_t = \sqrt{R_0^2 + X_0^2}$$

$$= \sqrt{\frac{R^2 + \omega^2\left[L - C\left(\omega^2 L^2 + R^2\right)\right]^2}{\left[\omega^2 C^2 R^2 + \left(\omega^2 LC - 1\right)^2\right]^2}}$$

$$= \frac{R^2 + \omega^2 L^2}{\sqrt{R^2 + \omega^2\left(CR^2 + \omega^2 L^2 C - L\right)}} \qquad ...(1)$$

The impedance is maximum when the denominator is minimum, *i.e.*

When $CR^2 + \omega^2 L^2 C - L = 0$ ...(2)

The frequency corresponding to this condition is called the resonant frequency. Hence,

$$\omega_r^2 = \frac{L - CR^2}{L^2 C}$$

$$= \frac{L}{L^2 C} - \frac{CR^2}{L^2 C}$$

$$= \frac{1}{LC} - \frac{R^2}{L^2}$$

or, $$\omega_r = \sqrt{\frac{1}{LC} - \frac{R^2}{L^2}}$$

or, $$f_r = \frac{1}{2\pi}\sqrt{\frac{1}{LC} - \frac{R^2}{L^2}} \qquad ...(3)$$

(i) *Advantages of CE configuration over other connections :*

(a) *High voltage gain.* In a common emitter connection, I_c

is the output current and I_B is the input current. In this circuit arrangement, collector current is given by :

$$L_C = \beta I_B + I_{CEO}$$

As the value of β is very large, therefore, output current I_C is much more than the input current I_B. Hence, the current gain in CE arrangement is very high. It may range from 20 to 500.

(b) *High voltage and power gain.* Due to high-current gain, the common emitter circuit has the highest voltage and power gain of three transistor connections. This is the major reason for using the transistor in this circuit arrangement.

(c) *Moderate output to input impedance ratio.* In a common emitter circuit, the ratio of output impedance to input impedance is small (about 50). This makes the CE arrangement an ideal one for coupling between various transistor stages. However, in other transistor connections, the ratio of output impedance to input impedance is very large and hence coupling becomes highly inefficient due to gross mismatching.

(j) *Zener breakdown :* It is the reverse voltage at which pn junction breaks down with sudden rise in reverse current. This situation arises when the reverse bias on a crystal diode is increased, a critical voltage, called breakdown voltage is reached and at this stage, the reverse current increases sharply to a high value. The breakdown region is the knee of the reverse characteristic as shown in Fig. 5.

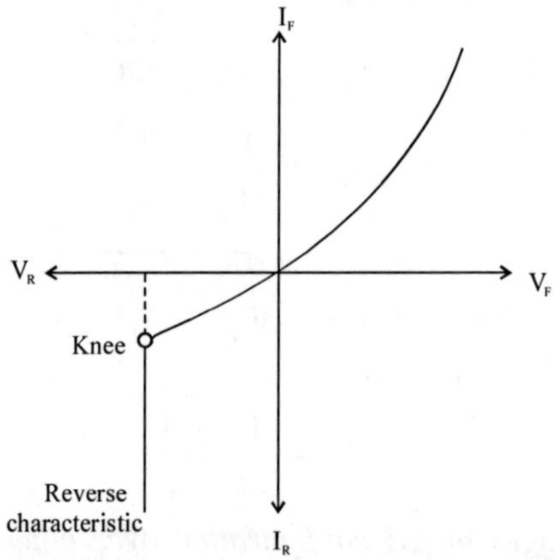

Fig. 5

American scientist C. Zener first gave the satisfactory explanation of this breakdown voltage which is therefore called *Zener Voltage* and the sudden increase in current is known as Zener Current.

The breakdown or Zener voltage depends upon the amount of doping. If the diode is heavily doped, depletion layer will be thin and consequently the breakdown of the junction will occur at a lower reverse voltage. On the other hand, a lightly doped diode has a higher breakdown voltage. When an ordinary crystal diode is properly doped so that it has a sharp breakdown voltage, it is called a *Zener diode*.

(k) *FET*

The FET is called an unipolar transistor because, in a FET, there is only one type of carrier, *i.e.* holes in p-type channel and electrons in n-type channel. Hence, FET is a three terminal semiconductor device in which current conduction is by one type of carrier *i.e.* electrons or holes.

(l) *Richardson's equation for thermionic emission :* Please refer to the answer of Q. No. 1(f) of 1996 (S).

(m) *Fall of gain of RC amplifier at low frequency :* Please refer to the answer of Q. No. 3(k) of 1994 (A).

1993 (A)

Group-A

1. (i) The unit of frequency is cycle/sec or Hertz (Hz).

(ii) 1 farad is the capacitance of a conductor to which 1 coulomb of positive charge is given to raise its potential by 1 volt.

(iii) The minimum energy needed to remove an electron from the surface can be written as

$$E_0 = h\upsilon_0$$

and is known as the *work function* of the metal.

(iv) The plate resistance of a triode is the ratio of change in plate voltage to the change in plate current at constant grid voltage. It is given by

$$r_p = \left(\frac{\Delta V_p}{\Delta I_p}\right)_{V_g}$$

where, r_p is the plate resistance,
ΔV_p is the change in plate voltage,
and ΔI_p is the change in plate current.

(v) Peak inverse voltage is the maximum reverse voltage that can be applied to the p-n junction without damage to the junction. If the reverse voltage across a diode exceeds this value, the reverse current increases sharply and breaks down the junction due to excessive heat.

(vi) *Class-A amplifier* : The amplifier in which the input bias voltage is such that the collector current flows for the full cycle of the input signal is called a class-A amplifier.

(vii) Two 0.1 μF capacitors will be connected in parallel to give 0.2 μF capacitance because the combined capacitance of parallel connection of capacitors is given by

$$C = C_1 + C_2$$
$$= 0.1 + 0.1 = 0.2 \ \mu F$$

(viii) In a N-type semiconductor, the majority carriers are the electrons and the minority carriers are the protons.

(ix) *Electromagnetic induction :* It is the process of linking a current in a coil by changing the magnetic flux linked through it.

Faraday found that whenever there is a relative motion between a coil and a magnet, a current is generated in the coil. This current is the induced current and the source of emf producing this current is called the induced emf. The process of inducing a current in a conductor by the charge of magnetic flux linked through it is known as *electromagnetic induction.*

2. (i) The work function of thoriated tungsten is *lower* than that of pure tungsten.
   (ii) If a diode is heavily doped, the depletion layer width will be *less*.
   (iii) More ac component in the rectifier output means *greater* ripple factor.
   (iv) A crystal diode can withstand reverse voltage much *higher* than vacuum diodes.
   (v) In full-wave rectification, if the input frequency is 50 Hz, then the output has a frequency of 100 Hz.
   (vi) A triode is normally operated with the grid at *negative* potential.
   (vii) A Hartley oscillator uses *positive* feedback.
   (viii) If the value of α of a transistor is 0.9, then its β is *9*.
   (ix) In a transistor, the base is *lightly* doped.
   (x) Negative feedback *increases* the bandwidth of an amplifier.

Group-B

3. (i) Thermionic emitters are heated in vacuum to emit electrons. If the *emitter* or *cathode* were heated to the required temperature in open air, it would burn up because of the presence of oxygen in the air.
   (ii) The Zener diode is always reverse connected across the supply because it develops a sharp breakdown voltage, called Zener voltage V_Z. At this breakdown voltage, the reverse current increases sharply to a high value. In this region, the curve is almost vertical (Fig. 1). It means that voltage across Zener diode is constant at V_Z even though the current through it changes. So, in the breakdown region, an ideal Zener may be represented by a battery of voltage V_Z as shown in Fig. 2. This indicates that a Zener diode can be used as a voltage stabiliser.

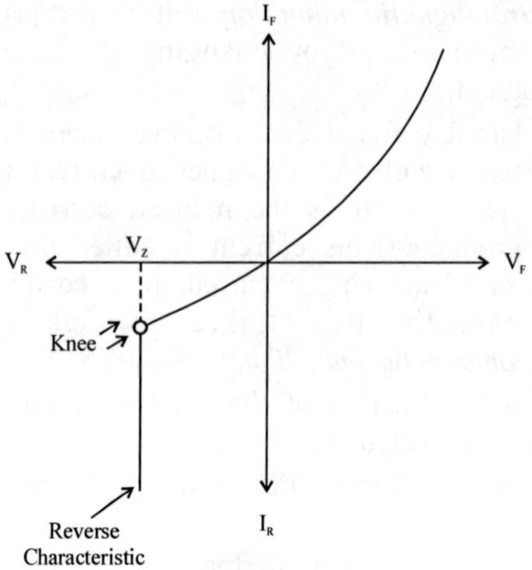

Fig. 1

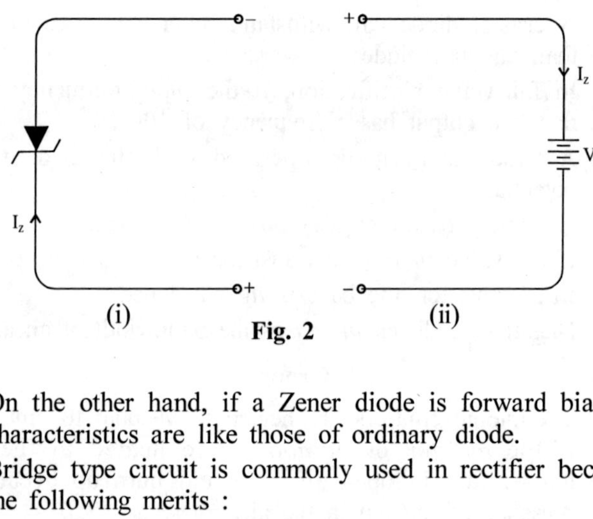

Fig. 2

On the other hand, if a Zener diode is forward biased, its characteristics are like those of ordinary diode.

(iii) Bridge type circuit is commonly used in rectifier because of the following merits :

(a) The output is twice that of the centre-tap circuit for the same secondary voltage.

(b) The PIV is one-half that of the centre-tap circuit.

(iv) In a vacuum tube amplifier, the output is obtained across the load R_L connected in the plate circuit. If the plate is placed at a positive potential, the electrons emitted from the indirectly heated cathode are attracted by the plate and hence a plate current flows through the load resistor R_L. As the value of R_L is quite high, a large voltage drop is produced across the load.

Moreover, a weak signal voltage produces a large change in plate current. This, in turn, produces a large voltage drop across R_L. Thus, a weak signal applied in the grid circuit (Fig. 3) appears in the amplified form in the plate circuit. In this way, the vacuum tube does the work of amplification.

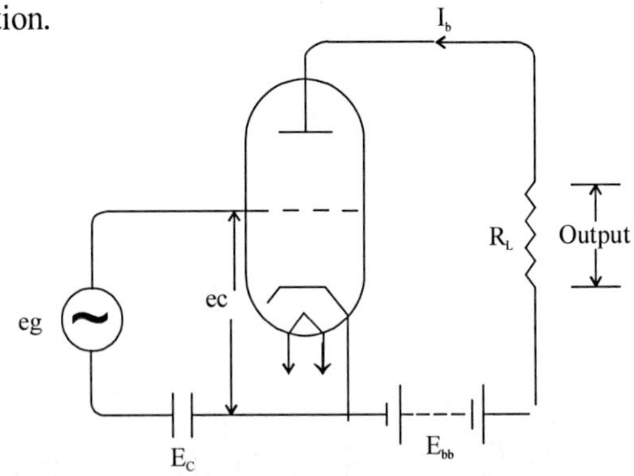

Fig. 3

(v) The characteristic curve of a tunnel diode :

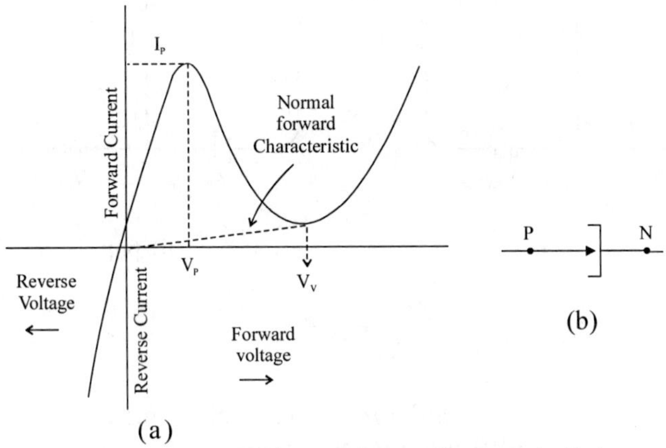

Fig. 4. (a) V~I Characteristics of a tunnel diode
(b) Symbol for funnel diode

(vi) *Richardoson's law for thermionic emission :* Please refer to the answer of Q. No. 3(b) of 1994 (A).

(vii) $L = 2 \text{ mH} = 2 \times 10^{-3}$ H

$f_r = 10$ KHz $= 10 \times 10^3$ Hz $= 10^4$ Hz

$$f_r = \frac{1}{2\pi\sqrt{LC}}$$

$$f_r^2 = \frac{1}{4\pi^2 LC}$$

$$\text{or,} \qquad C = \frac{1}{4\pi^2 f_r^2 L} = \frac{1}{4 \times 9.87 \times 10^8 \times 2 \times 10^{-3}}$$

$$= \frac{10^{-5}}{8 \times 9.87} = 0.01266 \times 10^{-5} \text{ farad}$$

$$= 0.1266 \ \mu\text{f}$$

(viii) *Common base connection of transistor :*

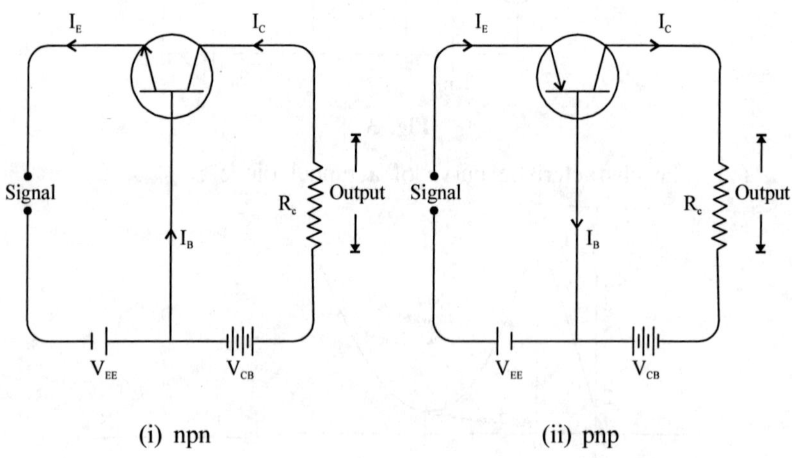

Fig. 5

(ix) *Conditions of oscillation :* The following three conditions must be fullfilled for producing sustained or undamped oscillations.

(a) There should be d.c. source for supplying energy which is to be converted to a.c. oscillations. The amount of energy supplied should be such as to compensate the losses in the tank circuit as well as the a.c. energy removed from the circuit by the load.

(b) The frequency of additional energy supplied should be same as that of oscillation in the tank circuit.

(c) The energy supplied should be in phase with the oscillations set up in the tank circuit. This is usually achieved by providing positive feedback.

Besides the above, the voltage gain with feedback is given by

$$A_f = \frac{A}{1 - mA} = \frac{V_0}{V_i} \qquad ...(1)$$

where, A is the voltage gain without feedback, m is the feedback factor, V_o and V_i are the output and input voltages respectively, m is the feedback factor.

When the loop gain $|mA| \to 1$, $A_f \to \infty$. Evidently, if input voltage V_i is Zero, there is output voltage, *i.e.*, the circuit becomes an oscillator. Thus, the two basic conditions for oscillations to occur in a feedback oscillator are :

(a) The loop gain $|mA| = 1$

(b) The feedback is +ve.

This satisfies the three conditions mentioned above. In all practical oscillators, however, $|mA| > 1$.

(x) *Advantages of negative feedback in amplifiers :* The advantages of negative feedback in amplifiers are gain stability, reduction of non-linear distortion, improvement of frequency response, increase of circuit stability and increase in input impedance and decrease in output impedance.

(a) *Gain stability.* An important advantage of negative feedback is that the resultant gain of the amplifier can be made independent of transistor parameters or the supply voltage variations.

$$A_{fb} = \frac{A}{1 + mA} \qquad ...(1)$$

For negative feedback in an amplifier to be effective, the designer deliberately makes the product *Am* much greater than unity. Therefore, in the above relation, 1 can be neglected as compared to *Am* and the expression becomes :

$$A_{fb} = \frac{A}{Am} = \frac{1}{m} \qquad ...(2)$$

It may be seen that the gain now depends only upon feedback fraction '*m*', *i.e.* characteristics of feedback circuit. As feedback circuit is usually a voltage divider (a resistive network), therefore, it is unaffected by changes in temperature, variations in transistor parameters and frequency. Hence, the gain of the amplifier is extremely stable.

(b) *Reduction of non-linear distortion :* A large signal stage has non-linear distortion because its voltage gain changes at various points in the cycle. The negative feedback reduces the non-linear distortion in large signal amplifiers. It can be proved mathematically that

$$D_{fb} = \frac{D}{1+mA} \qquad ...(3)$$

where, D = distortion in amplifier without feedback,

D_{fb} = distortion in amplifier with feedback.

It is clear that by applying negative feedback to an amplifier, distortion is reduced by a factor $(1+mA)$ feedback since feedback is being given from the output. This tends to oppose the increase in amplification and maintain it stable. The same is true if the output voltage decreases. As a result of which, the circuit stability is appreciably increased.

(c) *Increase in input impedance and decrease in output impedance :* The negative feedback increases the input impedance and decreases the output impedance of amplifier. Such a change is profitable in practice as the amplifier can then serve the purpose of impedance matching.

With negative feedback, the input impedance becomes

$$Z'_{in} = Z_{in}\,(1+mA) \qquad ...(4)$$

where, Z'_{in} is the input impedance with feedback, Z_{in} is the input impedance without feedback. It is thus clear that by applying negative feedback, the input impedance of the amplifier is increased by a factor $(1 + mA)$, where 'm' is the feedback factor, 'A' is the gain of the amplifier. Since 'mA' is much larger than unity, therefore, input impedance is increased considerably.

The output impedance with feedback is given by

$$Z'_{out} = \frac{Z_{out}}{1+mA} \qquad ...(5)$$

where, Z'_{out} = output impedance with negative feedback,

Z_{out} = output impedance without feedback.

(xi) Please refer to the answer of Q. No. 3(i) of 1994 (A) [According to 1993 syllabus].

(xii) $I_E = 10 \text{ mA} = 10 \times 10^{-3}$ A

$I_B = 50\ \mu\text{A} = 50 \times 10^{-6}$ A

$\alpha = ?$

$I_E = I_B + I_C$

or, $I_C = I_E - I_B = (10 \times 10^{-3} - 50 \times 10^{-6})$

$= 10^{-2} - 5 \times 10^{-5}$

$= 10^{-2}\,[1 - 5 \times 10^{-3}]$

$= 10^{-2}\,[1 - 0.005] = 10^{-2} \times (0.995)$ A

$$\alpha = \frac{I_C}{I_E} = \frac{0.995 \times 10^{-2}}{10^{-2}} = 0.995$$

SOME MORE CHSE (ORISSA) QUESTION PAPERS

1989 (A)

APPLIED ELECTRONICS

PAPER I

Full Marks : 75 Time : 3 Hours

Questions Cover Both 1988 & 1989 Syllabus

The figures in the right-hand margin indicate marks

*Answer **All** questions from groups A, B and **Four** from group C*

Group A

1. Answer very briefly, any *seven* of the following : (1 × 7)
   (a) Give the circuit symbol of a PNP transistor.
   (b) Write the expression for impedance of a series LCR circuit.
   (c) What is the frequency of radio waves whose wavelength is 10 mtr?
   (d) What is the unit of self-inductance?
   (e) What is the value of resistance with the following coloured bands?
   Brown, Black, Yellow, Silver.
   (f) One electronvolt is equal to how many Joules?
   (g) Write the relation among the tube constants of a triode.
   (h) What is the phase relation between applied voltage and current in a pure inductor?
   (i) What is the ripple frequency of a full-wave rectifier?
2. Fill in the blanks with appropriate answer from the bracket for any *eight* of the following : (1 × 8)
   (a) A FET is a — controlled device. (voltage, current)
   (b) With negative feedback, the voltage gain of an amplifier ____________ (increases, decreases)
   (c) Dynatron effect is present in __________ (Tetrode, Pentode)

(d) The voltage and power gain of ____________ amplifier is maximum. (CE, CB, CC)

(e) Inductive reactance __________________ with increase of frequency. (increases, decreases)

(f) _________________ feedback is used in oscillator. (Positive, Negative)

(g) Peak to Peak voltage of 220 V (RMS) AC is nearly ______. (308 V, 616 V)

(h) The temperature coefficient of resistance of a semiconductor is _______________. (positive, negative)

(i) A capacitive filter is used in _________________ load of a rectifier. (series with, parallel across)

(j) The phase difference between the output and input in a common emitter amplifier is _____________. (0°, 180°)

Group B

3. Answer briefly, any *five* of the following : (2 × 5)

(a) Show graphically how the impedance of a series LCR circuit varies with frequency.

(b) The plate current of a triode changes from 10 mA to 5 mA when the grid voltage changes from — 1 V to — 3 V at a plate voltage of 200 V. Calculate gm.

(c) How do you classify amplifiers according to operating point?

(d) Mention four essential tools used by a radio mechanic.

(e) A transistor has I_c = 9.5 mA and I_E = 10 mA. What is its β?

(f) What is the tuned frequency of a parallel LC circuit with L = 0.1 H and C = 0.001 μF?

(g) A resistor of 10 KΩ is connected across a voltage source of 100 V. What should be the recommended wattage of the resistor?

4. Answer briefly, any *five* : (2 × 5)

(a) Can you use a power choke in RF circuit? Justify your answer.

(b) What is meant by maximum plate dissipation as applied to a vacuum tube?

(c) Draw the VI characteristics of a Junction diode and explain its different parts.

(d) What is Zener Knee?

(e) A dynamo produces alternating current. Can it be called an oscillator? Justify your answer.

(f) Why heat sinks are used in transistor circuits?

(g) Can you interchange the collector and emitter of a PNP transistor? Justify.

Group C

(10 × 4)

5. Draw the Plate Characteristic Curve of a vacuum diode. Explain its different parts. What is Schottky effect?
6. What is meant by a 'resonant' circuit? Compare the characteristics of a series and parallel LCR circuit.

   (a) What are the safety precautions adopted by a radio mechanic?

   (b) Give the layout of a typical radio repairing shop.
7. What is meant by 'intrinsic' and 'extrinsic' semiconductor?

   Discuss how N-type semiconductor is developed. Mention the characteristic properties of a semiconductor.
8. Draw the circuit diagram of a single stage R-C coupled amplifier. Calculate its voltage gain. Discuss its frequency response.
9. Draw the circuit diagram of a Hartley oscillator and explain its working.

   Compare its performance with a Collpitt oscillator.
10. Draw the circuit diagram of a bridge-type rectifier. Explain its working. Compare its performance with a fullwave rectifier using two diodes.

    Show how a filter is used to eliminate ripples and a Zener diode for voltage regulation.
11. Write short notes on any *two* :

    (a) Load Line

    (b) FET

    (c) Transistor biasing by voltage division.

    (d) Mobility of charge carriers in an extrinsic semiconductor.

    (e) Frequency drift of an oscillator.

1989 (S)

APPLIED ELECTRONICS

PAPER I

Full Marks : 75 Time : 3 Hours

(According to 1988 & 1989 Syllabus)

The figures in the right-hand margin indicate marks
*Answer **All** questions from Groups A, B and **Four** from Group C*
(Questions cover both 1988 & 1989 syllabus)

Group A

1. Answer, 'very briefly', any *seven* of the following : (1 × 7)
   (a) Give the circuit symbol of an NPN transistor.
   (b) What is the frequency of radiowave whose wavelength is 15 metres?
   (c) What is the value of resistance with the following colour bands?
   Red, Red, Yellow, Gold.
   (d) What is the forbidden band gap in silicon expressed in ev?
   (e) What is the phase relation between applied voltage and Current in a pure capacitor?
   (f) What is the special use of a Thyristor?
   (g) What is the majority charge carrier in N-type semiconductor?
   (h) Show how reverse biasing is applied to a PN Junction diode.
   (i) Give the units of mutual inductance.
2. Answer *any* eight of the following : (1 × 8)
   Fill in the blanks choosing appropriate answers from the words given in the bracket.
   (a) A bipolar transistor is a ______________ controlled device. (current, voltage)
   (b) The bandwidth of an amplifier ______________ if there is negative feedback. (increases, decreases)

(c) Negative resistance is seen in a ______________. (Zener diode, tunnel diode)

(d) The voltage gain of a common collector amplifier is ________ than one. (more, less)

(e) Push-Pull amplifier is used for ______________ amplification. (voltage, power)

(f) Capacitive reactance ______________ with increase of frequency. (increases, decreases)

(g) Distortion is maximum in ____________ type of amplifier. (class A, class B, class C)

(h) The phase difference between the output and input in a CB amplifier is ______________. (0°, 180°)

(i) A Crystal is used in an oscillator to maintain ____________ steady. (gain frequency)

(j) When the RMS value of voltage is 200 V, the peak value is nearly ____________. (280 V, 240 V)

Group B

3. Answer any *five* briefly : (2 × 5)

(a) Show graphically how the impedance of a parallel LCR circuit varies with frequency.

(b) The plate current of a triode changes from 5 mA to 10 mA when the plate voltage changes from 100 V to 200 V at a grid voltage of — 2 V. Calculate r_p.

(c) Mention any *four* test equipments used in a radio repairing shop.

(d) A transistor has I_B = 5 mA, and I_c = 95 mA. What is its α?

(e) What is the impedance of a series LCR circuit at 20 KHz when

$$L = 0.01 \text{ H}, C = 100 \text{ pf, and } R = 100 \ \Omega?$$

(f) 10 mA current flows through a resistance of 100 KΩ. What is the recommended wattage of the resistor?

(g) With sufficient feedback, the voltage gain of a feedback amplifier is independent of tube or transistor parameter. Is the statement true? Justify your answer.

4. Answer any *five* briefly : (2 × 5)

(a) What are the common cathode materials? Which is most suitable? Why?

(b) Why n-p-n transistor is more suitable for high frequency use?
(c) What is Zener impedance?
(d) What are the different biasing methods used in transistor circuits? Which of them is most commonly used?
(e) Mention the advantages of a push-pull amplifier.
(f) What are the reasons of frequency drift of an oscillator?
(g) Why the gain of RC coupled amplifier falls-off at high frequency?

Group C

(10 × 4)

5. Draw the characteristic curves of a triode and explain how the tube constants can be calculated from them.
6. Draw the vector diagram for current and voltage in different sections of a series LCR circuit. Hence calculate the impedance, phase angle and the resonant frequency in this case.
7. Draw a circuit for drawing the input and output characteristics of a transistor in CE configuration.

   Draw the input and output characteristic curves and explain their different sections.
8. Draw the circuit diagrams of transistor amplifier in three common modes. Which of them is most commonly used? Why?
9. Draw the circuit diagram of a Collpitt oscillator. Explain its working. Compare its performance with a Hartley oscillator.
10. Write short notes on any *two* :
    (a) Tunnel diode.
    (b) Tube constants.
    (c) Tolerance and current rating of a resistor.
    (d) MOSFET.
11. Draw the circuit diagram of a fullwave rectifier. Explain its working. What is the peak inverse voltage in the circuit?

    Show how a filter is connected across the output to eliminate ripples.
12. (a) Suggest suitable wiring and arrangement of equipments on the work table of a radio mechanic.

    (b) You have a multimeter, a VTVM and an oscilloscope. For what type of measurement, you will uses each of them?

1988 (A)

Full Marks : 75 Time : 3 Hours

Answer *All* questions from Group A and Group B, and *Four* questions from Group C

The figure in the right-hand margin indicate marks

Group A

1. Answer the following : (1 × 8)
   (a) Give the value of the resistance with the following colour code :
   Red, Orange, Violet.
   (b) Give the colour code of a 1.9 Ω resistance.
   (c) Give the units of power of frequency.
   (d) Give the circuit symbol of a π filter.
   (e) What are the atomic weights of Silicon and Germanium?
   (f) Draw the output waveform of a half-wave rectifier working with a sinusoidal input.
   (g) The internal resistance of a voltage source is 0.05 Ω. How much voltage is dropped across this resistance when current through it is 2 A?
   (h) Define self-inductance?

2. Fill in the blanks with appropriate terms from the brackets : (1 × 7)
   (a) Gain of 3 db corresponds to power gain of __________. (2, 3, 4)
   (b) A transistor is a _______________ controlled device. (current, voltage)
   (c) A reverse biased diode has ________ resistance. (high, low)
   (d) The gain of an emitter follower is __________. (less than one, one, higher than one)
   (e) In a rectifier a capacitor filter is used in _________ with the load. (series, parallel)
   (f) The reactance of an inductor ______________ with increasing frequency. (increases, decreases)

(g) In a series resonant circuit, when capacitor is quardrupled the resonant frequency ____________. (doubles, is halved, remains unchanged)

Group B

3. Answer briefly any *five* questions : (2 × 5)
   (a) Name four important tools used in a radio repairing shop.
   (b) Discuss the origin of reverse saturation current in a semiconductor diode.
   (c) Why do you bias a transistor in an amplifier?
   (d) What is transistor α? What is its usual magnitude?
   (e) Why is speaker coupled to an audio amplifier through a transformer?
   (f) Why is an indirectly heated cathode preferred to a directly heated one?
   (g) A class *C* amplifier is more efficient than a class *B* amplifier, why?
4. Answer briefly any *five* : (2 × 5)
   (a) What is zener breakdown?
   (b) What is the reactance of a 5 mH inductor to 2 volt 1 KHz alternating signal?
   (c) Discuss the effect of the coupling capacitor on the low frequency gain of an RC coupled amplifier.
   (d) Suppose that only 2 per cent of the electrons injected into the base recombine with base-holes. If 1 million electrons enter the emitter in 1 μs, how many electrons will come out of the base lead in this period?
   (e) A transistor has $\beta = 400$. What is the base current if collector current is 50 mA?
   (f) Why does a crystal controlled oscillator give output of stable frequency?
   (g) Mention the advantages of oxide coated filaments.

Group C

5. Distinguish between class *A* and class *B* operation of amplifier. With a neat circuit diagram explain the operation of a single-stage class *A* amplifier. (10)
6. What is rectification? How does a full-wave rectifier rectify the ac signal? Discuss the effect of a capacitor filter on the output of such a circuit. (10)

7. Analyse the phenomenon of resonance in a parallel resonant circuit. (6)

   A capacitor of 0.5 μF is connected in parallel with a 5 mH inductor and 100 Ω resistor. Find the resonant frequency f_o of the circuit. What is the impedance of the circuit to a signal of frequency $2f_o$? (4)

8. Discuss the construction and working principles of a vacuum diode and a crystal diode. Point out the advantages of the latter over the former. (10)

9. Draw the circuit diagram of a Colpitt oscillator and explain its operation. In what way it is different from a Hartley oscillator? (10)

10. Discuss about the different types of capacitors and resistors used in a radio circuit.

11. Write notes on any *two* : (10)
    (a) Cathode-ray oscilloscope
    (b) Multimeter
    (c) Tuned amplifier
    (d) Layout of a radio repairing shop.

1988 (S)

APPLIED ELECTRONICS

PAPER I

Full Marks : 75 — Time : 3 Hours

*Answer **All** questions from Group A and Group B and Answer Four questions from Group C*

The figures in the right-hand margin indicate marks

Group A

1. Answer the following : (1 × 8)
   - (a) Give the colour code of a 4Ω resistor.
   - (b) Give the units of impedance and electric charge.
   - (c) What is the value of the resistance with the following colour code?

     Brown, Black, Green, Silver.
   - (d) Give the circuit symbol of an iron-core transformer.
   - (e) What are the maximum ambient operating temperatures of Silicon and Germanium?
   - (f) What is the wavelength of a 15 KHz radio-wave?
   - (g) Define mutual inductance.
   - (h) A voltage source of 9 volts and internal resistance 0.4Ω is connected to a 5Ω load. What is the load current?

2. Fill in the blanks with appropriate words from the brackets : (1 × 7)
   - (a) A vacuum tube is a ________________ controlled device (voltage, current).
   - (b) Silicon doped with Boron results in ____________ type semiconductor (*n*, *p*).
   - (c) A gain of 6 db corresponds to voltage ratio of ________ (2, 3, 4)
   - (d) A tuned amplifier has ___________ bandwidth. (large, small)

(e) In a rectifier an inductor filter is used in ________________ with the load. (parallel, series)

(f) The reactance of a capacitor is ____________ proportional to the signal frequency. (directly, inversely)

(g) In stepdown transformer the primary current is __________ than the secondary current. (none, less)

Group B

3. Answer briefly any *five* : (2 × 5)

(a) What is the significance of the peak inverse voltage of a diode?

(b) What is the bandwidth of a double stage amplifier formed by coupling two identical stages of bandwidth 10 KHz?

(c) Why is a VTVM preferable to an ordinary voltmeter?

(d) Why does a Hartley oscillator use an RF choke?

(e) A transistor has I_c = 100 mA and I_B = 0.5 mA. What is its α?

(f) An inductor filter in a halfwave rectifier is hardly preferred. Why?

(g) Give a list of measuring and test equipments used in radio repair.

4. Answer any *five*. Each answer should be in about 3 to 5 sentences : (2 × 5)

(a) Explain the term 'Gain' of an amplifier.

(b) A series resonant circuit uses an inductor of 5 mH, a capacitor of 50 μF and a resistance of 100Ω. Find its resonant frequency.

(c) Discuss the effect of internal capacitances of a transistor on the high frequency gain of an amplifier.

(d) What is the power dissipation in a forward biased diode if diode voltage is 0.7 V and the current is 100 mA?

(e) In a transistor the emitter current is 6 mA, and the collector current is 5.75 mA, what is the transistor β?

(f) The base of a transistor is narrow and lightly doped. Why?

(g) Mention the advantages of a semiconductor diode over a vacuum diode.

Group C

5. Draw the characteristics of a zener diode and explain its nature. Discuss its use in a voltage regulator. (10)

6. Draw the circuit diagram of a push-pull power amplifier and explain its operation. (10)
7. Discuss the principle of oscillation. Draw the diagram of a Hartley oscillator and explain the functions of different elements in the circuit. (10)
8. What do you mean by *n* and *p* type semiconductors? Explain, how they are prepared. (10)
9. What is meant by feedback? Discuss the effect of negative feedback on the performance of an amplifier. (10)
10. Give the circuit diagrams of transistor amplifiers in *CB, CE* and *CC* modes. modes. Which of them is usually used? Why? (10)
11. Write notes on any *two* : (10)
    (a) Crystal oscillator.
    (b) Equivalent circuit of a transistor.
    (c) Parallel resonant circuit.
    (d) Load line.

1985 (S)

Full Marks : 75 Time : 3 Hours

*Answer **All** questions from Group A and any **Ten** bits from Group B and any **Four** questions from Group C*

Group A

[Marks : 15]

Each bit carries *one* mark. Give your answer by a diagram or in numericals or at best in *one* sentence as the case may be.

1. (i) What is the relation between resistance and conductance?
   (ii) What is the meaning of a semiconductor?
   (iii) Write down the resistance from the following colour code. Red, Red, Red, Silver
   (iv) Write the colour code for the following resistance :
   33 KΩ
   (v) How transformers are used?
   (vi) Draw the circuit symbol for a gang condenser.
   (vii) Write down the units of inductance and conductance.
   (viii) How do you prepare a variable inductor?
2. (i) How *N*-type semiconductors are made?
   (ii) What is peak reverse voltage (PRV) of a solid state diode?
   (iii) What is the advantage of a common emitter transistors?
   (iv) A radio station broadcasts at 900KHz. What is its wave-length?
   (v) How many junctions and terminals are there in a *N-P-N* transistor?
   (vi) How *L-R-C* combination be used for tuning circuits?
   (vii) If only a single diode is used in a rectifier circuit, what type of output is expected?

Group B

[Marks : 20]

Each bit carries *two* marks. Answer any *ten* bits. Answer each bit in 3 to 5 sentences whenever necessary.

3. (i) Why filter circuits are used in electronics?

(ii) A Voltage vosin 2 πft is being applied to a pure inductor of inductance L. What is the current in the circuit and draw voltage and current versus time from $t = 0$ to $t = 1/f$.

(iii) Draw circuit symbols for :

(a) variable capacitor

(b) transformer

(c) fuse wire

(d) tapped resistor.

(iv) Explain how L-R combination can work as low pass filter?

(v) Describe a Zener diode.

(vi) Calculate revere resistance in MΩ for a junction diode operating at a reverse voltage of 25V generating a reverse current of 5μA.

(vii) Explain the significance of common emitter configuration of a transistor.

(viii) Explain why the current gain is less than unity in common base transistor amplifier.

(ix) Explain why a crystal oscillator gives steady oscillations.

(x) Define time constants for L-R and L-C circuits.

(xi) Describe the construction of a P-N-P transistor.

(xii) Discuss the methods used to identify the leads of a transistor.

Group C

[Marks : 40]

4. Explain the working of a simplified DC voltmeter at different ranges. Give circuit diagram wherever necessary.

5. (a) A microammeter has a resistance of 2 KΩ and can take maximum current of 50 μA. What is the series resistance connected to it when it is used as a volt-meter to read 1 kV?

(b) What is the shunt resistance for the above meter to be used as ammeter to read 10 mA?

6. Give the circuit diagram of a series of resonant circuit. Show the impedance of the circuit by means of vector diagram. What happens to the impedance and current of such a circuit at resonance.

7. Explain the functions of a Zener diode. How it can be used as a voltage regulator?

8. Explain the working of a N-P-N type transistor through neat circuit diagram. Show the emitter and collector characteristic curves.

9. What are the different types of coupling used in amplifier circuits? Describe a *R-C* coupled transistor amplifier giving an expression for its gain.

10. Writes notes :
    (a) Hartley oscillator
    (b) Full-wave rectifier.

2000

ELECTRONICS – I

Full Marks : 75 — Time : 3 Hours

*Answer **All** questions from Group A and Group B and any **Four** questions from Group C*

Group A and Group B questions should be answered serially and continuously

The figures in the right-hand margin indicate marks

No electronic gadgets are allowed into the Examination Hall

Group A

Answer very briefly (1 × 7)

1. (a) Draw the circuit symbol of battery.

Ans. Circuit symbol of battery :

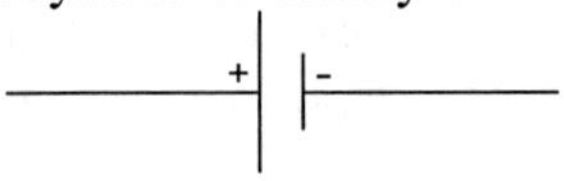

(b) Give the colour code for the resistor

$0.32\ \Omega \pm 10\%$

Ans. Colour Code for 0.32 + 10% :

The colour code for 3 is Orange

The colour code for 2 is Red

The colour code for 0.01 is Silver

The colour code for 10% tolerance is Silver

So, the colour code for the resistor 0.32 + 10% is Orange, Red, Silver, Silver.

(c) Mention the expression for the time constant of an LR circuit.

Ans. Time constant of LR circuit is given by

$$\lambda = \frac{L}{R}$$

(d) Whose work function is more — tungsten or thoriated tungesten ?

Ans. Work function of tungsten is 4.52 eV whereas the work function of thoriated-tungsten is 2.63 eV.

So, the work function of tungsten is more than thoriated-tungsten.

(e) In which of the three standard transistor configurations, input impedance is highest ?

Ans. Input impedance is highest in case of common collector transistor configuration.

(f) What happens to the voltage gain of an amplifier if negative feedback is introduced?

Ans. If negative feedback is introduced, the voltage gain of an amplifier is reduced. This is because

$$A_f = \frac{A}{1+\alpha A}$$

where, A is the gain of the amplifier without feedback and A_f, gain of the amplifier with feedback. α is the feedback factor.

(g) Write down the expression for frequency of oscillation of a Hartley oscillator.

The frequency of osscillation of a Hartley oscillator is

$$f_0 = \frac{1}{2\pi\sqrt{C\left(L_1 + L_2 + 2M\right)}}$$

Ans. Whereas, M is the mutual inductance between the two sections, L_1 and L_2. C is the capacitance of the tank circuit.

2. Choose the *correct* answer from those given in the bracket :

(1 × 8)

(a) Which of the following expressions for a transistor is incorrect?

$$\left[\beta = \frac{\alpha}{1-\alpha},\ \gamma = \frac{1}{1-\alpha},\ \beta = \frac{\alpha}{1+\alpha}\right]$$

Ans.
$$\beta = \frac{\alpha}{1+\alpha}$$

(b) Which of the following is used in PA system ?

(RF tuned amplifier, push-pull amplifier)

Ans. Push-pull amplifier.

(c) What is the nature of series LCR circuit at the resonant frequency? (Resistive, capacitive, inductive)

Ans. Resistive

(d) What is the resistance of a 230 V, 1000 W heater coil?

(52.9Ω, 51.3Ω, 58.0Ω)

Ans. P = 1000 W, V = 230 V

$$I = \frac{P}{V} = \frac{1000}{230} A$$

$$R = \frac{V}{I} = \frac{230 \times 230}{1000} = \frac{23 \times 23}{10} = \frac{529}{10} = 52.9 \text{ ohm}$$

(e) What is angle of conduction in a class-B amplifier ?
(360°, 180°, 90°)

Ans. 180°

(f) What happens to the bandwidth of amplifier on cascading?
(Increases, decreases, does not change)

Ans. decreases

(g) Which of the following combination is better matched?
(CE → CE, CE → CB, CE → CC)

Ans. CE → CE

(h) Which of the following filters is more effective in eliminating ripples in a rectifier? (π-filter, L-filter, capacitive filter)

Ans. π–filter

Group B

Answer briefly

3. Answer any *ten* of the following : (2 × 10)

(a) Calculate the minimum power rating of a resistor of 5 KΩ to be used with a voltage source of 100 V.

Ans. $P = VI = V \times \dfrac{V}{R} = \dfrac{V^2}{R}$

$$= \frac{(100)^2}{5 \times 10^3} = \frac{10^4}{5 \times 10^3} = \frac{10}{5} = 2$$

(b) What is thermionic emission? Discuss briefly.

Ans. ***Thermionic emission.*** The process of electron emission from a metal surface by supplying thermal energy to it is known as *Thermionic emission.*

At ordinary temperatures, the energy possessed by free electrons in the metal is inadequate to cause them to escape from the surface. When heat is applied to the metal, some of heat energy is converted into kinetic energy, causing accelerated motion of free electrons. When the temperature rises sufficiently, these electrons acquire additional energy equal to the work function of the metal. Consequently, they overcome the restraining surface barrier and leave the metal surface.

Metals with lower work function will require less additional energy and therefore, will emit electrons at lower temperatures. The amount of thermionic emission increases rapidly as the emitter temperature is raised. Actually, high temperatures are needed to cause thermionic emission.

(c) When the speed of electric fan in your home goes down, the mechanic usually replaces the capacitor. Why?

Ans. A capacitor allows a.c. to flow through it. When the speed of electric fan goes down, it indicates that proper amount of a.c. is not passing through the capacitor. So, the mechanic replaces the capacitor with a correct one.

(d) In a common base arrangement, the collector current is 0.9 mA and $\alpha = 0.9$. Calculate the base current.

Ans. $I_C = 0.9$ mA, $\alpha = 0.9$, $I_B = ?$

$$\alpha = \frac{I_C}{I_E} \quad \text{or } I_E = \frac{I_C}{\alpha} = \frac{0.9 \text{ mA}}{0.9} = 1\,\text{mA}$$

Again, $I_E = I_B + I_C$

So, $I_B = I_E - I_C = (1 - 0.9)\,\text{mA} = 0.1$ mA

(e) Give the classification of amplifiers according to various types of cascading. Give examples.

Ans.

Name of coupling	*Name of multistage amplifier*
R-C coupling	R-C coupled amplifier
Transformer coupling	Transformer coupled amplifier
Direct coupling	Direct coupled amplifier

(f) What is 'Zener Knee'? Explain its significance.

Ans. *Zener Knee* : It is the reverse voltage where the reverse current increases sparply to a high value. It is also called the breakdown voltage.

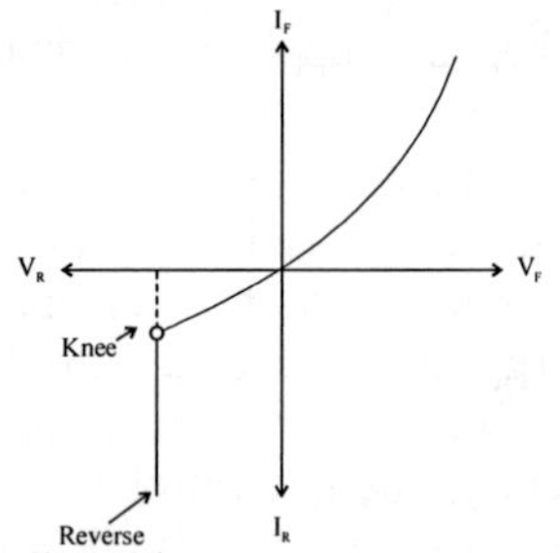

Fig. 1. Characteristic of Zener diode

Explanation. When the reverse bias on a crystal diode is increased, a critical voltage, called *breakdown voltage* or Knee voltage or Zener voltage is reached where the reverse current increases sharply to a high value. This breakdown voltage is the knee of the reverse characteristic as shown in fig. 1.

The satisfactory explanation of this breakdown of the junction was first given by the American scientist C. Zener.

As the applied reverse voltage in increased beyond the Zener voltage, current continues to increase rapidly. This results in further decrease of diode resistance. Consequently, the voltage drop (V = IR) across the Zener diode tends to remain constant at the level of Zener voltage. If the applied reverse voltage is brought down to the level of Zener voltage, current starts decreasing and again the voltage drop across the diode is maintained equal to Zener voltage. Hence, the breakdown effect is reversable, *i.e.*, the current through the diode could be varied over a large range without damaging it provided the power dissipation in the diode is limited to the specified rating.

(g) Impedance matching is easy to achieve in a transformer coupled amplifier but not so for a RC coupled amplifier. Justify.

Ans. *Impedance matching* : In a transformer coupled amplifier, it is easy to make the inductive reactance of primary equal to the output impedance of the transistor and reactance of secondary equal to the input impedance of next stage.

In multistage R-C coupled amplifier, impedance matching is poor. This can be explained as under :

(a) The output impedance of a common emitter or common base amplifier is high and the input impedance is low. Hence, when output of one stage is coupled to the input of the next stage, there is considerable mismatch and power is lost in the transformer.

(b) Again, the output impedance of R-C coupled amplifier is several hundred ohms whereas that of a speaker is only a few ohms, there is thus a mismatch. As a result, little power will be transferred to the speaker.

(h) A negative feedback amplifier is provided with a voltage feedback factor 0.05. Calculate the percentage reduction of the voltage gain if the initial voltage gain is 50.

Ans. $\alpha = 0.05$

$A = 50$

$$A_f = \frac{A}{1+\alpha A} = \frac{50}{1+(0.05\times 50)} = \frac{50}{1+2.5}$$

$$= \frac{50}{35} = \frac{500}{35} = \frac{100}{7}$$

% reduction in voltage gain

$$= \frac{50 - \frac{100}{7}}{50} \times 100$$

$$= \left(\frac{350 - 100}{50 \times 7} \right) \times 100$$

$$= \frac{250}{7} \times 2 = \frac{500}{7} = 71.4\%$$

(i) The zero signal collector dissipation in a power amplifier is maximum. Justify.

Ans. The objective of power amplifier is to achieve maximum possible power output. Thus, the collector efficiency is maximum.

The ratio of a.c. output power to the zero signal power (*i.e.*, d.c. power) supplied by the battery of a power amplifier is called the collector efficiency.

The amplifier converts d.c. power from the battery into a.c. output power. The greater the collector efficiency, larger is the a.c. power output.

Since the collector efficiency is maximum, the a.c. power output is also maximum. Due to flow of large amount of current through the transistor in a power amplifier, heat is developed during operation. The ability of a power transistor to dissipate heat without affecting its performance is known as power dissipation capability. So, maximum power dissipation in a transistor occurs when no signal power is supplied by the battery of a power amplifier.

(j) What is piezo-electric effect? Explain.

Ans. *Piezo-electric effect* : When an a.c. voltage is applied across certain crystalline materials like Rochelle salt, quartz and tourmaline etc., they vibrate at the frequency of the applied voltage. Conversely, when they are compressed or placed under mechanical strain to vibrate, they produce an a.c. voltage. Such an effect is called the piezo-electric effect. The crystals which exhibit piezo-electric effect are called piezo-electric crystals.

Frequency of Crystal : Each crystal has a natural frequency like a pendulum. The natural frequency 'f' of a crystal is given by

$$f = \frac{K}{t} \qquad \text{...(1)}$$

where K is the constant that depends upon the cut and t is the thickness of the crystal.

Hence, $f \propto \frac{1}{t}$ ...(2)

i.e., frequency increases as the thickness decreases. However, extremely thin crystals may break out of vibrations. This puts a limit to the frequency obtainable. In practice, frequencies between 25 KHz to 5 MHz have been obtained with crystals.

(k) Draw the output characteristics of a transistor in *CE* configuration and discuss its features. Explain how output resistance is computed using it.

Ans. *Output Characteristics of a CE transistor configuration :* It is the Curve between collector current I_C and collector-emitter voltage V_{CE} at constant base current I_B.

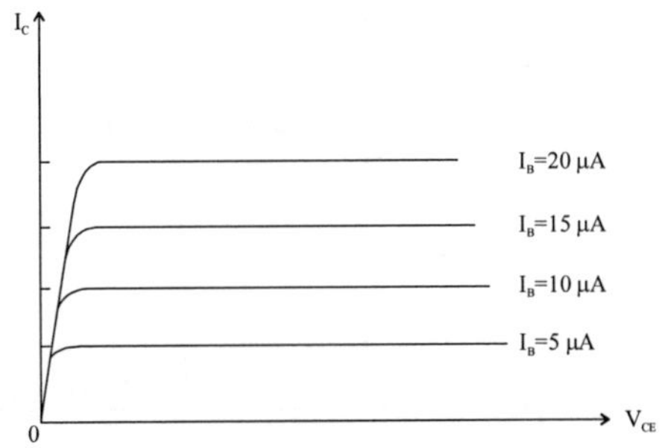

Fig. 2

Features

(i) The collector current I_C varies with V_{CE} for V_{CE} between 0 and 1 V only. After this, collector current becomes almost constant and independent of V_{CE}. This value of V_{CE} upto which collector current I_C changes is called the *Knee Voltage*. The transistors are always operated in the region above knee voltage.

(ii) Above Knee Voltage, I_C is almost constant. However, a small increase in I_C with increasing V_{CE} is caused by the collector depletion layer getting wider and capturing a few more majority carriers before electron-hole combinations occur in the base area.

(iii) For any value of V_{CE} above Knee Voltage, the collector current I_C is approximately equal to βI_B

(iv) The output characteristics have noticeable slope.

Output resistance

It is the ratio of change in collector-emitter voltage (ΔV_{CE}) to the change in collector current (ΔI_C) at constant I_B, *i.e.*,

$$\gamma_0 = \left(\frac{\Delta V_{CE}}{\Delta I_C}\right)_{I_B}$$

(l) How do you get positive feedback in an RC phase shift oscillator? Explain.

Ans. *Positive feedback in a RC phase shift oscillator :*

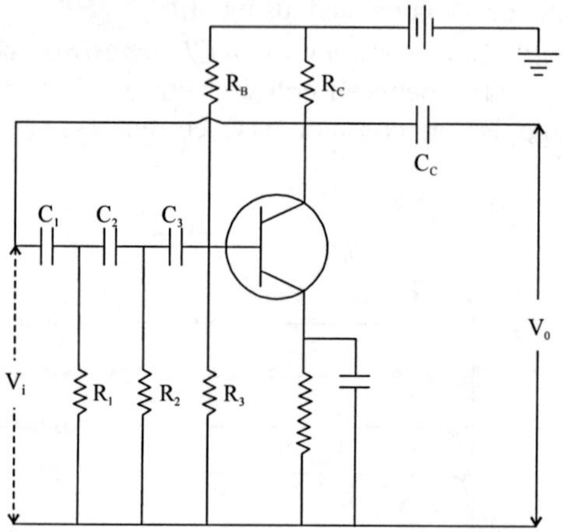

Fig. 3. Phase shift oscillator

A suitable RC network connected between the output and the input circuits of a common emitter amplifier can be designed to maintain positive feedback (fig. 3). This network is designed to introduce a phase-shift of 180°. In addition, the amplifier itself introduces a phase change of 180°. Thus, a total phase change of 360° can be attained to maintain positive feedback.

In order to achieve 180° phase-shift, three identical RC sections are generally used, each shifting the signal phase by 60° at the frequency of oscillation. Since the capacitive reactance of the network changes with frequency, 180° phase-shift condition is satisfied only at one frequency.

(m) Discuss the distinctions between an AF choke and a RF choke.

Ans. *Audio frequency choke (AFC)* : It works with frequencies ranging from 20 cycles/sec to 20 kilocycles/sec. So, greater inductance is required for low frequency circuits. The effectiveness of a choke coil depends upon its reactance. Therefore, iron-core choke coils with higher inductance are used as AFC.

Radio frequency choke (RFC) : It works with frequencies ranging from 20 KHz to 500 MHz. So, low value of inductance is needed for radio frequency circuits. An air core coil with low value of inductance is employed as RFC.

(n) Derive the expression for average d.c. through the load in a fullwave rectifier.

Ans. *Average d.c. through the load in a fullwave rectifier :* The average d.c. is that steady current which sends the same charge through a circuit as is done by a.c. in a half-cycle ($t = T/2$, $\omega t = \pi$).

Mathematically,

$$q = \int_0^q dq = \int_0^{T/2} I\,dt = \int_0^{T/2} I_m \,Sin\, \omega t\, dt$$

$$= I_m \left[-\frac{Cos\,\omega t}{\omega} \right]_0^{T/2}$$

$$= -\frac{I_m}{\omega}\left[Cos\,\pi - \cos 0\right]$$

$$= \frac{2I_m}{\omega} \qquad ...(1)$$

But $$q = \left(\bar{I}\right)_{half} \times \frac{T}{2} \qquad ...(2)$$

Where $\left(\bar{I}\right)$ is the average value of a.c. for the half-cycle.

From (1) and (2), we get,

$$\left(\bar{I}\right)_{half} \times \frac{T}{2} = \frac{2I_m}{\omega}$$

$$\left(\bar{I}\right)_{half} = \frac{4I_m}{\omega T} = \frac{4I_m}{2\pi}$$

$$= \frac{2I_m}{\pi} \qquad ...(3)$$

However, the average d.c. for a full-cycle becomes zero, since

$$\bar{I} = \frac{\int_0^T I_m \sin \omega t \, dt}{\int_0^T dt} = \frac{I_m \left[-\frac{\cos \omega t}{\omega} \right]_0^T}{T}$$

$$\bar{I} = \frac{-I_m}{\omega T} (\cos 2\pi - \cos 0)$$

$$= 0 \qquad ...(4)$$

Group C

Answer any *four* questions

4. Deprive a series LCR circuit with an a.c. source. Discuss the variation of current in the circuit when frequency is changed.
   Derive the necessary expressions. (2 + 8)
5. Explain (i) intrinsic semiconductor and (ii) extrinsic semiconductor. What are the different types of extrinsic semiconductor? How are they formed? Discuss. (10)
6. Draw the circuit diagram of a bridge rectifier with a filter and explain its operation.
   Derive an expression for its efficiency. (3 + 5 + 2)
7. Describe the 'Feedback resistor' method of biasing of a transistor amplifier. Explain its merits and demerits. (8 + 2)
8. Draw the circuit diagram of a single stage RC coupled amplifier. Calculate its voltage and power gain in midfrequency range. Why does the gain drops off for low and high frequencies? (2 + 6 + 2)
9. Mention the conditions necessary for maintenance of oscillation in an electronic oscillator.
   With a neat circuit diagram, explain the working of Colpitt oscillator. (3 + 7)
10. Write notes on any *two* : (5 × 2)
    (a) Dynatron effect
    (b) Colour code for capacitor
    (c) Self and mutual inductance.

❑ ❑ ❑